THE
Jefferson BROTHERS

Also by Joanne L. Yeck

*"At a Place Called Buckingham"...
Historic Sketches of Buckingham County, Virginia*

THE JEFFERSON BROTHERS

Joanne L. Yeck

Printed in the United States of America
Copyright © 2012 by Joanne L. Yeck

Cover Image:
Thomas Jefferson Silhouette by John Marshal. (Library of Congress)

All rights reserved. Printed in the United States of America. No part of this book may be reproduced in any manner whatsoever without written permission except in the case of brief quotations embodied in critical articles and reviews.

Library of Congress Cataloging-In-Publication Data

Yeck, Joanne L.
The Jefferson Brothers /by Joanne L. Yeck
   p. cm.
Includes bibliographical references and index.
ISBN 978-0-9839898-1-3
1. Jefferson, Randolph, 1755-1815. 2. Jefferson, Thomas, 1743-1826. 3. Snowden (Va.: Plantation) 4. Monticello (Va.: Plantation) 5. Shadwell (Va.: Plantation) 6. Plantation life—Virginia—Buckingham County—History. 7. Plantation life—Virginia—Albemarle County—History. 8. Slavery—Virginia—Buckingham County—Biography. 9. Jefferson, Peter, 1708-1757. 10. Jefferson, Jane, 1720-1776. 11. College of William and Mary—Virginia—Williamsburg—History—18thcentury.
I. Joanne L. Yeck, 1954 – II. Title.
E332.25 .Y43 2012
973.4'6'092

Designed by Carole Ohl
Printed by Greyden Press, LLC
Published by Slate River Press Ltd.
http://slateriverpress.wordpress.com
slate.river.press@gmail.com

CONTENTS

PART ONE: THE JEFFERSON FAMILY

1. . . . . . Peter Jefferson, Gent.
2. . . . . . Jane Jefferson's Shadwell
3. . . . . . The Early Education of Randolph Jefferson

PART TWO: RANDOLPH JEFFERSON

4. . . . . . A World beyond the Piedmont: Williamsburg and the College of William and Mary
5. . . . . . Randolph Jefferson, Patriot
6. . . . . . Randolph Jefferson, Planter
7. . . . . . Matters of Money
8. . . . . . Snowden: A Plantation in Buckingham County
9. . . . . . The Children of Randolph Jefferson

PART THREE: THE JEFFERSON BROTHERS

10. . . . . The Jefferson Servants
11. . . . . The Patient Randolph Jefferson
12. . . . . The Second Mrs. Randolph Jefferson
13. . . . . "My Brother Died This Morning"
14. . . . . The Jefferson Brothers: A Study in Contrasts

Acknowledgments
Selected Bibliography
Notes
Index
About the Author

For
Randolph Jefferson
and his many descendants

Peter and Jane (Randolph) Jefferson were married for seventeen years. Jane was frequently pregnant, delivering ten children into mid-18th century Colonial Virginia. Only two of their four sons survived to adulthood. The elder, Thomas, grew up to influence a generation and help shape a nation. The younger, Randolph, led a simple life as a planter on the banks of the James River. A great deal is known about Thomas, and his life's work will continue to be the subject of discussion for decades to come. By contrast, very little is known about Randolph whose story sheds a much, much smaller light on his neighborhood in northern Buckingham County, Virginia.

In the summer of 1757, Peter Jefferson suddenly died. Tom, then fourteen years old, was nearly a man in 18th-century terms. Randolph, only two, was yet unformed and would grow up fatherless. Despite their extremely divergent paths in life, both of the Jefferson brothers would be significantly shaped by Peter's bequest of two quite separate land holdings: the Rivanna lands and the Fluvanna lands of Albemarle County, Virginia.

Part One
THE JEFFERSON FAMILY

CHAPTER ONE

Peter Jefferson, Gent.

No society is so precious as that of one's own family.
Thomas Jefferson to Randolph Jefferson, 1789

Over the decades, the life of Peter Jefferson has been preserved more as legend than as fact. The long shadow cast by his prodigiously talented son, Thomas, has obscured and lessened Peter's pioneering accomplishments. Stepping back to Colonial Virginia, Peter Jefferson emerges as an influential figure in the development of Virginia's Piedmont, with a distinct vision for the James River's Horseshoe Bend.

Peter Jefferson was born on February 29, 1708, south of the James River in what was then part of Henrico County.[1] His parents were Mary (Field) and Thomas Jefferson, II, who set an example of civil service which Peter would follow.[2] Thomas was a Gentleman Justice for Henrico for nearly twenty years, a Captain in the Militia, and for at least two years, held the office of High Sheriff in the county. He also aided in the building of a church in Bristol Parish, long referred to as "Jefferson's Church."[3] He owned a racing mare and fraternized with the local gentry, with whom the Jeffersons would long intermarry.

Prior to Thomas Jefferson's death, Peter began developing his father's land to the west, which lay on Fine Creek. By 1728, he had spent £3.10 on a house and had invested an additional £3.2 on two tobacco houses. These expenses were "charged to Mr. Thomas Jefferson's Acc't with Peter."[4] The young Virginian was also eager to establish his own homestead and, shortly after he came of age, Peter acquired 322 acres of "new" land. The land grant, issued in 1730, came from His Majesty King George II of England and was signed by Governor William Gooch. Not far from Fine Creek, the tract lay in the newly established Goochland County, on the south side of the James River, nestled among the French Huguenot refugees who had settled land around Manakin Town in the early 1700s. For a generation, these French Protestants had successfully cultivated the area, infusing the surroundings with their Gallic sensibilities.[5]

The area appealed to several of the younger Jeffersons. Two of Peter's sisters married men established in this part of Goochland. Martha Jefferson wed Bennet Goode, who operated Jude's Ferry from his land on the south bank of the James River to Col. John Fleming's land on the north bank.[6] Mary Jefferson married Thomas Turpin, who eventually served as one of Peter Jefferson's five executors.

Peter had barely struck out on his own when his father died, leaving him additional land south of the James River. The last will of Thomas Jefferson, II was recorded in Henrico County in April of 1731. It reads in part:

> ITEM. I give unto my son PETER JEFFERSON and to his Heirs forever all my land on Fine Creek and on Manakin Creek, . . . I also give unto my son, Peter and to his Heirs forever, two Negro's Farding and Pompey I also give unto my Son PETER my Chest and Wearing Cloathes with the Cloth and Trimming that is in the Chest, my Cane, Six Silver Spoons, which I bought of Turpin, two Horses named Norman and Squirrell, my Trooping Arms and Gunn I had of Joseph Wilkinson, Two Feather Beds, Ruggs and Blankets, one Suit of Curtains and Vallains, a Diaper Table Cloth and six Napkins, two Iron Potts and Hooks, one large and one small, A brass Kettle I had of Thomas Edwards, the Couch standing in the Hall and the Two Tables standing there, six Leather Chairs, Half my Stock of Cattle, Sheep and Hogs, on condition my Son PETER live to be Twenty One years old. . . .[7]

PETER JEFFERSON: SURVEYOR GOOCHLAND COUNTY

When Goochland County surveyor, William Mayo, died in 1744, Peter Jefferson was named his successor. Mayo had served the county since its formation in 1728, took an interest in Jefferson when he settled in Goochland, and may have taught him the rudiments of surveying. The Goochland Court Order Book for the December Court of 1744 preserves details of Jefferson's appointment and his commission received from the President and Masters of the College of William and Mary. As one of his first orders of business, Jefferson was directed to run the new county line between Goochland and Albemarle. He was to meet with such persons appointed by the Albemarle County Court. There, Joshua Fry had been appointed Surveyor with Peter Jefferson as his assistant. Anne Mayo, William's widow, was ordered to deliver unto Peter Jefferson, Gent. the books of plats and the original entry books and all other papers belonging to the office of Surveyor. Apparently, Anne was reluctant to surrender the papers. At Goochland County's February Court of 1744/45 (OS), Jefferson claimed that "Anne Mayo Executrix of the last Will and testament of William Mayo deced hath refused to Deliver the Books belonging to the Surveyors Office Pursuant to an order made at December Court last," and requested an order demanding that Mrs. Mayo comply.

[Silvio A. Bedini, "William Mayo (1684-1744) Surveyor of the Virginia Piedmont," *Professional Surveyor Magazine*, Parts I & II (February 2000, March 2000)]

Peter, having reached his majority prior to his father's death, accepted his father's appointment as Executor. Initially, he and his brother, Field, continued to operate the family's ferry across the James River at Osbornes (a.k.a. "Jefferson's Landing").[8] This familiarity with, and perhaps affection for, life on the river would later affect at least two of Peter's real estate investments.

Ultimately, neither Jefferson brother elected to stay at Osbornes. Field removed southward to Lunenburg County, Virginia, where he served as a Vestryman in Cumberland Parish and as a Justice of the Peace.[9] Peter returned to Fine Creek and Goochland, which became his training ground as a public servant. There, his associates and neighbors around Fine Creek, Beaverdam, and Dover Church included local heavyweights such as the Randolphs, Flemings, and Woodsons.[10]

Numerous Goochland County records survive, documenting Peter Jefferson's contribution to the development and maintenance of this increasingly settled section of Virginia frontier. In 1734, he was appointed Surveyor of the Road, from the Mountains to Lickinghole Creek, and was recommended, along with his friend, William Randolph, and brother-in-law, Thomas Turpin, as a Gentleman Justice.[11] The Goochland Court Order Book for the October Court of 1734 reads:

> Justices recommended Upon consideration of the want of a sufficient number of Justices for the dispatching the business of the Court & the preservation of the peace administration of justice in the severall (sic) parts of the county the following gentlemen to witt Dudley Digges, Charles Lewis, William Randolph, John Netherland, George Carrington, Peter Jefferson, Thomas Dickins, & Thomas Turpin are recommended to the Honble. William Gooch Esqr. his Majesty's Lieut. Governour as person proper to be added to the commission of the peace.[12]

Soon, Peter found himself serving as the foreman on a Grand Jury.[13] By 1737, he was Sheriff of Goochland County. The security bond for his appointment as Sheriff was provided by Isham Randolph and William Randolph, Jr., indicating solid Randolph family support in establishing Peter's position of authority in the county. As Magistrate, Surveyor, and Sheriff, even if Jefferson was not a significant planter, his "earned" status made him an appropriate mate for Isham Randolph's daughter, Jane.[14]

Becoming a large-scale planter in the increasingly crowded Goochland was difficult in the 1730s, so Peter looked further west to the next frontier. In May of 1736, he struck a distinctive deal with his good friend, William Randolph, who deeded him 200 acres of land in exchange for "Henry Weatherborne's biggest bowl of Arrack punch."[15] The agreement concerned a tract near the foot of the Blue Ridge Mountains to which Randolph held the patent. This desirable land was located considerably west of Fine Creek, in the part of Goochland that would soon become Albemarle County. There, Jefferson could establish himself as an important planter, build a home, and a life of his own.

A basic dwelling house was constructed, and by the fall of 1739, Peter was ready to marry Jane Randolph. She was nineteen years old; he was a mature thirty-one. Their marriage bond was recorded in Goochland, and they were likely wed at her family home, Dungeness.[16]

If the newlyweds remained in Goochland, they did not tarry long. Jane was immediately pregnant, and their daughter, Jane, Jr., was born on June 27, 1740, not at Fine Creek or at Dungeness, but at the Jeffersons' rustic new home in the west.[17] Lying along the Three Notch'd Road on the north side of the Rivanna River; the Jeffersons called the farm Shadwell, after Jane's home parish in England.[18] Compared to Fine Creek, with its more established community, Shadwell was isolated and remote. Ellen (Randolph) Coolidge, Peter Jefferson's great-granddaughter, described her ancestors' homeland as "thinly peopled and densely wooded."[19] It was indeed a daunting place to deliver a first baby in an era when many women went home to mother for their confinement. Two more Jefferson children, Mary and Thomas, would be successfully delivered at Shadwell while it was still part of Goochland County.

Meanwhile, Virginia's swelling population pushed westward along the James, its westward extension known as the Fluvanna and the Rivanna rivers, and, in 1744, county lines in the Piedmont were redrawn. The vast county of Goochland was sub-divided and, as a result, Peter Jefferson became a founding father of a new entity called Albemarle County. Once cut from Goochland, a

Albemarle Courthouse was established on Scott land where Toiter Creek enters the James River. This rendering of the Edward Scott dwelling house at Scott's Ferry, drawn by architect Floyd Johnson, represents a typical 18th-century dwelling in central Virginia. The original house at Snowden may have resembled the Scott home. (Courtesy of Scottsville Museum)

new county government was necessarily established, and the experienced thirty-six-year-old was primed to step into Albemarle's top offices.

It is no surprise that most of the county's first officials had long-cemented relationships in Goochland's government. They included Joshua Fry, who was first to be sworn in as Chief Magistrate and County Lieutenant ("Commander of the Plantations" and head of the Militia). His friend and surveying partner, Peter Jefferson, was sworn in alongside him as Magistrate. They, in turn, heard the oaths of Allen Howard, Gent. and William Cabell, Gent., both former Goochland Magistrates, as well as Joseph Thompson and Thomas Bellew. The group comprised Albemarle's first Gentleman Justices. Fry also served as Chief Surveyor, with Jefferson as his Assistant Surveyor.[20]

These men then established Albemarle's new county seat. Located on the James River at "Scott's Ferry," the chosen spot was both central and convenient to transportation on the river.[21] The landing, soon to be home to a courthouse, jail, and various services, was situated on the land of Daniel Scott, which had been patented by his father, Edward Scott, a former Goochland Magistrate, Sheriff, and Burgess.[22] In 1745, the new county government went to work planning the building of the courthouse and improving transportation routes to the area. The territory was raw and demanding. John Hammond Moore describes some of the challenges:

> The essential fact of Albemarle life during these years is the creation of a prosperous, stable community wrested from virgin frontier. The county's early citizens were spared bloody confrontation with Indian tribes and never suffered the attacks from England's enemies that made day-to-day existence so hazardous in other colonies and in the western counties. . . . But the men and women who took up the arduous task of establishing new homes, farms, and plantations in the foothills of the Blue Ridge had to contend with the loneliness of isolation, separation from loved ones, and the rigors that wilderness living imposes upon all who dare to push forward into its depths.[23]

Apparently, Peter Jefferson was just the man to transform this virgin frontier into a planter's paradise, with Shadwell an early oasis in the wilderness.[24] At least, over the years, that is how he has been characterized. A blend of near super-human strength and elegant style, he embodied the paradox of the Virginia countryside. Cast as a classic frontier hero, legends clung to him like fog to the Blue Ridge. In the 1960s, Virginia Moore concisely repeated the oft-told details concerning the man who looms like a mythological giant of the Virginia wilderness behind his famous son, Thomas:

> Fry's fellow judge, surveying assistant and close friend Peter Jefferson was another man who wore tall boots. As a Goochland Magistrate and Sheriff, he had married Jane, daughter of Isham Randolph of Dungeness, and . . . took her to live at newly-built Shadwell on the Rivanna (River Anna), north branch of the James. Huge-framed Peter was noted for his strength. Once he carried a mule on his back; and standing between two hogsheads of tobacco could

do the impossible: head them up. Self-educated, he knew where the values lay. On his shelves, side by side with the Book of Common Prayer, stood Pope, Swift, Addison, Steele, Shakespeare. The February day he rode twenty-odd miles to the Scott house for the swearing-in, three little children were running around Shadwell; a pair of daughters and a two-year-old son named Thomas.[25]

Moore and many others, from James Parton writing in 1874, to Fawn Brodie publishing a century later, have reworked Thomas Jefferson's early biographer, Henry S. Randall, in regards to Peter Jefferson. In 1858, Randall preserved the already mythic Jefferson family narrative:

> Many well attested facts and anecdotes are yet extant of the life of the father of Thomas Jefferson. . . . They all show that he was no ordinary man. He owed none of his success to good fortune or ingratiating manners. He was a man of gigantic stature and strength – plain and averse to display – he was grave, taciturn, slow to make, and not over prompt to accept, advances. He was one of those calmly and almost sternly self-relying men, who lean on none – who desire help from none. And he certainly had both muscles and mind which could be trusted! He could simultaneously "head-up" (raise from their sides to an upright position) two hogsheads of tobacco weighing nearly a thousand pounds apiece! He once directed three able-bodied slaves to pull down a ruinous shed by means of a rope. After they had again and again made the effort, he bade them stand aside, seized the rope, and dragged down the structure in an instant.... [A]s a surveyor through the savage wilderness, after his assistants had given out from famine and fatigue, subsisting on the raw flesh of game, and even of his carrying mules, when other food failed, sleeping in a hollow tree amidst howling and screeching beasts of prey, and thus undauntedly pushing on until his task was accomplished.[26]

Incredibly, in 1901, Field family genealogist, Frederick Clifton Peirce, lifted Randall's prose about Peter virtually verbatim and attributed it to Peter's father,

Thomas Jefferson, II, making this minor, but so significant, adjustment in the opening: "Many well attested facts and anecdotes show that Capt. Thomas Jefferson was no ordinary man."[27]

By the time Henry Randall collected these stories, 100 years had already passed since Peter Jefferson's death, allowing plenty of time and space for embellishment. There is no way to know if either father or son were true Goliaths. In fact, stories like these were commonly applied to many pioneering men. As the sire of the monumental Thomas Jefferson, stories of Peter Jefferson's physical prowess grew to match the impressive intellectual prowess of his first-born son, until these anecdotes of a rural Hercules approached folkloric magnitude. It is remarkable that in the over 150 years since Randall published these virtual tall tales that they have not been critically evaluated. Repeated over and over as literal, they have been passed down unquestioned.

Even allowing for exaggeration, Peter Jefferson was probably a large man and a fit one, and he may have inherited a strong physic and a "gigantic" stature from his father. However, contrary to Jefferson family tradition, Peter definitely was *not* a plain man, adverse to display. He sported a silver-hilted sword, equipped with a silver-hilted cutlass, wearing silver spurs on his boots – all of which he owned at his death. A pair of brass-barreled pistols completed this image of a well-heeled frontiersman.

The plain facts are that Peter did ride fearlessly along mountain precipices and did call Native American Chiefs his friends. He was ready, willing, and able to tackle a host of wilderness challenges while on his surveying adventures – including bear, adverse weather, and very possibly near starvation. Over the coming years, these myriad stories were undoubtedly told to his wide-eyed children, who in turn told them to their equally wide-eyed children and grandchildren, taking on a life of their own as oral traditions tend to do.[28]

∼

Despite the fact that Peter Jefferson was present at the inception of Albemarle County, his intimate, day-to-day involvement with the brand new county was ultimately short-lived. In 1745, he uprooted his young family from Shadwell, moving their residence to Tuckahoe, the estate of the late William Randolph, where he would act as guardian for William's three orphaned children. This gesture fulfilled the last wish of his friend and his wife's first cousin, which read:

"Whereas I have appointed by my will that my dear only son Thomas Mann Randolph should have a private education given him in my house at Tuckahoe, my will is that my dear and loving friend Mr. Peter Jefferson to move down with his family to my Tuckahoe house and remain there till my son comes of age with whom my dear son and his sisters shall live."[29] William Randolph's generosity had resulted in the establishment of Shadwell. Now it was Peter Jefferson's turn to oblige his friend's family and respect William Randolph's dying request.

Located in Goochland County, down river from Albemarle and much nearer to Richmond, Tuckahoe offered an elegant abode compared to Jefferson's frontier farm. The plantation's genteel atmosphere was both familiar and comfortable to Jane Jefferson. Living there guaranteed that her oldest children would benefit from graceful, more established living, as opposed to the relatively primitive Albemarle. The Jeffersons remained at Tuckahoe for at least six years. There, among other duties, Peter served on Goochland's St. James Northam Parish Vestry, to which he was appointed on January 18, 1747, replacing Benjamin Cocke, Gentleman.[30]

The distance from Tuckahoe to Albemarle County, however, did not deter Peter's plans to settle permanently there. He continued his involvement in the

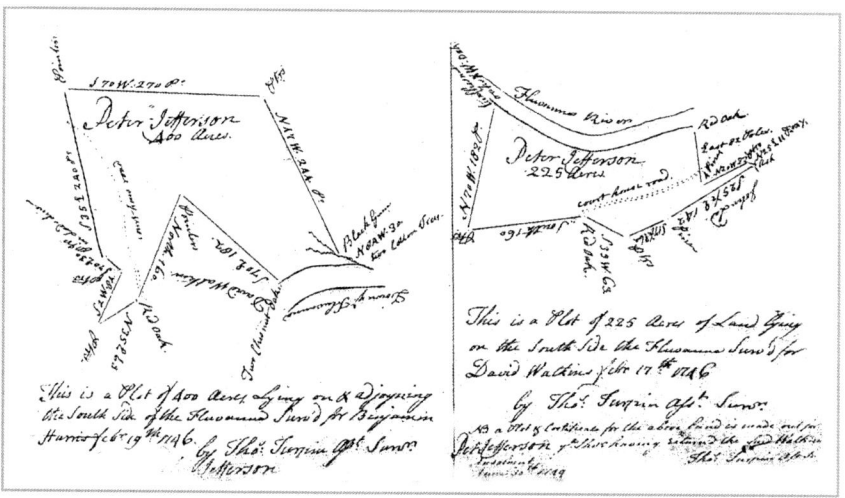

Peter Jefferson's holdings were surveyed by his friend and brother-in-law, Thomas Turpin. In 1757, Jefferson chose Turpin as one of the executors for his estate. (ALBEMARLE COUNTY PLAT BOOK)

county government, his surveying adventures with Joshua Fry, and his plan to create a plantation along the Fluvanna River directly opposite the new county seat. With Shadwell as his base camp, Jefferson set about acquiring land along the south side of the Fluvanna River. As Virginia Moore observed, Edward Scott wisely chose a mix of acreage, his land providing him a double guarantee: "rich lowlands for crops, and sites elevated enough to escape the river's periodic rampage."[31] Peter Jefferson would make similar choices when acquiring his land by purchase and patent, mirroring Scott on the south side of the river.

Exactly how and when Jefferson conceived of his plan to develop his Fluvanna plantation across from Albemarle Courthouse is not known. It may have been soon after the county seat was established in 1744. By early 1746, he was acquiring land south of the river, beginning with two parcels, totaling 625 acres, which had been previously patented by others.[32]

Ultimately, Jefferson was interested in the spot directly across from Scott's Ferry, where Daniel Scott established a ferry on the north side of the river to serve the new Albemarle Courthouse. The ferry bridged the primitive roads then being constructed, designed to connect vast stretches of Albemarle to the north and to the south.

In colonial days, a ferry wasn't merely a business; it was "public business" and the business of the Crown of England. In February of 1745, under King George II, a statute was recorded into the Laws of Virginia which included Daniel Scott's permission to operate a ferry. A man or a horse, they paid three pence apiece to cross the Fluvanna River.[33] Additionally, on July 25, 1745, Albemarle County granted Daniel Scott an ordinary license.[34] At the very least, his tavern and ferry would be bustling on court days.

The precise location of the ferry landing at Albemarle Courthouse is not currently known. The north side landing was described as near the mouth of Totier Creek. The south side landing was on Noble Ladd's property at the tip of the Horseshoe Bend, adjacent his brother, John Ladd.[35] If Peter Jefferson wanted the ferry landing on the south side, he was going to have to buy it from the Ladd family, who had owned this property since the 1720s.[36]

In March 1740/41, Amos "Lad" of Goochland divided 400 acres in the Horseshoe Bend between his son, Noble, and Col. John Bolling. Noble paid his father £10 for 100 acres.[37] By 1745, Amos Ladd was dead, and Noble Ladd

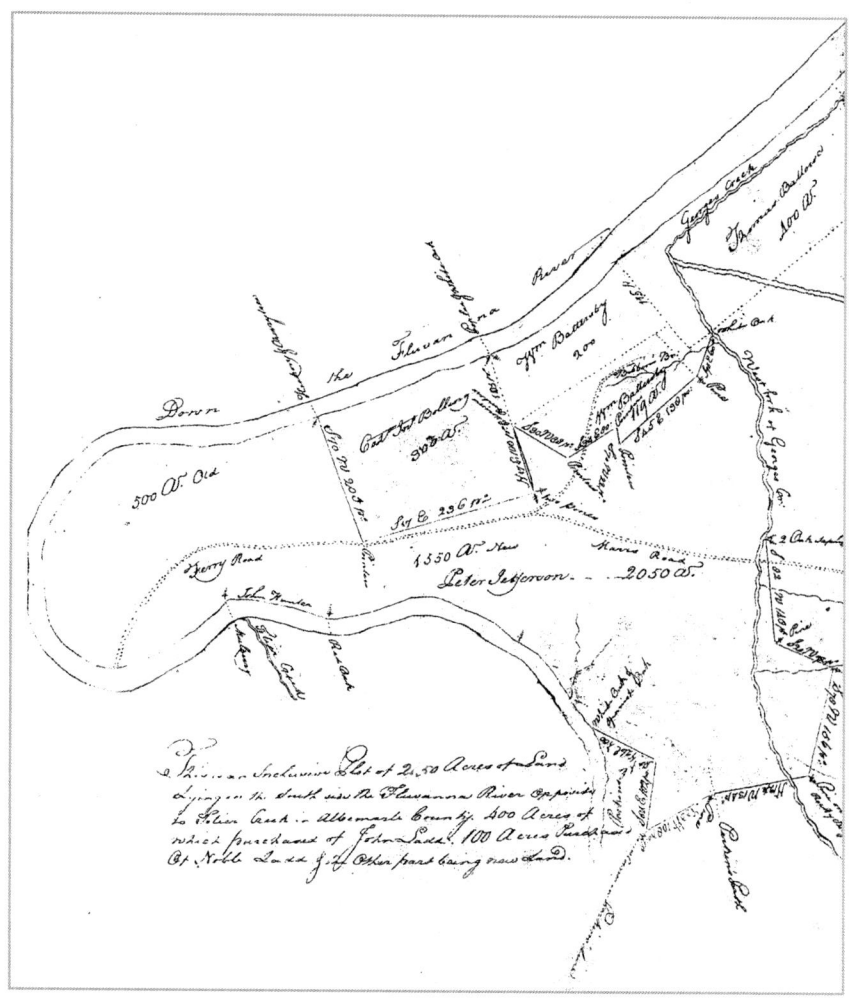

This undated survey (c.1754) shows the fully assembled Snowdon plantation, consisting of 2,050 acres. Peter Jefferson's "Fluvanna Lands" included 1,550 acres of "new" land and 500 acres of "old" land he purchased from John and Noble Ladd. Jefferson assembled his plantation of high and low grounds, well-timbered and immensely fertile, along the western extension of the James River. The Ferry Road ran through the center of the tract. His neighbors, Col. John Bolling, William Battersby, Thomas Ballow, and Hardin Perkins were all in place. (Albemarle County Plat Book)

was named his Executor.[38] Then, in 1750, Noble sold the land to Jefferson for £105, a tidy profit over the £10 he paid his father for the property. Of course, in the intervening ten years, the tract's desirability as a ferry landing had vastly increased its value and Noble Ladd had himself improved the property. The deed between Jefferson and Ladd describes it as the plantation tract whereon Noble Ladd "lately dwelt," indicating that Ladd had also improved the property with a dwelling.[39] Soon after, his brother, John Ladd, sold Peter an additional, adjacent 400 acres.[40] Peter had reached his goal; he now owned the tip of the Horseshoe Bend, including the ferry landing.[41]

Mr. Jefferson's Landing

At some unknown point in time, Col. Peter Jefferson named his Fluvanna River plantation Snowdon. Family tradition holds that it remembered his family seat in Wales.[42] Once the land was cleared for planting, the view from the bluff was stunning. Situated at the river's Horseshoe Bend, the water wrapped Jefferson's land in a ribbon of blue. The man who owned this strategic point 400 feet above the county seat commanded the crowning point at the heart of a vast and fertile Albemarle County.

The Ladds' 500 acres added extremely fertile lowlands to the plantation, and with a road running from the river south through the property, Snowdon was poised to serve the developing southern section of Albemarle. This significant addition to Peter Jefferson's farm established him and all the successive proprietors of Snowdon as the owners of the south side landing of Scott's Ferry, cementing a relationship with the north bank and, later, with Scottsville, which would continue until the ferry closed in 1907 in deference to the new woodenplanked bridge.[43] Additionally, Noble Ladd's land came with the ordinary, soon to be known as the "Snowdon Ferry House."[44] In short, Peter had successfully recreated the "Jefferson's Landing" of his childhood, potentially a very lucrative commercial site.

When John Hammond Moore described taverns, like the one at Snowdon, he quipped that while Peter Jefferson made an income from his property, he likely never had the misfortune to stay the night there.

Albemarle's colonial ordinaries, although ostensibly controlled by the county court, appear to have been rather dismal, often filthy affairs far removed from romantic pictures we have today of cozy rural inns complete with jovial host and voluptuous serving maids. If the justices themselves occasionally had been forced to seek meals or lodging in them, perhaps conditions would have improved in short order. Constant complaints indicate these public facilities left much to be desired; whenever possible, travelers preferred to spend the night in private homes. The usual meal consisted of eggs, bacon, and hoecake washed down with either peach brandy or whiskey, both of local origin. Beds often were crawling with bugs and vermin. Both Peter Jefferson and William Cabell owned taverns, although these structures were not within present-day Albemarle nor were they operated by these gentlemen.[45]

The south side landing of Scott's Ferry was at Snowdon from 1745 until the ferry closed in 1907. Folk artist, Lewis Miller, depicted the ferry crossing the James River from Buchanan to Pattonsburg. (LEWIS MILLER, "SKETCHBOOK OF LANDSCAPES IN THE STATE OF VIRGINIA, 1853-1867," ABBY ALDRICH ROCKEFELLER FOLK ART MUSEUM, THE COLONIAL WILLIAMSBURG FOUNDATION, WILLIAMSBURG, VA., GIFT OF DR. AND MRS. RICHARD M. KAIN IN MEMORY OF GEORGE HAY KAIN)

Ten years earlier, both Cabell and Jefferson had helped determine regulations for ordinaries in the county. When the second court for the newly-formed Albemarle met in March of 1745, Peter Jefferson, Magistrate, was present at Edward Scott's house when the judges discussed the plans for the new courthouse and the necessary building of a nearby ordinary – to lodge, feed, and provide liquor for travelers coming to the county seat. The men set the prices in detail: "For a Diet, twelve pence. For a Servant Diet, six pence. For one night's Lodging, seven pence, half penny. Indian corn by the Gill, four pence."[46] According to Virginia Moore, "If the food was monotonous, the choice of liquids on the price list had more range: West Indian rum, French brandy, peach brandy, Madeira wine, Virginia cask or bottled beer, and 'good Virginia cyder.'"[47]

In 1750, it would soon be more than a ferry that connected Peter Jefferson of Snowdon and Daniel Scott of Scott's Ferry. The eldest son of Edward Scott (1695-1738), Daniel Scott (1725-1754) had inherited his father's lands on the north side of the Fluvanna River and successfully established and operated his ferry and tavern for several years, initially dealing with Noble Ladd and William Battersby on the south side. There is every reason to believe that when Peter Jefferson took over the south-side landing, the two men were already collaborators, rather than competitors; for on November 28, 1751, Daniel Scott married Anne Randolph, Peter Jefferson's sister-in-law.[48]

At Isham Randolph's death in 1742, Peter Jefferson had been named as one of five guardians for Randolph's minor children, including Anne (a.k.a. Anna), who was about eleven years old at the time.[49] During the years the Jeffersons lived at Tuckahoe (1746-1751), they were physically close to the Isham Randolph home at Dungeness, where his widow, Jane (Rogers) Randolph, and youngest Randolph children remained; the proximity would have facilitated frequent family visits.[50] During these years, Peter and Jane Jefferson grew especially close to young Anne. It may even have been through the Jeffersons that she met Daniel Scott. Not yet twenty-one years old in the fall of 1751, Anne asked Peter's consent to marry, which he gave as her guardian, signing the couple's marriage bond as Anne's nearest male relative.[51]

After their marriage, the Scotts settled on the river at Albemarle Courthouse. About the same time, the Jeffersons returned to Shadwell. It is possible the men had long envisioned a co-operative ferry business, with one branch of the family owning and operating the ferry on the north bank of the river and the other branch providing services on the south bank, imitating the Branch and Jefferson family arrangement at Osbornes. For the next couple of years, they had ample opportunity to continue any plans to develop the area around Scott's Ferry.

Then, late in 1753 or early in 1754, Daniel Scott died, leaving Anne a childless widow. Only about twenty-two years old, Anne returned east to Goochland County.[52] Her mother was still alive to see Anne settled once again, marrying Jonathan Pleasants on June 14, 1759. Jane Randolph may have favored Anne, leaving her a gold watch. Following Jonathan Pleasants' death in 1765, Anne married his cousin, James Pleasants, and they were the parents of James Pleasants, Jr. who became a United States Senator and Governor of Virginia.[53]

Despite Anne's return to Dungeness, the affection between her and the Jefferson family was abiding and was formally acknowledged in October of 1755, when not one but two Jefferson babies honored Aunt Anne. When Jane and Peter's baby girl was accompanied by a twin boy, the Jeffersons conceived of the clever and charming gesture of splitting Anne's name between the twins, calling them Anna Scott and Randolph Jefferson.

Peter Jefferson knew the value of river commerce; however, he did not see himself personally operating the ferry house and decided to raise his family further north, away from the immediate vicinity of Scott's Ferry. As a result, Snowdon's riverfront property was leased to a man named Richard Murray, who paid Jefferson £4 per year for four acres, a house, and an ordinary.[54] One day, Peter imagined, his adult son would inherit this lucrative property. From the vantage point of 1750, that fortunate son was Thomas. How different the world might have been, if Peter Jefferson had lived long enough to see his elder son turn twenty-one and establish him at Snowdon.

Life at Shadwell

Having assembled the Fluvanna plantation, Peter Jefferson provided himself with a steady rental income which could be used to support the ongoing improvement of his farm at Shadwell or fund other new land investments. In 1750, his family still resided at Tuckahoe, and Peter Jefferson embarked on the "remodeling" of the dwelling at Shadwell in anticipation of their return to Albemarle when Thomas Mann Randolph reached his majority.[55]

By 1753, when the Jeffersons eventually returned to Shadwell, Peter Jefferson was in his early forties and the father of a still expanding family. Six children now needed to be accommodated in Shadwell's dwelling house and two more would be born there within two years. Jefferson wasted no time improving the farm and expanding his home. Structures had no doubt deteriorated somewhat during their absence at Tuckahoe. His accounts include multiple entries for building materials, acquired from carpenter John Biswell, who was long associated with Jefferson and the grist mill at Shadwell. From "posts for porches" to "hewing sills for Dwelling house," improvements were made on the mansion house; the stable was moved; roofs were repaired on three small houses; the hen house was repaired. He also placed a significant order with joiner Francis West, ordering new furnishings for the dwelling's interior.[56]

The farm had a typical complement of buildings and implements for agriculture, livestock, and modest "industry." Barns stored tobacco and grain; stables housed the horses. Vegetable gardens fed the family, and flower gardens provided beauty. Orchards gave forth fruit. Unimproved woodlands provided timber and wild game. There was a grist mill and a brewing operation. Livestock included cattle, hogs, and sheep. In the 1990s, archeology at Shadwell revealed that fences and gates marked off the main house. The separate kitchen sat in the yard, and slave quarters ran east to west on the ridge top.[57]

Historians and archeologists have long speculated about Shadwell's exact floor plan and have not reached a consensus. In 2010, Susan Kern concluded, "The total length of the house cannot be determined from what remains archaeologically."[58] However, based on the general limits of the foundation, typical floor plans of the day, and the historical evidence concerning Shadwell's contents, particularly as presented in the 1757 inventory of Peter Jefferson's estate, it is possible to imagine the shape of life at Shadwell. It is life at Shadwell, not at Tuckahoe, which would shape Peter Jefferson's younger son, Randolph.[59]

Jefferson family history describes the house as a story-and-a-half-dwelling, which was *de rigueur* for 18th-century planters in central Virginia. Like other homes in the region, when the family expanded, the core house was likewise expanded, frequently in more than one direction. Randolph Jefferson was born in and grew up in an elaborated version of the original dwelling. Early biographer of Thomas Jefferson, Henry S. Randall, describes it as follows:

PETER JEFFERSON'S LIBRARY

Peter Jefferson's fairly extensive library of books and periodicals also provided a basis for the education for his children. Subjects represented included history, literature, religion, and natural philosophy, as well as law. Peter's tastes were broad and included the poets Jonathan Swift (1667-1745), Joseph Addison (1672-1719), Alexander Pope (1688-1744), and William Shakespeare (1564-1616); the latter being his favorite. From Swift's satire to Shakespeare's sonnets, did Peter Jefferson commit verse to memory and recite it in the Jefferson household, perhaps still delivered with traces of lyrical Welsh tones? The collected English periodicals included Joseph Addison and Richard Steele's daily *Spectator* and the *Tatler*, to which both Swift and Addison contributed. The stated aim of the London daily *Spectator* was "to enliven morality with wit, and to temper wit with morality . . . to bring philosophy out of the closets and libraries, schools and colleges, to dwell in clubs and assemblies, at tea-tables and coffeehouses." In the mid-18th century, this aspiration to sophistication extended to Jane and Peter Jefferson at the frontier edge of Virginia. Assuming that the Jefferson library was intact until the house burned in 1770, Randolph Jefferson had access to his father's collection until he was fourteen years old and may have been attracted to the practical subjects most applicable to the planter's life that lay ahead of him, drawn to titles such as *The London and Country Brewer* (1744) and Stephen Switzer's *Practical Husbandman and Planter*, which included horticultural illustrations demonstrating such agrarian arts as "how to graft a tree." His father's map collection and surveying tools, which likewise represented applied sciences, may have intrigued him.

It was of a story and a half in height; had the four spacious ground rooms and hall, with garret chambers above, common in those structures . . . and also had the usual huge outside chimneys, planted against each gable like Gothic buttresses, but massive enough, had such been their use, to support the walls of a cathedral, instead of those of a low, wooden cottage. In that house was born Thomas Jefferson. . . .[60]

If this vision is correct, based on family tradition passed down to the 1850s, it likely represents the expanded house, Shadwell as it existed at Peter Jefferson's death. His estate inventory, which lists the contents of the house in 1757, contains plenty of furnishings to fill a dwelling this size. The inventory not only lists the contents of the house, but also specifically ascribes Peter Jefferson's office furnishings to an unheated room. There he worked at a desk made of cherry wood, surrounded by his maps, surveying tools, and a library of about fifty books, an extensive collection compared to others in the county.[61]

In 2005, Susan Kern envisioned the first house with "two or three rooms on the ground floor and the main entrance may have been directly into one of these rooms instead of into a passage that buffered inner living spaces from the outdoors. The porch added in the 1750s united the old and new parts of the house and framed the entrance that now opened to the passage flanked by imposing rooms."[62] In 2010, she pictured the expanded house with four rooms with a significant middle passageway.[63] Hallways, like this one, provided more than access to main rooms, they flexibly functioned as another room. Open doors at both ends created a breezeway, a perfect place for a game of cards, to linger with an after-dinner drink, or enjoy a cooler place to sleep.

While the precise layout of the Shadwell dwelling house is ultimately left to the imagination, Peter Jefferson's household inventory reveals that life on the plantation emulated Virginia Tidewater gentry and English manners and that this was the standard of living provided for the Jefferson children at Shadwell in the 1750s. Growing up with fine appointments no doubt influenced Randolph Jefferson's eventual aspirations for Snowdon, which was a virtual *tabula rasa* when he came into his inheritance.

The genteel Jeffersons subscribed to the famous Virginia hospitality, and were well-equipped to entertain and clearly valued this aspect of gracious living.

Jane Jefferson owned three tea tables and a tea service, a mix of cups and saucers from China and "white stone tea ware." Jane was equally prepared for coffee drinkers, with a silver coffee service and a set of coffee cups. This was Jane's most valuable possession. Her "one silver coffee pot teapot & milk pot" was valued at £17.10.[64]

Table linens, silver, wine glasses, and a silver-plated dinner service graced their board. They owned chairs to accommodate as many as twenty persons for dinner, seated at a collection of tables. In 1750, Peter commissioned several small oval tables from Frances West, marking his plans for domestic expansion. The Jeffersons' spiked punch was served with a silver ladle. A large looking glass in a cherry frame, with candlesticks, provided beautifully reflected illumination.[65]

Guests were no doubt plentiful at Shadwell, both relations and friends. The plantation's proximity to the Three Notch'd Road, the main east-west thoroughfare from Richmond to the Valley, and Old Fredericksburg Road, running north and south, guaranteed traffic and visitors. At Snowdon, the Ferry Road promised similar traffic and a steady opportunity to entertain. Shadwell visitors included the peripatetic Anglican minister Rev. Robert Rose (1704-1751), who, like Jefferson, was a land speculator and surveyor, as well as the illustrious Joshua Fry (c.1700-1754).[66]

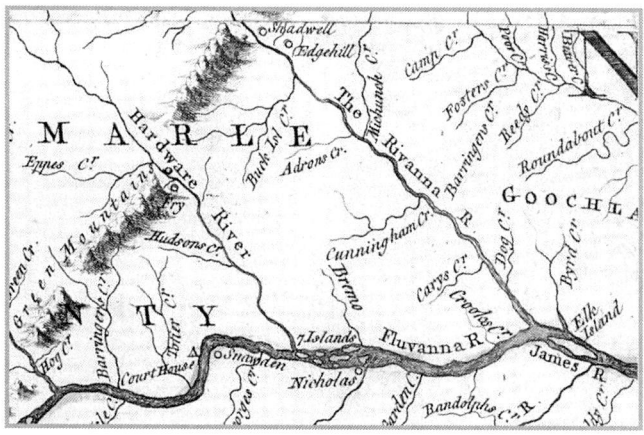

The Fry-Jefferson map (1751) locates both of Peter Jefferson's plantations, Shadwell and Snowden. (COURTESY OF THE LIBRARY OF VIRGINIA)

Joshua Fry was a surveyor, pioneer, and professor of mathematics and natural philosophy at the College of William and Mary. His partnership with Peter Jefferson on the "Fry-Jefferson Map" is their best known association; however, they were connected over many years through surveying, land speculation, and county government. Fry's home lay south of Shadwell, about halfway to Albemarle Courthouse at Scott's Ferry, its location encouraging visits by Jefferson as he rode down on county business.[67]

On October 1, 1755, the Jeffersons' youngest children were born in this expanded, bustling mansion house; they were the twins, Anna Scott and Randolph. Thomas, twelve years their senior, had been born in Shadwell's ruder, smaller house, constructed in the 1730s. Later, Thomas would envision a lifestyle far grander than he'd known as a boy, even at Tuckahoe. Randolph, on the other hand, was content to recreate the "frontier elegance" that had been his primary experience.

During the mid-1750s, Peter Jefferson's world steadily expanded professionally as well as personally. As County-Lieutenant, Commander-in-Chief of the Albemarle Militia, Peter had earned the title of Colonel. In 1754, he replaced Joshua Fry as the surveyor of Albemarle and, in 1756, he became a Burgess to the General Assembly, representing the county in Williamsburg.[68]

Then, in the summer of 1757, just as Peter Jefferson was busy reinforcing and expanding his world, something happened to compromise the health of this apparently exceptionally vital man. Beginning the last week in June his neighbor and friend, Dr. Thomas Walker, made frequent visits to Shadwell. Walker called on the household fourteen times that summer.[69] On July 13, 1757, it became clear to Jefferson that he might not survive, and he wrote his lengthy and very precise last will. A little over a month later, on August 17th, Peter Jefferson was dead. He was forty-nine years old. Like his father before him, he was stopped in the prime of life.

Though he assigned personal slaves to all of his children, Jefferson's last will stated that his land was to be equally divided between his two sons, Thomas and Randolph. He bequeathed a few individual slaves to his children. These included Myrtilla's son Peter, whom he gave to Randolph and his heirs forever. Additionally, Jefferson gave "all my Slaves not herein otherwise disposed of to be equally divided between my two sons Thomas and Randolph at such time as my son Thomas shall attain the age of twenty one years." Thomas was given his choice of the Rivanna lands, which included Shadwell and what would become Monticello, and the Fluvanna lands which included Snowdon. Shadwell, the Jefferson homeplace, would provide for Jane during her lifetime.[70]

When Thomas chose the Rivanna lands as his own, he necessarily had to wait until his mother's death to come into full possession of the property. He leased the farm from her until her death, which came in 1776. Thomas' choice of the Jefferson homeplace left the less developed and more remote Fluvanna lands to his much younger brother, Randolph. A distant working farm, Snowdon would have to wait many years for Randolph to grow up and become its squire.

While Snowdon awaited Randolph Jefferson's maturity, a change took place that would deeply effect the potential of the plantation's location. In 1761, the land in Albemarle County which lay south of the Fluvanna River was cut out to form the new Buckingham County. As a result, Albemarle's county seat was moved to Charlottesville, a more central location given Albemarle's new boundaries. Folks began to call Scott's Ferry "Fool's Corner." Once at the heart of an enormous county, it was now relegated to the southeast corner, and Buckingham's county seat was located a distant twenty-three miles south of the Snowdon tract.[71]

The farm's proximity to the river and Scott's Ferry was still desirable; however, the fact that the Scott land was no longer home to Albemarle's county seat and at center of a much larger Albemarle necessarily impacted the growth of what would eventually become Scottsville. Snowdon, which had once been a very convenient ferry ride from the county seat with its bustling activity, was now remotely located in relation to both Albemarle's and Buckingham's courthouses.

Chapter Two

Jane Jefferson's Shadwell

*The happiest moments of my life have been the few which
I have past at home in the bosom of my family.*
Thomas Jefferson to Francis Willis, Jr., 1790

Following Peter Jefferson's death in 1757, Master Thomas Jefferson was, for the most part, away at school and Jane ran a household filled with daughters – Jane, Jr., Mary, Elizabeth, Martha, Lucy, and little Anna Scott. The sole male exception was Jane's youngest, toddler Randolph.[1]

There is no known portrait or other image of Randolph Jefferson. One wonders if he resembled his older brother, who had their father's height. Was his hair reddish or did he take after his mother's Randolph family, which tended towards dark hair and dark eyes? His father was remembered as a "grave and taciturn" giant, while Ellen (Randolph) Coolidge described Randolph Jefferson's mother as "mild and peaceful by nature, a person of sweet temper and gentle manners."[2] Where in this spectrum did Randolph's nature fall?

Most of the Jefferson children had nicknames. Thomas was known as Tom. Mary was "Polly," Martha was "Patsy," and Anna Scott was affectionately called "Nancy." Was Randolph ever called Randy or Ranny? Thomas strictly referred to him as "my brother," RJ, and Randolph. His niece called him "Uncle Randolph." Likely pronounced "Randuff," his name would not have been fully enunciated, which was common at that time. Interestingly, when Thomas' slave, Isaac Jefferson, dictated his memoirs, Master Randolph was transcribed as "Mass Randall."

Randolph barely knew his father, leaving his mother, Jane (Randolph) Jefferson, as his sole parental influence. Details of her life and of the life at Shadwell have remained obscure for several reasons. Thomas Jefferson wrote virtually nothing concerning his mother or his childhood home. As a result, some of his biographers have concluded that his relationship with her was, at best, conflicted. Also, it has been assumed that Thomas destroyed correspondence with his mother, though there may never have been voluminous correspondence between them. This was compounded by a devastating fire at Shadwell in 1770, which destroyed both Peter and Jane Jefferson's collected papers. The family data she left behind in "Jane Jefferson – Her Booke," which is dated Sept. 6th (1772), was written after the fire, perhaps reconstructing what had been lost.[3]

What little we do know about Jane's life indicates it was both typical and atypical of an affluent woman raising a family in Albemarle County in the mid-1700s. She was born Jane Randolph, on February 9, 1721 (OS), in London, England. Her father, Isham Randolph, who had been educated at the College of William and Mary in Virginia, was a mariner. He commanded his first vessel, a merchant ship called the *Henrietta*, at the age of twenty-five. Following the death of his first wife, Isham Randolph lived in England for many years. There he married Jane Rogers, and their daughter, Jane, was the second of their ten children. Upon their relocation to Isham Randolph's native Virginia, their daughter, Mary, was born in Williamsburg in October of 1725. It is unknown if Jane resembled her father or her mother or, like her roots, combined a bit of both, distinctively entwining England and Virginia. At least one of her brothers, William Randolph, chose to live in England, indicating he was at home there.[4]

Once back in Virginia, Isham Randolph settled down as a planter, serving Goochland County as Magistrate, rising to the rank of Colonel, and eventually acting as Adjutant General of the Colony. His epitaph summarizes his contribution to early Virginia:

> Sacred to the Memory of Col. Isham Randolph of Dungeness in Goochland County Adjutant General of this Colony
> He was the third son of William Randolph and Mary his wife. The distinguishing qualities of the Gentleman he possessed in an eminent degree: To justice probity & honour so firmly attached That no view

of secular interest or Worldly advantage, no discouraging frowns of fortune could alter his steady purpose of heart. By an easy compliance and obliging deportment he knew no enemy, but gained Many friends, thus in his life meriting an universal esteem. He died as universally lamented Nov'r, 1742 age 57

Gentle Reader go & do likewise.[5]

Isham Randolph (1685-1742) sat for this portrait in London (c. 1724), before returning to Virginia to live at his home, Dungeness. (ARTIST UNKNOWN, COURTESY OF VIRGINIA HISTORICAL SOCIETY)

Jane's mother, Jane (Rogers) Randolph, may not have been held in such universal affection. According to Ellen Coolidge, Jane was "a stern strict lady of the old school, and feared and little loved by her children."⁶ Her daughter, Jane, matured to be a very different sort of woman, and while there are no descriptions of her during her lifetime, Thomas Jefferson's biographer, Henry S. Randall, neatly summed up family tradition concerning her softer personality:

> She was an agreeable, intelligent woman, as well educated as the other Virginia ladies of the day, of her own elevated rank in society – but that by no means implying any very profound acquirements – and like most of the daughters of the Ancient Dominion, of every rank, in the olden time, she was a notable housekeeper. She possessed a most amiable and affectionate disposition, a lively, cheerful temper, and a great fund of humor. She was fond of writing, particularly letters, and wrote readily and well.⁷

Once in Virginia, Jane's home was at Dungeness in Goochland County. The plantation included "extensive gardens enclosed with brick walls, specialized plantation buildings including a coach house, mill house, well house, and hen house, horses and a chariot for riding." Surviving portraits of the Randolph family, including one of her father, reflect the family's status and wealth.⁸

Jane Randolph married Peter Jefferson in the fall of 1739. They may have lived briefly at Peter's home at Fine Creek in Goochland (later Powhatan County); however, historians disagree about this. The nineteen-year-old bride immediately became pregnant. Their marriage bond is dated October 3, 1739 and Jane, Jr. was born at Shadwell on June 27, 1740. That is eight months and twenty-four days from marriage to delivery.

Peter likely took his bride to a rather basic dwelling on his Rivanna lands, in northern Albemarle County. The couple named the plantation Shadwell, after Jane's birthplace, Shadwell Parish, Tower Hamlets, London, announcing Jane's genteel background and the decidedly English influence she would bring to the Virginia frontier.

Jane's married life was divided into three parts: the early years at Shadwell, the years spent at Tuckahoe, and the return to Shadwell. Following the birth of

The dwelling house at Tuckahoe, seen from the northwest. (COURTESY OF VIRGINIA DEPARTMENT OF HISTORIC RESOURCES)

Jane's fourth child in Albemarle, in November 1744, the Jeffersons made their home at Tuckahoe, William Randolph's estate, which provided an environment that no doubt resonated strongly with Jane's own heritage and cultivation. During the years the Jeffersons spent in residence there, especially the oldest children were influenced by Tuckahoe's many refinements. Once back at Shadwell, Jane gave birth to her last two children, Randolph and his twin sister, Anna Scott "Nancy" Jefferson, born on October 1, 1755. Along with their sister Lucy, Shadwell would provide a singular influence for the youngest Jefferson children.

The accounts for Peter Jefferson's estate give us a small, but informative, window into domestic life at Shadwell during the years between Peter's death in 1757 and Jane's death in 1776. Through the details of recorded expenditures – from satin slippers for young Jane to stockings for Elizabeth to a gown for Anna Scott – the quality of life at Shadwell is revealed. The house and its environs may have lacked the grandeur of Randolph-owned Tidewater homes, still the Jefferson children enjoyed the sophisticated appointments and finery that Jane had known as a girl.[9]

When Peter Jefferson died in the summer of 1757, the woman he called his "Dear and Well Beloved Wife," was thirty-six years old and left with eight minor children in her care.[10] The birth and death of two sons at Tuckahoe, Peter Field (16 October 1748 - 29 November 1748) and an unnamed boy (born and died on 9 March 1750), created a gap of nearly six and one half years between the first closely grouped set of children and the second set, which included Lucy and the twins. As a result, the youngest three Jefferson children established a bond unrelated to their older siblings.

In the summer of 1757, Thomas was already fourteen and one-half years old; virtually a generation separated the only Jefferson boys. Tom was preparing to go into the world, growing up very much apart from his only surviving brother. Conversely, young Jane and Mary were old enough to "mother" the not yet two-year-old twins, if they were so inclined. However, Randolph and Anna Scott likely remembered little of Mary's years at Shadwell. In January of 1760, she married John Bolling and left her home; the twins were only four years old.

As for sister Elizabeth, too little is known about the exact nature of her intellectual deficiencies to imagine her impact on her siblings, but she was an ongoing burden for her mother and a constant concern for the household. Depending on Elizabeth's temperament, she may have been a perpetual child, a welcome playmate, who never grew up. The young ones, of course, did grow up and eventually left home, while Elizabeth did not develop independence. She was, however, well-dressed, becoming a young lady of her station, and cared for by her family and attended by her personal servant, Cate.[11] The fact that Elizabeth was provided with an acceptable social wardrobe suggests that she was capable of being taken on outings and receiving guests in the Jefferson

THE CHILDREN
OF PETER AND JANE JEFFERSON

Jane Jefferson, Jr.: 27 June 1740 - 1 October 1765, b. *Shadwell*

Mary "Polly" Jefferson: 1 October 1741 - d. bef. 21 January 1824, b. *Shadwell*, m. John Bolling on 24 January 1760

Thomas Jefferson: 13 April 1743 - 4 July 1826, b. *Shadwell*, m. Martha (Wayles) Skelton on 1 January 1772

Elizabeth Jefferson: 4 November 1744 - abt. 24 February 1774, b. *Shadwell*

Martha "Patsy" Jefferson: 29 May 1746 - 3 September 1811, b. *Tuckahoe*, m. Dabney Carr on 20 July 1765

Lucy Jefferson: 10 October 1752 - 26 May 1810, b. *Tuckahoe*, m. Charles L. Lewis in 1769

Randolph Jefferson: 1 October 1755 - 7 August 1815 b. *Shadwell*, m. Anne Lewis abt. 30 July 1781

Anna Scott "Nancy" Jefferson: 1 October 1755 - 6 July 1828, b. *Shadwell*, m. Hastings Marks in October of 1787 home. If she had been confined to her room, unable to interact socially, there would have been little need for her to be well-dressed.

Martha, the youngest of the first set of Jefferson children, was nine years old when the twins were born. She did not marry until the summer of 1765; as a result, Randolph and Anna Scott spent over nine years with this sister. Her marriage to Thomas' best friend, Dabney Carr, guaranteed her a close association with the Thomas Jeffersons and Monticello. Martha's happiness with Dabney Carr was comparatively brief. A widow by 1773, she lived for long periods of time at Monticello, where her six children were influenced by and in turn influenced the culture on the mountain top.[12]

Of the first cluster of Jefferson children, Jane, the eldest, potentially had the largest consistent impact on her three youngest siblings. Jane never married and remained at Shadwell until her death in 1765. She was Thomas' favorite and, years later, Sarah Randolph wrote of Jane's death in the twenty-eighth year of her life:

> . . . The eldest of her family, and a woman who, from the noble qualities of her head and heart, had ever commanded their love and admiration, her death was a great blow to them all, but was felt by none so keenly as by Jefferson himself. The loss of such a sister to such a brother was irreparable; his grief for her was deep and constant. . . .[13]

Biographer Henry Randall noted that Thomas Jefferson "ever regarded her as fully his own equal in understanding, and there was a depth, earnestness, purity and simplicity in her high nature which made an impression on his mind which was never effaced."[14] Jane's apparent strong intellect, as well as her passion for nature and music, may have been absorbing and she may have had little interest in the smaller Jefferson children and babies. The fact that she did not marry at an early age, leave home, and begin having children of her own, may indicate that she was not particularly domestically inclined.

Clearly, Jane, Jr. enlivened the household and added complexity to life at Shadwell. Her sweet voice accompanied Thomas' fiddle. Randall wrote, "many a soft summer twilight, on the wooded banks of the Rivanna, [one] heard their voices, accompanied by the notes of his violin, thus ascending together."[15]

Jane Jefferson and her daughters were accomplished in domestic arts, including spinning. As each girl matured, she received a wheel of her own. Martha Jefferson was sixteen years old when she received hers; Jane Jefferson, Jr.'s wheel was among the possessions inventoried at her death. ("Spinning," by Thomas Eakins, 1881)

She was also a well-appointed young lady. Her wardrobe included a dozen frocks made of silk, chintz, and calico, as well as Virginia cloth. Wigs and hair pieces were at her disposal. She owned satin slippers for parties and dancing. Gold and silver rings adorned her fingers, while a collection of hats kept the hot Virginia sun off of her face and shoulders. One of Shadwell's eight dressing tables served her, and Jane could admire the results in a "dressing glass." Unlike her brother, Thomas, who left home to be educated beyond the limits of Shadwell, Jane was likely primarily educated at home, augmented with dancing and music lessons.[16]

At the time of her death, young Jane owned six books of her own, in addition to her Bible, and had the use of her mother's small collection and her father's library, more than half of which encompassed four periodicals collected in nineteen volumes including the *Tatler*, the British magazine of collected articles concerning popular culture and literature. There were also eighteen volumes of history, maps to study, and religious books to read.[17]

Jane probably remained at home until her death in the fall of 1765. The fact that she died on the tenth birthday of the twins surely added to the devastation of the day. This time the youngest children were old enough to feel the full impact of a death in the house. One of Shadwell's brightest lights had been extinguished.

~

Growing up somewhat apart from the oldest children, the three youngest Jeffersons experienced many milestones in common. It is not surprising that the significant, mutual experiences of Lucy, Anna Scott, and Randolph Jefferson resulted in loyalty and interconnectedness among the three siblings. Importantly, they were not part of the Jeffersons' years at Tuckahoe, which molded the older children. Their father's death in 1757 was the primary factor which shaped their very early lives at Shadwell. Soon, their brother Thomas departed, first to board at Rev. Maury's School, and then to Williamsburg and the College of William and Mary in 1760. Tom's absence certainly created a void, leaving the five-year-old Randolph in a world of females.

Lucy, the oldest of the second cluster of Jefferson children was born at Tuckahoe on October 10, 1752. Almost immediately, the Jeffersons returned to Shadwell, making it Lucy's only home. In 1769, she married her first cousin, Charles Lilburne Lewis. Her first surviving child, born about 1773, was named Randolph Lewis, carrying the surname of both his grandmothers, Jane (Randolph) Jefferson and Mary (Randolph) Lewis. His name inevitably also acknowledged Lucy's brother, with whom she had a long and affectionate relationship. In 1781, when Randolph Jefferson married Charles Lilburne Lewis' sister, Anne, the bond between the families was further strengthened. This was followed by the inter-marriages of several of their children and grandchildren, creating a Jefferson-Randolph-Lewis "dynasty."

Likewise, Anna Scott Jefferson, Randolph's twin sister, remained close to Lucy and for most of her life relied heavily on the kindness and generosity of her siblings. When Jane Jefferson died in 1776, Anna Scott was twenty-one years old and still single. Her marriage may have been delayed for numerous reasons, particularly the lack of an adult to guide her search for a suitable husband. Her father had been dead for nearly two decades, and her older brother, who might have acted as a surrogate father, was occupied in realms beyond domestic concerns at Shadwell. An uncle, either a Jefferson or a Randolph, might have helped her find a suitable match, but none lived nearby. Her late marriage to Hastings Marks in 1787 was eventually facilitated by the Lewis family. Early widowhood meant that she spent many years living with her siblings, including both of her brothers.

In her younger days, there was widespread affection for Anna Scott. Called Nancy by the family, their correspondence speaks kindly of her, and to her, tolerant of whatever her limitations may have been and valuing her strengths. On April 22, 1786, Thomas Jefferson closed a letter to her, "Pray remember me to my sisters Carr, and Bolling, to Mr. Bolling and their families and be assured of the sincerity with which I am my dear Nancy your affectionate brother."[18]

During the early months of 1787, the still unmarried Anna Scott lived with her widowed sister Martha (Jefferson) Carr. On January 2, 1787, Martha wrote to Thomas Jefferson that they were enjoying good health and that Nancy had "grown very fat." Described as "rolly-polly," Nancy Jefferson may have provided a very comfortable lap for young nephews and nieces.[19]

Indeed, when Thomas Jefferson's daughter, Maria, was very young and did not wish to join her father and sister, Martha, in France, she wrote, "I had rather stay with Aunt Eppes. Aunt Carr, Aunt Nancy and Cousin Polly Carr are here."[20] Since young "Polly" Jefferson had other aunts in the vicinity whom she did not name, it might be concluded that "Aunt Nancy" was a favorite.

Anna Scott was honored with lasting legacies, as well. Eventually, Lucy (Jefferson) Lewis' children would include Anna Marks Lewis, and Randolph and his wife, Anne (Lewis) Jefferson, named their only daughter Anna Scott. She, too, would be known as Nancy.

CHAPTER THREE

The Early Education of Randolph Jefferson

The bulk of mankind are school boys thro' life.
Thomas Jefferson, 1784

All of the Jefferson children received a combination of education at home and instruction from hired tutors. Like the older ones, the younger siblings had access to Peter Jefferson's library of books and maps. While he did not live to pass his interest in history and poetry on to them directly, his collection remained intact until 1770, when it burned with the house. Without Peter Jefferson to quote Addison, Swift, Pope, and Shakespeare, did their mother, their sister, Jane, or their brother, Tom, recite for them in their father's stead?[1]

Sometime in the spring of 1762, Jane Jefferson decided to send young Master Randolph to live with her sister, Mary Lewis, at Buck Island, demonstrating her confidence in the operation of the Lewis household. Buck Island was far enough away for Randolph to experience life on another plantation, yet close enough so that the six-year-old boy was still near his home and living with family members, rather than strangers.

Mary Randolph, Jane's next sister, was born in Virginia on October 15, 1725, making her five years Jane's junior. She wed Charles Lewis, Jr. on July 15, 1746, in Goochland County, probably at her home, Dungeness. That summer, the Jeffersons were living nearby at Tuckahoe, so Jane easily could have joined in the celebration of her sister's marriage. Their mother, Jane (Rogers) Randolph, had been a widow since 1742. Isham Randolph's last will provided Mary with a dowry of £200 at her marriage.[2]

Already established in Albemarle, Charles Lewis, Jr. was in his mid-twenties when he became Peter Jefferson's brother-in-law. Charles lived on the Rivanna River, south of Shadwell, at Buck Island Creek, and was improving one of his father's land patents. Making his mark in the new county of Albemarle, like Peter Jefferson, Charles became one of its first Magistrates.[3]

It is not known how closely Peter Jefferson and Charles Lewis were associated; however, they shared roots in Goochland County. Jefferson was thirteen years Lewis' senior and was living away from Shadwell when the Lewises began married life.[4] Nor is it known how close the Randolph sisters were. In Charles' will, he described Mary as his "affectionate and virtuous wife." The mother of eight children, she long out-lived her husband, remaining at their Buck Island home until her death on October 13, 1803.

Charles Lewis was about nineteen years old when he went west to his father's land, which still lay in Goochland County's uncultivated frontier. In 1731, Charles Lewis, Sr. of The Byrd had patented 1,200 acres of land on the Rivanna River, near the mouth of Buck Island Creek.[5] The land was divided by the river, with 800 acres lying to the north (ultimately located in Albemarle's Fredericksville Parish) while the other 400 acres lay to the south (ultimately located in Albemarle's St Anne's Parish).[6] Rev. Edgar Woods described how the creek, which watered the Lewis plantation, got its name:

> Buck Island Creek was so designated from the beginning. It is a mistake to write it Buckeyeland, as if derived from the deer-eyed tree. The name was taken from an island in the Rivanna opposite its mouth, and as in the case of so many objects of natural scenery, was suggested by the great numbers of deer found everywhere in the country.[7]

For many years, Charles Jr. held a deed of tenancy from his father, but not full title to the land. As a "tenant," he was a successful planter of the land allotted him, and he acquired hundreds more acres. His crops included tobacco, corn, rye, oats, and wheat. He eventually received the land outright from his father, and at the time of his death in 1782, Charles owned fifty-seven slaves, eleven horses, livestock, significant farming equipment, as well as industrial equipment: a cotton gin, cotton wheels, a flax wheel, looms, and heckles, which processed hemp and flax. His personal property was valued at £3,466.[8]

Achieving the military rank of Colonel, Charles Lewis was a politically active planter and served as Thomas Jefferson's alternate at the second session of the Virginia Convention in 1775. Beginning in December, he "sat as a regular member for Albemarle, and he continued in his seat in the Convention of May 6, 1776."[9] A man of revolutionary spirit and a fervent Presbyterian, Charles also stood firmly behind the separation of Church and State in Virginia. He would become a vocal dissenter against the domination of the Anglican Church. The Lewis library included various religious books, and this Calvinist atmosphere probably resulted in an orderly household and may have influenced Randolph Jefferson's religious beliefs.[10] At his death, Charles Lewis would name his wife's nephew and "his friend," Thomas Jefferson, as one of the executors of his will. His other executors reflected the tight Lewis family group; they included his sons-in-law, Charles Hudson and Bennet Henderson, as well as sons, Charles Lilburne and Isham Lewis.[11]

Author Boynton Merrill suggests that the Lewis dwelling house at Buck Island was typical of the early Albemarle dwellings, likely a hewn log house covered by weatherboards.[12] However, no specific description of it has survived. The exterior may have been basic, but this was a well-appointed home. At the time of Charles' death, there were seven feather beds and furniture, numerous books with a bookcase to keep them in, an accompanying desk at which to write, chests of drawers, leather bottomed chairs, as well as spinning wheels and looms which may have been used both by the Lewis women and their slaves.[13]

Mary Lewis was well-equipped for gracious entertaining, owning numerous tables and tablecloths to cover them. There were pewter plates and dishes, as well as Queen's China plates for fancy occasions and earthenware dishes for every day. Candlesticks and a pair of snuffers were part of her dining service. Silver teaspoons accompanied her coffee cups and saucers. She had both a copper coffee pot and a copper tea kettle. Her chafing dish was worth £4; there were six custard cups, and a Delft bowl was valued at five shillings. Despite Buck Island's isolation on the western frontier, like her sister, Jane, at Shadwell, Mary (Randolph) Lewis lived in relative elegance which reflected her upbringing at

Dungeness. In short, Randolph Jefferson's country cousins were living in style in a home very comparable to Shadwell and made perfect surrogate parents for the fatherless boy.[14]

Following Peter Jefferson's death in 1757, it is quite possible that Jane increasingly relied upon the Lewises. Charles Lewis helped her with the preparations for Peter's funeral, at least to the extent that he obtained two shillings/six pence worth of sugar used for punch and other beverages.[15]

The youngest Jefferson and Lewis children became especially close. The families were apparently compatible, and their association resulted in two intermarriages. The first was on September 12, 1769, when Charles Lilburne Lewis married his first cousin, Lucy Jefferson. The second would be the wedding of Randolph Jefferson and his cousin, Anne.

One of Jane Jefferson's motivations behind placing her young son in the Lewis household may have been to expose him to men and boys of various ages. With Thomas away at school, Randolph was the lone male in a household of females. By the time Randolph arrived at Buck Island, the Lewises had been married about twenty years and all of their children were born. Randolph could look up to Charles Lilburne Lewis, who was eight years his senior, while Isham, closer to Randolph's age, may have provided a schoolmate.

Few exact birthdates are known for the Lewis children; however, their father's last will may suggest birth order, especially for the girls. The boys, who are named first, may not be older than their sisters. The children are mentioned in the following order:

Charles Lilburne Lewis (b. 1747) m. Lucy Jefferson, 1769
Isham Lewis (d. 1790)
John Lewis (d. before 1782)
Mary Randolph Lewis m. Col Charles Lewis, of North Garden, son
 of Col. Robert Lewis of Belvoir and Jane Meriwether
Jane Lewis m. Charles Hudson
Elizabeth Lewis (b. 1752) m. Col. Bennett Henderson
Anne Lewis m. Randolph Jefferson, abt. 30 July 1781
Frances Lewis (b. 24 January 1759) m. John Thomas
Mildred Lewis m. Edward Moore[16]

At the Lewis home, Randolph's world certainly expanded. There he benefitted from a collection of surrogate brothers and sisters on a day-to-day basis, as well as surrogate parents, most importantly a father figure in Charles Lewis. This extremely tight-knit Lewis family would become the core of Randolph Jefferson's relations.[17] His father's early death, his time spent living at Buck Island, his mother's death before his twenty-first birthday, his only brother's distance from the neighborhood during his early adulthood, his sister Lucy's marriage to Charles Lilburne Lewis and, finally, his own intermarriage with Charles Lewis' daughter, Anne – all contributed to Randolph's deep alliance with the Lewis family, an alliance which would be reflected most profoundly in the marriages of his children.

In 1762, however, Randolph Jefferson was only six years old when he took his first significant step away from Shadwell and went to board at the home of Charles Lewis, Jr. The Lewis plantation lay six to seven miles southeast of

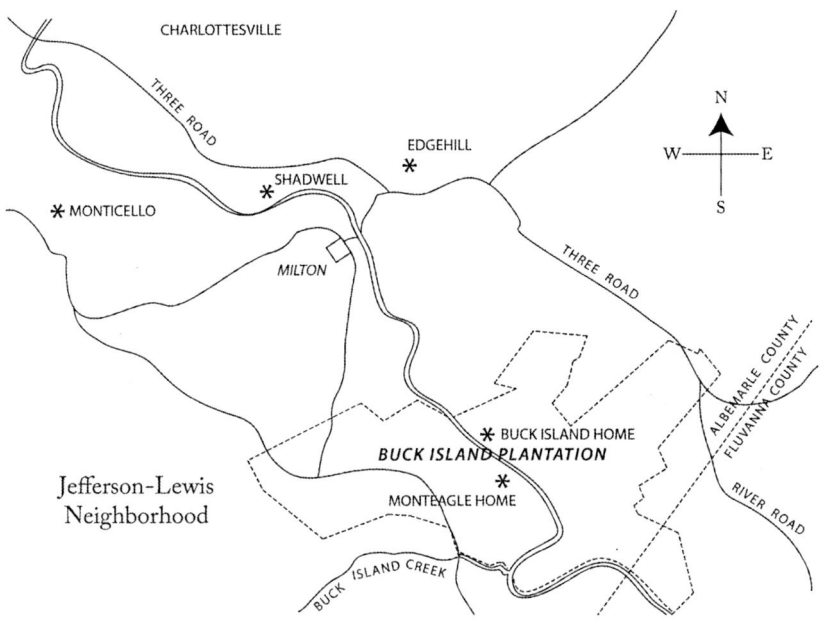

The distance between the Jefferson home at Shadwell and the Lewis home at Buck Island was just far enough to offer young Master Randolph Jefferson a new experience at his cousins' plantation. (BASED ON A MAP BY THOMAS E. CLARK)

Shadwell. The dwelling house, which sat on the north side of the Rivanna River, had been the home of Charles and Mary (Randolph) Lewis since their marriage in 1746.

These years at Buck Island marked the beginning of Randolph Jefferson's formal education, some of which may have been provided by a school teacher hired by the Lewises, coupled with classes conducted by Benjamin Sneed. The older Jefferson children had begun attending Sneed's English School at least by the fall of 1758, when Martha, age twelve, was his pupil. The younger children followed suit.

The Sneeds lived near the junction of the Three Notch'd Road and the road north to Orange County Courthouse. The school sat about two miles east of Shadwell's dwelling house, which was located just south of the Three Notch'd Road.[18] The Jefferson children probably walked to school, where they may have studied with others from the neighborhood. During the 18th and 19th centuries, many children, male and female, walked or rode on horseback long distances to school. While Shadwell and Sneed's farm may not seem close by modern standards, by 18th-century standards, they were neighbors. Forty years later, Martha (Jefferson) Randolph's son, Thomas Jefferson Randolph, continued the family tradition at Sneed's school; he walked from his home at Edgehill. In 1798, Martha wrote to her father concerning Jeffy's attendance:

> We have been all well but Jefferson who had declined rapidly for some time from a disorder which had baffled every attention and change of diet, the only remedy we ventured to try, but Mr. Sneed opening school and Jeffy being hurried out of bed every morning at sunrise and obliged after a breakfast of bread and milk to walk 2 miles to school: his spirits returned his complexion cleared up and I am in hope that disorder has left him entirely.[19]

Entries for the education of Jane Jefferson's children are scattered throughout the "family expenses" in the account books of Peter Jefferson's estate. They show both the payments to Benjamin Sneed and the monies concurrently paid to Major Charles Lewis for Randolph Jefferson's board.[20]

Oct 8, 1758
  To Benja Snead for Schooling Martha Jefferson   1. - .-
Feb 9, 1761
  By Mr. Benja. Snead pr acct. for (illegible)   2.8.9 [?]
Feb 9, 1761
  To Mr. Benja Snead for teaching the children   1.5.-
May 24, 1762
  To Maj^r Charles Lewis for boarding Master Randolph a year 6.-.-
Sept 25, 1762
  To pd. Ben Snead for teaching Miss Lucy Jefferson 3 months -.13.4
July 28, 1764
  By Boarding Randolph Jefferson one year   6. - . -
July 29, 1764
  By pd. Maj^r Charles Lewis in part for
  Master Randolph's Board   3. -. -
Aug 8, 1764
  By ball. pd. Mr. Benja Snead for 1 years schooling
  of Rand. Jefferson   1.1- . [?]
Jan 4, 1765
  By pd. Maj^r Cha for the Ball of Master Randolphs Board 3.- . -

These payments are open to interpretation, though they clearly reveal that Master Randolph probably spent part or all of three formative years living with the Lewis family. These may or may not have been consecutive years. Payments for his board stretch from 1762-1765; they might be in advance of his stay, in arrears, or both. Ultimately, the four payments, totaling £18, appear to add up to three full years of room and board away from home.[21]

At what point did Randolph study with Benjamin Sneed? As with many other Jefferson accounts, we know when Sneed was paid but not when he provided his service. The August 8, 1764 entry states that £1.1 is the balance due the schoolmaster, indicating a prior payment toward Randolph's account. Randolph was eight years old at the time. It is possible that the August 1764 settlement of Randolph's account covered his education in 1763-1764, when there was not an obvious payment to Charles Lewis. During the last part of

1763-1764, Randolph may have returned to Shadwell and studied at Sneed's English School. His brother, Thomas, could have been the reason for his return. During most of 1763, when Randolph was seven years old, Thomas was at home reading the law in preparation for his career as an attorney.[22] Jane Jefferson may have preferred to have Randolph at home when Thomas was there, bringing him back for part of 1763. In Thomas' absence, however, Jane encouraged the male companionship her younger son enjoyed with the Lewises.

It is difficult to imagine Randolph attending Benjamin Sneed's English School while boarding at the Lewis home, which lay five-to-seven miles to the south, unless Sneed stayed at or near Buck Island for a period of time, and there is no known evidence that any of the other Jefferson children came to board at the Lewises or attended school there.[23] Importantly, in 1764-1765, Sneed was a married man with children and a farm to manage. It is more plausible that the Lewises hired an unmarried schoolmaster for their children, and Randolph was included in that classroom. In *Cabells and Their Kin*, Alexander Brown describes this sort of plantation school where the Virginia gentry prepared their children for classical schools and college:

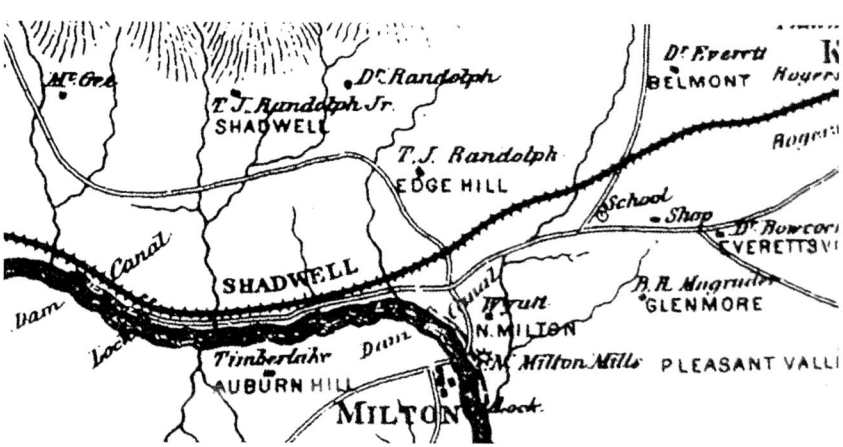

This map, created in 1875 by Green Peyton, shows the relationship between the Jefferson-related farms, Shadwell and Edgehill, and the English School established by Benjamin Sneed which lay to the southeast on the Three Notch'd Road. (COURTESY OF ALBEMARLE CHARLOTTESVILLE HISTORICAL SOCIETY)

It seems well to say here that the custom with the landed gentry of this region with their minor children, before the Revolution, was this: First one and then another of a circle of friends would employ a tutor, and take the young sons of the others as boarders.... From these private tutors, or from such classical schools as those of the Rev. Mr. Maury or the Rev. Mr. Douglas, the boys were sent to William and Mary College, or to England, or to Scotland, to complete their education.[24]

Whatever the arrangement in the 1760s, as a schoolmaster Benjamin Sneed was a fixture in Albemarle for decades and had a fundamental effect on hundreds of children.

~

Randolph was only six and one-half years old when the first payment for board went to Charles Lewis, and he continued at Buck Island beyond his ninth birthday in October of 1764, perhaps well into 1765. Though the exact timing of Randolph's return to Shadwell is unclear, he likely continued his studies at home. His mother and/or older sisters may have tutored him, though the single most important influence on the boy was probably the presence of his elder brother.[25]

Late in 1765, Thomas returned from Williamsburg to continue his intensive schedule of reading at Shadwell. Randolph, now ten years old and a veteran of both the Sneed School and Buck Island, could observe his mature brother's behavior with new eyes. Thomas, age twenty-two and over six feet tall, was a significant masculine presence as he focused on activities relating to his burgeoning law practice – conducting interviews, maintaining correspondence, keeping his fee and case books, preparing cases for trial, and traveling locally to various courthouses. The degree to which Thomas directly concerned himself with Randolph's education is unknown, but his presence at Shadwell certainly might have influenced their mother to bring Randolph home from the Lewis plantation.[26]

~

The next dramatic event in the lives of the Jefferson brothers was the burning of the dwelling house at Shadwell on February 1, 1770. The fire was reported in the *Virginia Gazette*, on February 22, 1770, when the paper printed the following: "We hear from Albemarle that about a fortnight ago the house of

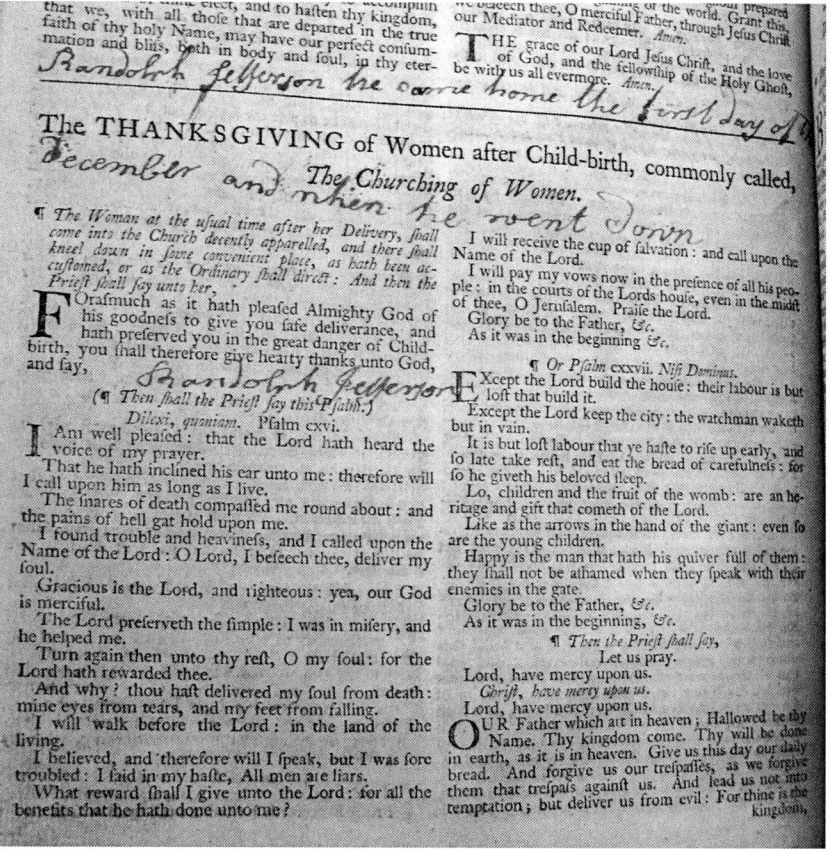

At some point in young Master Randolph's education, he may have misused the Jefferson family's Book of Common Prayer to practice his penmanship. Paper perhaps being scarce, someone wrote and rewrote "Randolph" and "Jefferson" in the top margin and between lines on several pages. He also may be the author of a few practice sentences, including several versions of, "Randolph Jefferson he came home the first of December." The phrase seems an odd thing to write about oneself, however. It is equally possible that his twin, Anna Scott, or his sister, Lucy, wrote this after being separated from Randolph while he was away at school. Additionally, Tucker M. Woodson's name was inappropriately written in Jane Jefferson's Booke in what appears to be a youthful hand. Woodson, born about 1744, married Martha Eppes Hudson, the sister of Charles Hudson who had married Jane Lewis, first cousin to the Jefferson children. If Jane Jefferson's Booke does not pre-date the Shadwell fire of 1770, these entries are particularly inexplicable. In December of 1771, Randolph Jefferson was in Williamsburg at William and Mary. At that time, he and Anna Scott were sixteen years old, and Lucy was already married. (JEFFERSON FAMILY BIBLE, SMALL SPECIAL COLLECTIONS, UNIVERSITY OF VIRGINIA LIBRARY)

Thomas Jefferson, Esq., in that county, was burnt to the ground, together with all his furniture, books, papers &c. but which that Gentleman sustains a very great loss. He was from home when the accident happened."[27]

Much later, in 1837, George Tucker wrote, "The house at Shadwell at which [Thomas Jefferson] lived with his mother, caught fire, while they were on a visit to a neighbour, and in the alarm and confusion of the slaves, almost everything in it was consumed."[28]

Family tradition holds that when Thomas asked one of the slaves if they had saved any of his books, the reply was, "No master, all burnt, but we save your fiddle."[29] This anecdote seems to indicate that the Shadwell servants understood how much Thomas cherished his violin. It may also reflect the values of the slaves. What importance would illiterate men put on books and accounts? Yet the value of a violin, and the beautiful music it made, was fully understood by all.

When Thomas Jefferson wrote of the house fire, he noted the great personal loss of his papers and books (the value of which he totaled at £200 sterling); however, he did not comment on how this event affected his mother or sisters, Elizabeth and Anna Scott, who were still living at home. Nor did he mention the effect on his brother Randolph, who was also presumably living at Shadwell and would not leave home for Williamsburg until October of 1771.

It is not known where the Jeffersons stayed while their new house was being built. They may have gone to Buck Island, the Lewises being their closest relations. Lucy had just married Charles Lilburne Lewis, further cementing the closeness of the two families. If they indeed spent at least some of this period at Buck Island, Randolph, who was now fifteen years old, had the opportunity to spend extended time with his cousin, Anne, perhaps laying the groundwork for the their marriage ten years hence. When the Jefferson family returned to Shadwell, it was to a smaller dwelling situated just west of the original building. Surviving bricks of the destroyed house were used in building the stairs to the new cellar.[30]

Thomas Jefferson Comes of Age

During Randolph's time at Buck Island, perhaps the most significant change at Shadwell was that Thomas Jefferson turned twenty-one, came into his inheritance, and gradually gained more influence over the decisions made in the household. One of Thomas' first and most important decisions was the division of Peter Jefferson's land between Randolph and himself. Thomas chose the Rivanna lands, placing the whole of Randolph's eventual inheritance at faraway Snowdon. While this would mean little to an eight-year-old boy, it was a decision which shaped the rest of Randolph's life.

John Harvie's role as Peter Jefferson's primary executor ended by 1765, transferring responsibility to Thomas Jefferson, who now shouldered the double duty of managing his own estate as well as the property of his mother, his dependent sister, Elizabeth Jefferson, and his minor siblings. At that time, Randolph was still a very long way from stepping into his responsibilities as proprietor of Snowdon.[31]

In 1764, when Thomas Jefferson turned twenty-one, he began the redistribution of slaves to satisfy his father's will. Ultimately, twenty-four slaves were assigned to Randolph. Quash, Nell, little Bellow/Bella, and, possibly, Betty, slaves originally living at Snowdon, left the farm and took up residence at Monticello. Initially, Crummel, Sanco, and Bellow may have stayed on at the Buckingham plantation, while other slaves were selected from the Rivanna lands in Albemarle and ultimately moved to Snowdon.[32]

It is not known who decided how the slaves were to be divided. Jane Jefferson, in her desire to keep families together, may have influenced the decision, and she no doubt wanted certain servants to remain in her household. As there was no dwelling house yet established at Snowdon, Randolph was not immediately in need of domestic servants at his farm. Some may have been assigned to him, but remained at Shadwell. His personal valet, Peter, had been with him virtually since birth and continued to serve him. The other individuals who eventually came to Snowdon were chosen from all five "Quarters" of Peter Jefferson's estate.[33] The number distributed to each son from each Quarter survives, but a list of names does not:

Quarter	Thomas	Randolph
I	11	10
II	2	3
III	6	4

IV	3	4
V	4	3
Total	26	24

While the assignment made specific slaves Randolph's property, these individuals didn't necessarily move to Snowdon immediately or all at once. Among those transferred to Snowdon before the mid-1770s was a woman named Hannah and her thirteen-year-old daughter, Fan. By this time, Martin Dawson, who had managed the farm for many years, had been replaced by an overseer named Isaac Bates.[34] Under his supervision, a shocking incident took place at Snowdon: Hannah's death allegedly caused by Bates' intemperate punishment.

In June of 1770, Thomas Jefferson, who was then practicing law in Albemarle County, brought suit against Isaac Bates. John Nicholas, Clerk of Albemarle County from 1749-1792 and one of Peter Jefferson's executors, gave his approval for Thomas to pursue a case against Bates as Randolph Jefferson's "next friend."[35] Thomas claimed that Bates was responsible for damages to Randolph's property, having killed Hannah "by a cruel whipping." In 1757, "Hannah & her Child Fan" had been valued together at £45. Likely still in her thirties, Hannah represented a significant investment. This lawsuit concerned the loss of Randolph's personal property; it was not viewed as a case of murder.

Only a few fragmentary pieces of evidence remain of the case, the loss of the Buckingham records having erased other details. On June 21, 1770, Thomas noted, "Jefferson v. Bates, Inclosed al. cap. to J. Nicholas," and on December 27, 1770, Thomas sent something concerning the case to the Sheriff of Buckingham County. As late as 1772, Thomas Jefferson's accounts for Randolph's property show that he had paid "Secretary's ticket in your suit vs. Bates 100 # tobo. @ 12/6."[36] It is not clear if Thomas Jefferson successfully collected damages for his minor brother; however, no credit appears among Randolph's accounts. Not much more is known about Isaac Bates, including how long he had been the overseer at Snowdon, or if he was immediately dismissed upon Hannah's death.

The Bates family was long established in central Virginia. Many were Quakers living in Goochland County during the early 18th century, where they intermarried with the Fleming and Woodson families, who were also members of The Religious Society of Friends. Ironically, a descendent of witnesses for peace would be accused of beating a woman to death.

In a 1719 will, John Bates II named his youngest son, Isaac Bates, leaving him the plantation on which he then lived. The family migrated west from Goochland County and, by 1747, Isaac Bates (1687-1752) and John Cannon owned land adjacent each other in Albemarle, in what would become the northeast quadrant of Buckingham County, where their neighbors included John Nicholas of Seven Islands. Isaac Bates married Elizabeth Cannon and they produced another Isaac Bates.[38]

Thomas Jefferson, John Nicholas, and the Bates family were associated at least by 1768, when John Nicholas had employed Thomas to defend plaintiff Isaac Bates against his neighbor, John Cannon. The nature of the dispute is unknown. Thomas took few cases in Buckingham County; in fact this may have been only his second to date. Ironically, he was defending the family he would soon bring charges against.[39]

One Bates family history states that Isaac Bates III (1750-1786), born in Albemarle, was the overseer at Snowdon. Other family histories suggest that Isaac Bates III was born in 1752, making him even younger when he was hired. If the Isaac Bates of Snowdon was between eighteen and twenty years old in 1770, his youth certainly could have added to his immoderate behavior. It is possible that he left Buckingham shortly after Hannah's death; he does not appear there on the 1773 or 1774 list of tythables.[40]

At fifteen, Randolph Jefferson had learned that property, especially human property, was a vulnerable investment. It is not known how close Randolph was to Hannah and Fan; however, he had lived with them all of his life, and Fan was near his age. Hannah's death potentially made a deep impression on Randolph. Over the years, he demonstrated a protective position toward his slaves. Hannah's death may have been formative in this attitude and may also have shaped Randolph's notions about plantation management, absentee landlords, and the selection of overseers.

Nothing more is known of Hannah's daughter, Fan, or the direct effect Hannah's death had on the community at Snowdon. Twelve years later, when the next surviving slave list was taken at Snowdon, there is no Fan or Fannie living on the farm.[38]

There is, though, a coda to Hannah's story. A few months later, in November of 1770, Jane Jefferson gave Randolph a slave girl named Rachael. It is possible that the gesture was, in part, to make up for the loss of Hannah. Rachael, the daughter of Little Sall, was a child, perhaps as young as two years old. Her fate is currently unknown, and she may never have lived at Snowdon. She is not listed among the slaves living there in 1782.[41]

Part Two
RANDOLPH JEFFERSON

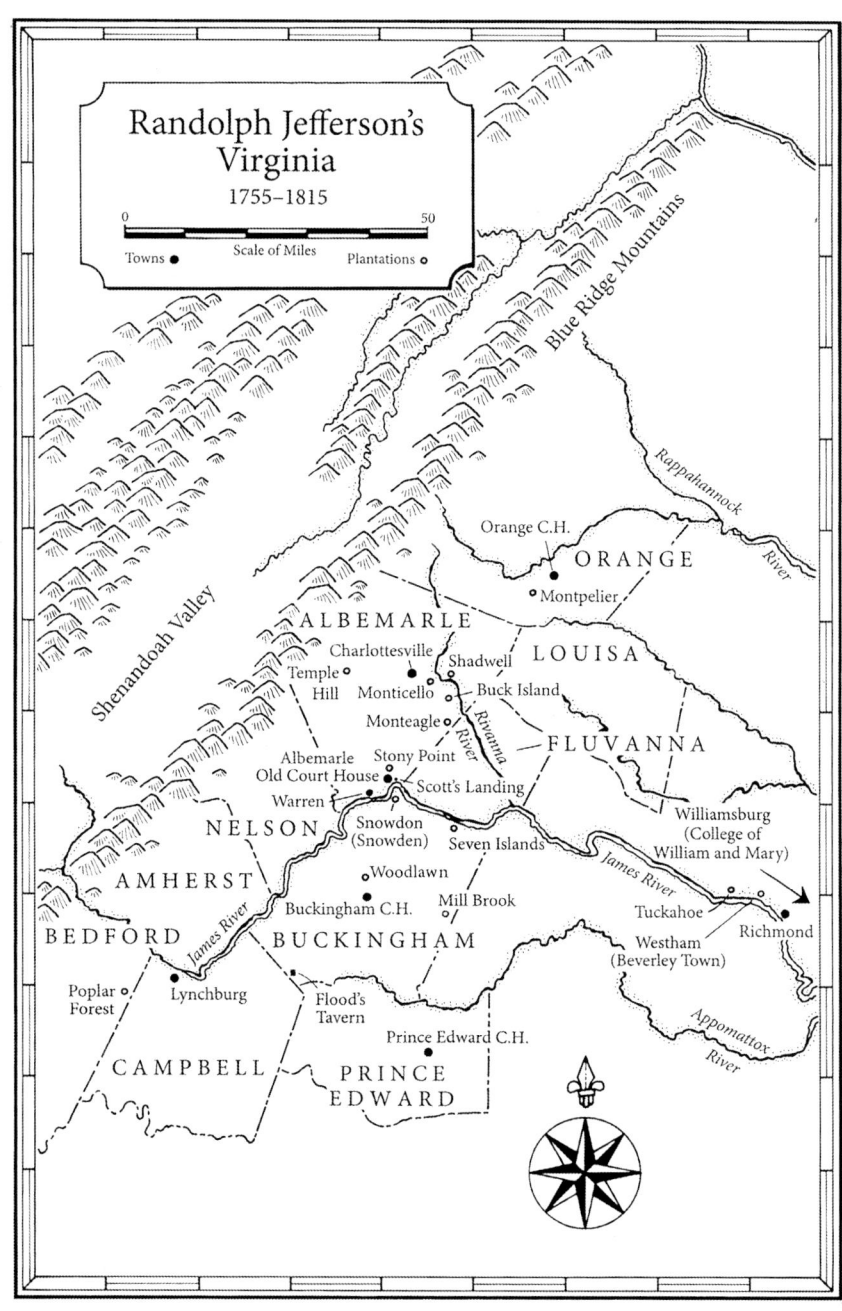

Map by Rick Britton

CHAPTER FOUR

A World beyond the Piedmont: Williamsburg and the College of William and Mary

The boys of the rising generation are to be the men of the next, and the sole guardians of the principles we deliver over to them.
Thomas Jefferson to Samuel Knox, 1810

In the late summer of 1771, Randolph Jefferson was nearly sixteen years old, an unofficial ward of his older brother, Thomas, who was twenty-eight years old and still a bachelor. Jefferson family tradition holds that Peter Jefferson's dying wish was that his elder son, Thomas, be classically educated. Indeed, his wish was granted. Thomas was first schooled at Tuckahoe, and beginning in 1752, Thomas attended Rev. William Douglas' classes nearby in Goochland County. Shortly after Randolph's birth at Shadwell, Thomas was sent away to Rev. James Maury's school; from Maury he learned the ancient languages.[1] This was followed by college at William and Mary (1760-1762) and the study of law with George Wythe until Thomas' admission to the Virginia bar in 1767. By contrast, Randolph Jefferson's only known educational opportunity outside his home had been three years of boarding at Buck Island, with his uncle, Charles Lewis, coupled with study at Benjamin Sneed's English School.

The decision to send Randolph to Williamsburg, to study at the royally chartered College of William and Mary, was certainly sanctioned by his brother and may have been essentially his idea.[2] A decade earlier, Thomas had appealed to his guardian, John Harvie, requesting funds and "permission" to go away to school. He laid out his good reasons for going to college and may have envisioned similar ones for sending his younger brother.

> Shadwell Jan, 14, 1760
> Sir –
> I was at Col° Peter Randolph's about a Fortnight ago, and my Schooling falling into Discourse, he said he thought it would be to my Advantage to go to the College, and was desirous I should go, as indeed I am my self for several Reasons. In the first place as long as I stay at the Mountain, the Loss of one fourth of my Time is inevitable, by Company's coming here and detaining me from School. And likewise my Absence will in a great Measure, put a Stop to so much Company, and by that Means lessen the Expenses of the Estate in House-Keeping. And on the other Hand by going to the College, I shall get a more universal Acquaintance, which may hereafter be serviceable to me; and I suppose I can pursue my studies in the Greek and Latin as well there as here, and likewise learn something of the Mathematics. I shall be glad of your opinion.
>
> Th:Jefferson[3]

As Thomas had justified the expense of additional education to John Harvie in 1760, he now had to justify to himself the investment in his brother, Randolph. In his autobiography, Thomas describes his experience in Williamsburg, especially the profound influence that Dr. William Small, professor of Natural Philosophy, had on him as a student. Thomas felt strongly about his mentor, admitting that this gentleman "probably fixed the destinies of my life."[4] Years later he wrote to William and Mary professor, Louis Hue Girardin, saying that Dr. Small was "to me as a father. To his enlightened and affectionate guidance of my studies while at college, I am indebted for everything."[5] Of course, Thomas was keenly aware that Randolph had lacked a father in his life and may have sincerely wished that his younger brother also would find such a mentor at William and Mary.

In years to come, Thomas was critical of conditions at the College as well as its system of education, and had many recommendations for its reform. In *Notes on the State of Virginia*, he wrote: "The buildings are of brick, sufficient for an indifferent accommodation of perhaps a hundred students." He further described the College and the hospital as "rude, misshapen piles, which, but that they have roofs, would be taken for brick kilns."[6]

In 1771, however, Thomas' own broadening experiences in Williamsburg were fresh in his memory, and he decided to fund a similar opportunity for Randolph. In his brother's case, Randolph would be enrolled in William and Mary's Grammar School.[7] His studies of Latin and Greek would be augmented by private tutelage in Mathematics and Natural Philosophy. An outgrowth of the Enlightenment, Natural Philosophy was more than the study of "nature," it embraced the rapidly changing field of the physical sciences, including an expanding set of "natural laws" realized by Galileo, Kepler, and Newton. By the time Randolph was ready to go to Williamsburg, Dr. Small had already returned to England, and it would be the much younger and newly arrived Rev. Thomas Gwatkin who would instruct Randolph in the ideas that had so shaped Thomas Jefferson's vision.

And so, in the early fall of 1771, Randolph Jefferson agreed to further his education at his brother's *alma mater* and to learn something of the larger world, concurring that to gain "a more universal Acquaintance" and experience away from Shadwell was a good idea. As it turned out, Randolph would not develop Thomas' scholarly orientation or his political ambitions, which might have been ignited in Williamsburg. In spite of this, Randolph's world certainly would be greatly expanded as a result of the eleven months he spent there.

Preparations for Randolph's trip to William and Mary began in earnest on October 1st, when Thomas gave him 15 shillings, 9 pence toward his expenses going to Williamsburg.[8] This also marked Randolph's sixteenth birthday and that of his twin sister, Anna Scott.

Thomas likely counseled his younger brother concerning the temptations of Williamsburg society, warning Randolph not to engage in idleness, horse racing, gambling, drinking, and other follies which could ensnare and tarnish young men. He could have shared lessons learned from his first year in Williamsburg when he had wasted precious time, indulging too much in society and not enough in his studies, as well as reminding Randolph that he had been sheltered from such corruption in the bucolic world of Albemarle. Of course, any advice Thomas gave to Randolph was in conversation and left no tracks.[9]

As Randolph turned sixteen, Thomas departed Shadwell for Williamsburg, a week ahead of his brother. He gave Randolph monies for his traveling expenses, lent their mother 5 shillings, 9 pence, and headed east with his servant, Jupiter. Their first night was spent in Louisa County, about twenty-five miles east of Charlottesville on the Three Notch'd Road. There, Thomas paid 1 shilling, 6 pence for his night's "entertainment" at Byrd's ordinary. The next day, Thomas crossed the James River at Tucker Woodson's ferry, just south of Goochland County Courthouse, and stopped at The Forest to see his intended, Martha (Wayles) Skelton. There, he gave the Wayles family servant, Ben, 5 shillings as a gratuity. By October 9th, Thomas had arrived in Williamsburg where he settled in and, among other activities, awaited his brother's arrival the following week.[10]

Randolph left Shadwell within a few days of Thomas' departure. He and Will Beck arrived in Williamsburg by October 13th, when Thomas paid Beck 5 shillings for "transporting" Randolph. Thomas had allowed Will nine days for traveling, "coming and returning" to Albemarle.[11]

Initially, Randolph was not alone in Williamsburg. For the next few months his brother, Thomas, would be living in the capital as well. During the coming weeks, Thomas was present for the meeting of the General Assembly, was re-elected to the House of Burgesses on November 29th, and remained in Williamsburg until mid-December, leaving just prior to his marriage to Martha (Wayles) Skelton on January 1, 1772.[12] During the autumn months, Thomas paid several calls to Martha's home at The Forest; however, he maintained a continuous residence in the capital, where he would have been available to Randolph. Following Thomas' marriage, the newlyweds returned to Williamsburg on April 9, 1772, where they remained until June 11, 1772, giving Randolph an opportunity to better know his sister-in-law.[13]

William Beck, who transported Randolph, had begun working for Thomas Jefferson in 1768 as a lime-burner, well-digger, and messenger, so Randolph had been acquainted with him for several years before they traveled to Williamsburg together.[14] Will likely helped Randolph get settled, and they established a mutual account with Cuthbert Hubbard, wigmaker, barber, and tavern keeper.

When Thomas settled the account with Hubbard on May 15, 1772, it totaled 22 shillings and 2 ½ pence. Randolph's expenses came to 7/2 ½; the rest, Thomas noted, was charged to Beck.[15]

Cuthbert Hubbard had taken a lease on "Anderson's Tavern," where Randolph and Will Beck probably lodged the night before Randolph moved into his quarters at William and Mary on October 14th. The tavern was a popular

Thomas Jefferson (1743-1826) by Charles Willson Peale, 1791. (COURTESY OF INDEPENDENCE NATIONAL HISTORICAL PARK)

place; George Washington mentioned it numerous times in his diaries between 1771 and 1774.[16] The ordinary was located east of the College, on Duke of Gloucester Street (Williamsburg's "Main Street"), could "entertain" ten or twelve gentleman, and offered "good Stabling &c. for their Horses."[17] When Hubbard took over the tavern three years earlier, he ran an advertisement in the *Virginia Gazette* letting Williamsburg know that he was now an innkeeper as well as a wigmaker.

> Williamsburg, March 7, 1771
> The Subscriber begs Leave to inform the Publick that he has taken the House lately occupied by Mr. *Robert Anderson*, where Gentlemen may be accommodated in the best Manner; and hopes, by his Care and Assiduity, that he shall be able to give Satisfaction to all who may please to favour him with their Custom.
> CUTHBERT HUBBARD.
> *N.B.* He still carries on his Business of Peruke making, Shaving, and Hair dressing, in a Shop nearly opposite Mr. *James Cocke's* Store.[18]

Once settled in at William and Mary, Randolph Jefferson, like his brother before him, boarded in what is today known as the Wren Building. Completed by 1700, burned in 1705, and immediately rebuilt, this single "U"-shaped structure was still meeting the College's needs decades later. The students both slept and ate there.[19] The total cost for Randolph's board for two terms was £12/5/6.

The College building also housed the staff, both white and black, who tended to the needs of the faculty and students, and functioned as an enormous household. The kitchen and servants' rooms were located in the basement. The housekeeper was central to the smooth functioning of the College and may have lived on the third floor near the boys.[20] She was in charge of the procurement and preparation of food, as well as the supervision of the cleaning of the building and the laundering of sheets. Finding and keeping a competent person to handle this job proved difficult over the years. During 1771-1772, Mrs. Margaret Garret held this position. The Bursar's books show she was paid £9 per quarter.[21]

During Randolph's tenure, resident nurse Phoebe Dwit not only cared for the sick, but made clothing for the servants and cleaned the dormitories and apartments. In December of 1771, it was decided that the room across from hers was to be the Infirmary, as soon as Mr. Nathanael Burwell left it.[22]

An unnamed hired stocking mender likely did not live in; she was paid £12 per year, provided she took care of her own "Lodging, Diet, Fire & Candles." The steward and gardener, James Nicholson, was in charge of the grounds and the building; he also served as the janitor and porter, locking the gates and doors at night.

Salaries for some the household employees at the College in 1770 were as follows:

Housekeeper, Mrs. Margaret Garrett	£ 30
Usher, James Emmerson	£ 75
Second Usher, James Marshall[23]	£ 40
Gardener, James Nicholson	£ 30
Janitor, James Nicholson	£ 5
Surveyor of Woodcutters, James Nicholson[24]	£ 5

Slaves belonging to William and Mary worked under these employees. Among them was a man named Winkfield, who was of particular service to Randolph Jefferson during 1771-1772.[25] In the fall of 1772, Thomas Jefferson paid Winkfield for personal services performed for Randolph, including "finding knives and forks, cleaning shoes, &c." Thomas gave Winkfield 22 shillings 6 pence, indicating that he may have taken care of Randolph for his entire stay. Interestingly, in Thomas' accounts, he identifies Winkfield as "Winkey," suggesting that he may have served Thomas during his student days a decade earlier.[26]

In an undated account of the "College Negroes," Winkfield heads the list and appears to be one of the few able-bodied servants at the College:

A List of Negroes at College
Winkfield, Daniell, Dick – almost a invalid, Pompey, Adam, Nedd, old Lucy – a invalid; old Kate, a invalid; Nanny a invalid; Effy – not much Better.
Negroes Hired outt
Lemon, James, Letty, Charlott, Frankey, Betty and two Gerrels
Mr. Bellini's three – Molly, Mass and Lucy.[27]

Accommodations at the College varied considerably; faculty and students lodged together on the building's third floor:

> Each professor had a private apartment of two plainly finished rooms, corresponding with the hall and chamber of a private dwelling. Leftover spaces were distributed among the "better Sort of the big Boys" living three or four to a room, while Grammar School students slept in the undivided dormitories over the hall and chapel. Thirteen dormer windows lit each of these barrack-like rooms, and in each case, a fireplace at the eastern end was the only source of heat. Curtains may have afforded some visual separation between individuals or groups, but compared to other students, those who occupied these common sleeping rooms enjoyed little privacy.[28]

There were those within and without the College who disapproved of the mixed boarding situation, particularly the housing of the younger Grammar School boys with the College students. The atmosphere, the critics said, was not always conducive to serious study.[29] The particulars of Randolph's exact accommodations are unknown; however, as the son of Peter and Jane (Randolph) Jefferson, at age sixteen, he certainly might have been classified as a "better sort of big boy."[30]

In this mixed living situation, Randolph had an opportunity to become friendly with students of various ages, Grammar School or College students or both. Within the group of scholars boarding during 1771-1772, there were

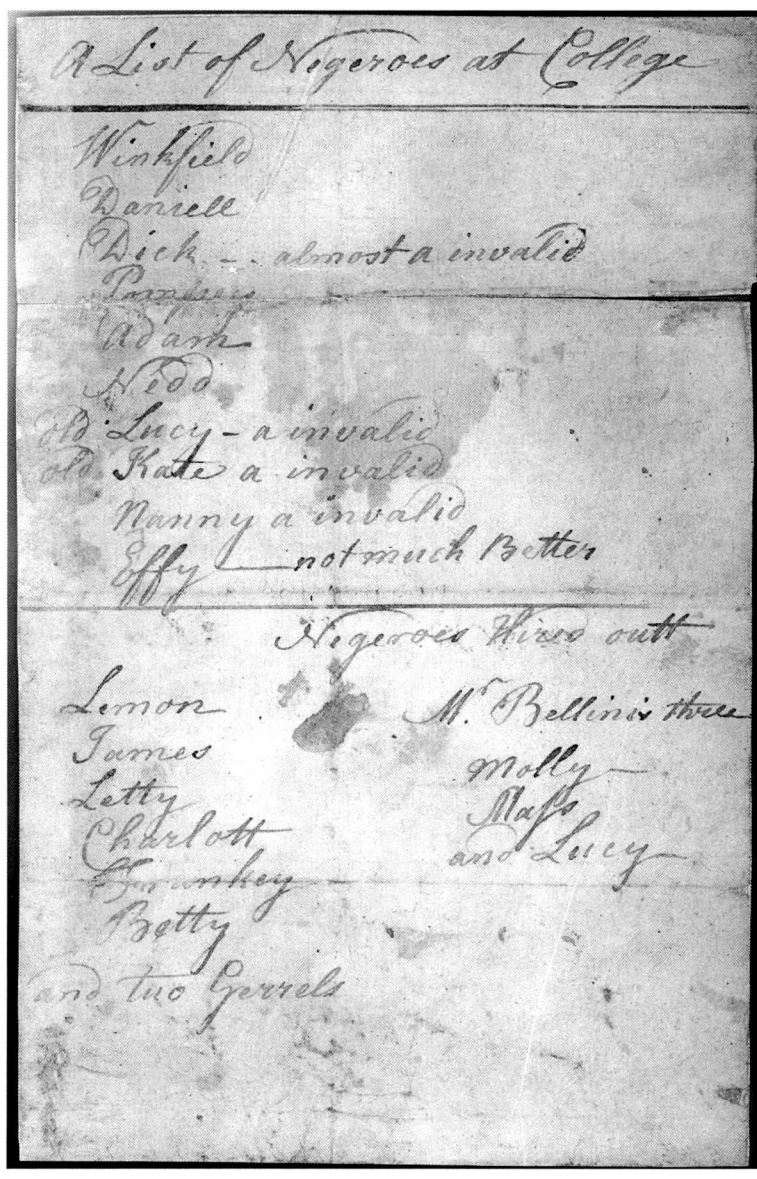

Winkfield, who was of particular service to Randolph Jefferson, heads the list of slaves owned by the College of William and Mary, c. 1780. (RECORDS RELATING TO THE OFFICE OF THE PRESIDENT, SPECIAL COLLECTIONS RESEARCH CENTER, SWEM LIBRARY, COLLEGE OF WILLIAM AND MARY)

several young men with particularly close ties to the Randolph and Jefferson families. Others would become entwined with them.

These students included three cousins, Peyton Randolph (enrolled 1771-1774) and two sons of Col. Peter Randolph, Beverley Randolph (enrolled 1771-1772) and Robert Randolph (who may have arrived in 1772, 1773-1776). Col. Archibald Cary, as the executor of Peter Randolph's estate, paid the boarding bills for Robert and Beverley. Beverley, who was a year or two older than Randolph Jefferson, boarded at William and Mary from March 1771 - March 1772, so he was there when Randolph arrived at school in October of 1771. Beverley's political star rose quickly; he became Governor of Virginia in 1788, at age thirty-four.[31]

St. George Tucker, son of Henry Tucker, Esq. of Bermuda, boarded at William and Mary from March 25, 1772 - December 10, 1772. Whether he and Randolph Jefferson became friendly during that school year is unknown. Their families, however, eventually would mingle. Tucker, born in Bermuda in 1752, was three years older than Randolph. His education was typical for his social station. Before arriving in Williamsburg, he attended a Bermudian grammar school, following instruction at home. Tucker enrolled at William and Mary specifically to read law with Thomas Jefferson's mentor, George Wythe, then, in 1778, married Frances Bland, the widow of John Randolph of Matoax. As a result of this marriage, Tucker became step-father to her sons, Richard, John and Theodorike Randolph, and became responsible for their education. In addition to his law practice, Tucker taught law at William and Mary. He eventually made his home in Williamsburg, where he is credited with constructing the town's first bathroom. His house remains one of the most admired in Williamsburg.[32]

Randolph Jefferson may have already known Robert Burton of Albemarle, who was a student in 1772 and elected writing master at the College in 1773, and Walker Maury (enrolled 1771-1775), son of Rev. James Maury, Thomas Jefferson's former teacher.[33] Walker Maury, who was three years Randolph's senior, came to Williamsburg from Louisa County. After leaving the College in 1775, Maury first taught in Orange County, where he offered instruction in English, Latin, and Greek. Board and tuition at his school was £25 per year.[34] After William and Mary's Grammar School closed in 1779, Walker Maury returned to Williamsburg and opened his own Grammar School.[35] Among his students there were the Randolph stepsons of classmate St. George Tucker, one of whom would eventually style himself John Randolph of Roanoke, the eccentric, anti-Jefferson politician.[36]

Additionally, three sons of the Hon. John Page, Esq., were at William and Mary during 1771-1772. They were first cousins to John Page (1744-1808), classmate and close friend of Thomas Jefferson, who had graduated from William and Mary in 1763. Brothers John and William Page ordered their caps and gowns in February of 1771, apparently finishing school before Randolph Jefferson arrived in the Fall; however, their younger brother, Carter (enrolled 1771-1776), continued to board in the College building during Randolph's time in Williamsburg.[37]

Born in 1758, Carter Page was slightly younger than Randolph Jefferson and may have first attended the Grammar School when he arrived in 1771. Upon leaving William and Mary, Carter Page married Mary "Polly" Cary, daughter of Archibald Cary, in 1783 at Tuckahoe, connecting him with the Randolph family. His second marriage, in 1799, was to Lucy Nelson, the daughter of Governor Thomas Nelson; at his death, he was Maj. Carter Page, and his widow received a pension for his services during the Revolutionary War.[38]

Any and all of these young gentlemen might have been among Randolph Jefferson's friends or even roomed with him. As boarders at the College, they all were in proximity of each other, taking breakfast, dinner, and supper in the College's Common Room, located on the second floor of the great building. Next to it was the Convocation Room, where the Board of Visitors met to direct the affairs of the college.

All classes were held in the building, as well. Classrooms, including the Grammar School, were located on the first floor, in the east range of the main building. At either end were the lecture rooms for Moral Philosophy (north) and Natural Philosophy (south). The Grammar School occupied a large space just north of the entry. A covered, cloistered walkway ("piazza") ran behind these rooms, providing a protected passage between the rear wings of the building.[39]

On the first floor, the Chapel and the Great Hall created the parallel sections of the building's "U." In the Great Hall, the boys attended concerts, dances, oratory, and stood for public examinations. The Chapel accommodated the Anglican services, a required component of every scholar's education.

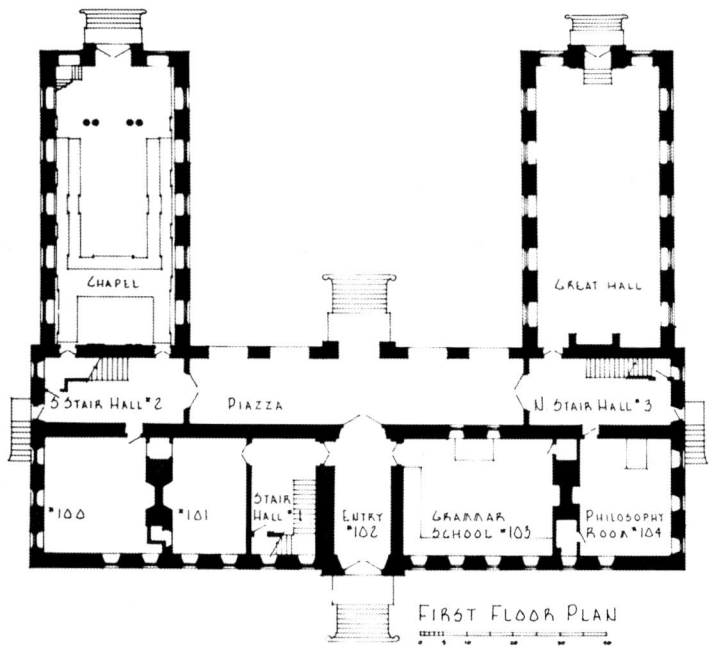

FIRST FLOOR PLAN

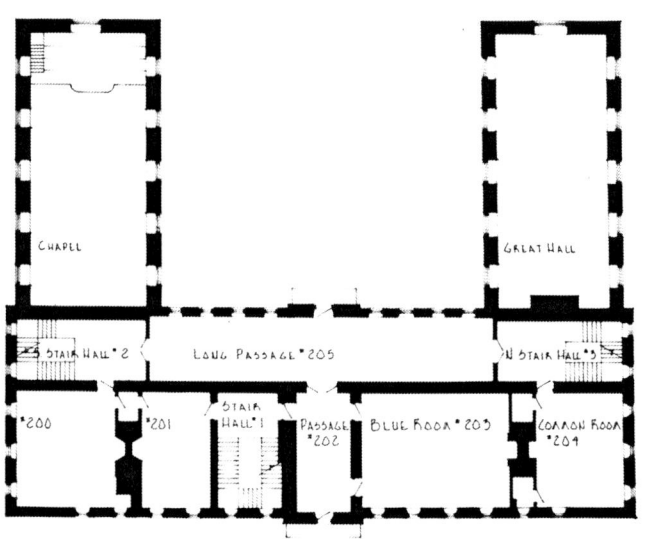

SECOND FLOOR PLAN

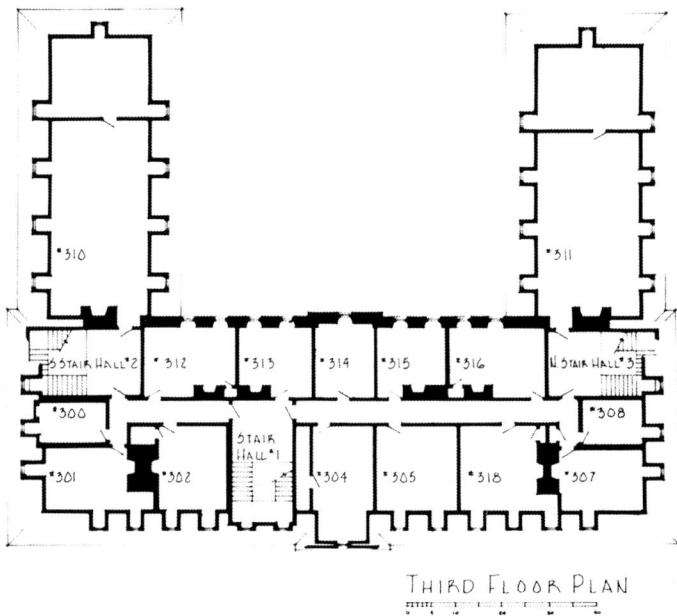

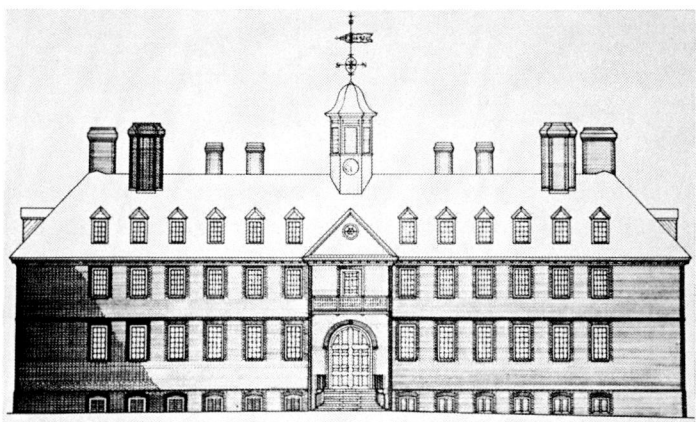

When Randolph Jefferson attended classes at the College of William and Mary, the Wren Building served both the Grammar School and College students. Many students, including Randolph, boarded in the building as well. (IMAGES COURTESY OF SPECIAL COLLECTIONS, JOHN D. ROCKEFELLER, JR. LIBRARY, THE COLONIAL WILLIAMSBURG FOUNDATION)

Daily prayers at 6 a.m. began the students' day in the summer months. Winter brought a slight reprieve, with prayers at 7 a.m. The day closed with evening prayers. This routine no doubt dramatically escalated Randolph Jefferson's devotions; in Albemarle formal worship services were infrequent due to sparse population and vast distances between churches. Additionally, his new tutor was an Oxford-educated, ordained minister, a sharp contrast to his previous teacher, Benjamin Sneed, an Albemarle local.[40]

The bustling atmosphere at the College and in the town of Williamsburg may have come as a bit of a shock to Randolph Jefferson, who had spent the first fifteen years of his life in the relative quiet of rural central Virginia. He was not alone in his adjustment. The town could be enticing, with its opportunities for gambling, drinking, and racing, as well as overwhelming. The faculty often struggled to maintain tight order among the young men. At nine o'clock each night a roll call was read in the Commons, the building's institutional-size "parlor." With the masters and the president present, each student was accounted for before retiring to his room or dormitory. Additionally, three times a week, ushers visited the students' rooms, spotting infractions and reporting on misconduct.

Discipline among the students was a persistent problem. Following Thomas Jefferson's time at the College, the situation continued to escalate. Author Mark R. Wenger summarized the situation:

> By the late 1760s, attendance at Commons had become so sporadic that the faculty was compelled to make participation mandatory for students and ushers. Similar efforts to enforce attendance at chapel reflected a declining compliance in that observance as well, but disciplinary problems were not limited to not appearing at communal functions. Several years before [Thomas] Jefferson's arrival, the faculty had decreed that no student should "presume to Lie, or curse or swear, or talk or do any Thing obscene, or quarrel and fight, or play at Cards or Dice, or set in to Drinking, or do any Thing else that is contrary to good Manners."[41]

John Murray 4th Earl of Dunmore (1732- 1809) by Sir Joshua Reynolds, 1765. Here, Lord Dunmore is depicted wearing the Highland Dress of the 3rd Regiment of Foot Guards prior to his term as the last Governor of Virginia. (COURTESY OF SCOTTISH NATIONAL PORTRAIT GALLERY, PURCHASED 1992 WITH CONTRIBUTIONS FROM THE ART FUND AND THE NATIONAL HERITAGE MEMORIAL FUND)

Conditions at William and Mary had reached a nadir in the 1760s, but by the time Randolph Jefferson arrived, the situation was improving and the students behaved in a relatively orderly fashion. Between 1769 and 1771, various new policies were put into place, including the Bursar's collection of funds and the insistence of prompt arrival of scholars at the beginning of terms. Expansion and building repairs were planned, both concurrent with Randolph's tenure. Lord Dunmore (John Murray, fourth Earl of Dunmore and the last Royal Governor of Virginia), in his capacity as rector of the Board of Visitors, embarked on an addition to the College building. He engaged Thomas Jefferson, who prepared a floor plan which would more than double the size of the building, turning the existing "U" shape into an enclosed quadrangle. The piazza would then line the four sides, creating a covered walkway for the inner court.[42]

As part of the reforms, changes in academic policy were also enacted. Students were now expected to arrive promptly when the term began, and academic procedures became more systematic, which included the establishment of examinations which Grammar School students were required to pass before they progressed to the College.[43]

Additionally, in 1770-1771, the collection of revenues for the College was examined and redesigned. New policies included requesting student payments for board in advance of their next term. Unsurprisingly, in debt-ridden Virginia, fathers and guardians of students were frequently delinquent in their payments. Beyond the problem of collecting fees, actual costs at William and Mary exceeded what students paid in tuition. Notably, Thomas Jefferson paid Randolph's bills promptly.

Still, there were persistent discipline problems during Randolph's year in school. On August 11, 1772, at the Meeting of President and the Masters of the College, which included Rev. Mr. John Camm, newly appointed president, Josiah Johnson, and Randolph Jefferson's tutor, Thomas Gwatkin, it was resolved that "a person be employ'd to attend constantly at the College, & take particular care that no Damage be done to the Buildings or Furniture; that if he should see any of the young Gentlemen committing any Waste, he give immediate Intelligence of the same to the President & Society who engage to support him;

that he be subject to any farther Directions, & that he be allow'd at the Rate of £30 Pr. Ann: for his Trouble."[44]

Thus, when Randolph Jefferson arrived in 1771, he stepped into a sea of sweeping change at William and Mary. The betterment of the College and its components was the goal, and Thomas' input would become a part of it that very year.

Over the course of his life, Thomas Jefferson enjoyed giving advice concerning education, both formal and informal, and one of the great achievements of his life would be his contribution to the conception and creation of the University of Virginia. Directing and advising his younger brother's advanced education was surely no exception. While there is no surviving description of Thomas' educational plan for Randolph, the results indicate that he devised a sort of accelerated Grammar School experience for his younger brother.

William and Mary's Grammar School provided Randolph with a natural progression from the education he had already received at Benjamin Sneed's English School. Its curriculum included lessons in catechism, advanced writing skills, Latin, and Greek; it was considered preparation for the College and was comparable to the curriculum at Rev. James Maury's school, where Thomas had studied before coming to Williamsburg as an "undergraduate."

Throughout the colonial era, the Grammar School represented the largest component of instruction at William and Mary. By 1771, it was under the direction of Rev. Josiah Johnson, who was a grammar master, and had been sent to Virginia from England by Bishop Terrick of London. The Grammar School utilized the best educators at the College. It was also innovative; Hugh Jones, renowned Professor of Mathematics and Philosophy, pioneered instructional methods there for teaching grammar.[45]

In addition to the coursework Randolph Jefferson tackled at the Grammar School level, he also studied privately with Rev. Mr. Thomas Gwatkin, who taught Mathematics and Natural Philosophy for the College. Advancing to a complete course of study at the College demanded mastery in the classical languages, something Randolph likely never accomplished. By adding Rev. Gwatkin's personal instruction to the Grammar School curriculum, Randolph benefitted from studies in Natural Philosophy and Mathematics, which reflected the College offerings, without competency in Latin and Greek.[46]

No correspondence survives from Randolph Jefferson during his time in Williamsburg. If he wrote to his mother, the letters were either lost or destroyed. However, the concerns of grammar school boys were fairly universal. A generation after Randolph attended William and Mary, young Harry Tucker wrote to his father, St. George Tucker.

> My dear papa,
> I now sit down to write you a few lines to show you my filial affection, and at the same time to let you know how we all are. Mama has given Brother Tudor and myself leave to go to a barbeque which the boys are to have at college on Saturday, and more-over they are to run footraces. I have often wished to go to many places with the boys, but have said to myself, Would my papa like it? Would he do a thing his papa had bid him not to do? And by asking that question I hardly want to go anywhere with them, since I am sure you do it only for our good. I am now reading Cicero which though very hard yet it is very pretty, and I am also reading that part of Virgil where the Trojans & Rutuli are engaged in battles: on the Rutulian side because Euryalus and Nisus two youths who were sent to Aeneas slew many of the Rutulian chiefs: who being enraged slew Euryalus, and Nisus slew himself seeing his friend dead. The Rutuli then cut off their heads and set them on long spears and then engaged in battle. I have one thing now to tell you, and that is that poor granny has had a very sore eye though it is getting better. Adieu my Dear papa and believe me to be your affectionate and dutiful son.
> Henry St. George Tucker.[47]

Had the fatherless Randolph Jefferson similarly invoked his brother, asking, "Would my brother like it?" Or did Randolph have some other paragon as his model of behavior?

A payment of £8 made to Rev. Johnson for Randolph's "entrance" in the Grammar School clearly indicates that he was at least initially enrolled there, primarily to acquire the ancient languages, the next necessary step in his formal education. Likewise, a payment of £10 made to Rev. Thomas Gwatkin clearly indicates Randolph's supplementary tutelage. These experiences may have been simultaneous or sequential.[48]

Rev. Thomas Gwatkin, the man who served as Randolph's counterpart to Dr. William Small, was born in 1741 in Herefordshire, England and was a comparative newcomer to Williamsburg. Educated at Jesus College, Oxford University, he matriculated there on July 16, 1763. Four years later, in 1767, he was ordained by the Bishop of London, Richard Terrick, who was also Chancellor of the College of William and Mary. Terrick nominated the twenty-eight-year-old Gwatkin to his post in Virginia. In January of 1770, Gwatkin arrived in Williamsburg, began by teaching Natural Philosophy and Mathematics and, in 1775, added Language to his lectures.[49]

Gwatkin was unmarried and a relatively youthful thirty years old when Randolph arrived in Williamsburg; he may have provided a new "acquaintance," as well as a tutor. He certainly could have functioned as an advisor for Randolph, standing in for Thomas Jefferson. Randolph was not his only private student. In 1774, Gwatkin tutored Governor Lord Dunmore's eldest son, Viscount Fincastle. Later, Dunmore presented him with a gold watch in appreciation for his work with the boy, who would become the 5th Earl of Dunmore.[50]

Gwatkin's curriculum was rigorous, reflecting the College's reforms toward a more structured course of study:

> Under Thomas Gwatkin, the progression of work in mathematics was carefully prescribed in four courses. First the student worked through six books of Euclid, hearing propositions explained on one day and demonstrating them at the next meeting. Then he studied plane trigonometry, including work in surveying and the use of logarithms; next came algebra. The third course stressed the properties of mechanical powers and the use of globes. It involved those aspects of physics that could be comprehended without a previous knowledge of solid geometry, conics, the elements of fractions and physical astronomy. Gwatkin thought that a diligent student could master the first three courses in two years, learning all the practical mathematics needed for measuring timber and the like. The entire series, he said, could be completed in three years. At William and Mary, then, each

week college students received two days' instructions in mathematics, one in natural philosophy, and three in the various phases of rhetoric, logic, and moral philosophy. No college in America placed more emphasis on mathematics and physics, those studies advocated by most progressive reformers.[51]

It is unknown where or how often Rev. Gwatkin and Randolph Jefferson met for their tutorials. Randolph might have attended Gwatkin's lectures as well. Gwatkin likely adapted his college-level curriculum to Randolph's background and abilities; however, he may have held firm to his position that his job was to teach reasoning, not practical "math and science." While Gwatkin taught some practical mathematics, including surveying principles, he was vehemently against abandoning a philosophical approach for a practical one. In August of 1771, just before Randolph Jefferson's arrival, Gwatkin had been quite vocal about his position.

Divinity Professor John Dixon had suggested a more practical curriculum for William and Mary, specifically one that would discard Euclid and teach surveying, navigation, and measuring. Gwatkin responded to Dixon's suggestion with a lengthy letter to the *Virginia Gazette* indicating his contempt for the suggestion that young Gentlemen might "receive the education of CARPENTERS and EXCISEMEN," rather than the fundamentals of "just reasoning."[52]

Indeed, the tutor who greeted Randolph Jefferson was no quiet academic, but a hotly involved, outspoken man of controversial opinions. He did not hesitate to take strong public stances and was not long in Williamsburg before he took a stand which endeared him to Virginians. In the spring of 1771, Gwatkin was one of four men who opposed a petition to appoint an American Bishop to the Episcopal Church. The General Assembly thanked them "for their steady and well-timed opposition to a scheme so detrimental to the interest of Society."[53]

As the decade progressed, however, Gwatkin's British loyalties caused increasing problems for him within the community. When he refused an invitation from Richard Henry Lee and Thomas Jefferson to write a defense of the Continental Congress, a gang of armed men came to the College, threatening harm to him if he did not change his mind. Gwatkin refused to comply, standing on his oath of allegiance to The Crown. The harassment continued, which he believed endangered his life and caused him life-long ill health.

So Gwatkin sought the protection of Lord Dunmore, still the Royal Governor of Virginia, which immediately lost the professor his post, salary, personal papers, personal library, and household furniture. Returning to England now seemed his only choice. He fled Williamsburg with Lady Dunmore and his pupil, Viscount Fincastle, first on the *Flowey*, then on the *Magdalen* to England on June 29, 1775.[54]

Upon Rev. Gwatkin's return to England, he earned an M.A. at Christ Church, Oxford. Though he never returned to Virginia, he maintained concern for his students there. In an undated letter written from Black Lion Inn, Water Lane, Fleet Street, London, Gwatkin wrote that he hoped "the troubles" in America would soon end and that he might be restored to his professorship. He desired that his pupils at William and Mary not be dispersed and lost to him and wished that Rev. John Bracken, minister of the Bruton Parish Church in Williamsburg, would take care of them. Bracken, unlike Gwatkin, was then "in

Old Bruton Church, Williamsburg, Virginia, in the Time of Lord Dunmore, by Alfred Wordsworth Thompson, 1893. (COURTESY OF THE METROPOLITAN MUSEUM OF ART)

favour with the ruling powers of the colony." Gwatkin suggested that Bracken receive one-half of his salary, paid from the £200 still owed him.[55]

During Randolph Jefferson's time in Williamsburg, however, the political waters were still fairly calm for Rev. Gwatkin. In 1772, he recommended a reading list for the so-called "Flat Hat Club," of which he also may have been a member or an advisor. Established in 1750, this secret society used the Latin letters F.H.C. and was known to the outside as the Flat Hat Club. It was America's first "fraternity." Thomas Jefferson joined while he was a student at William and Mary, as did Randolph's classmate, St. George Tucker. It is unknown if Randolph Jefferson was a member.[56]

By late summer of 1772, Randolph Jefferson decided to leave William and Mary. The circumstances of this decision are unknown. Will Beck returned to Williamsburg to escort Randolph home, leaving about September 11th and arriving at Shadwell in mid-September. The expenses of the trip were charged to Randolph's accounts, which Thomas Jefferson still kept as part of the Peter Jefferson estate.

> 7 Sept 1772: Credit W. Beck 7 day's himself and horse going to Wmsburgh. for R. Jefferson and charge Randolph, two days hire of W. Beck, to wit going from Mr. Wayles's (whither I sent him on my own business) to Wmsburg, and back again @ 5/pr. day, their ferriages 1/6 and hire of a horse from Ford 7 days @ 1/3 in all to R.J. 20/3.[57]

When Thomas Jefferson sent Will Beck to Williamsburg to collect his brother, he was at Monticello awaiting the birth of his daughter, Martha, which occurred on September 27th. By the time little "Patsy" arrived, Randolph was presumably ensconced with his mother and sisters at Shadwell and within two weeks, Thomas was off to Williamsburg again, arriving there on October 11th. While there, Thomas personally settled Randolph's accounts. He paid Winkfield at the College and also paid Rev. Mr. Thomas Gwatkin and Rev. Mr. Josiah Johnson for Randolph's schooling. Thomas remained in Williamsburg until November 2nd, when he was reimbursed by the College Bursar for an overpayment in Randolph's board.[58]

October 13: Pd. Winkey at College for R. Jefferson
... 22/6 as per acct. & rect.
October 28: Pd. Revd. Mr. Gwatkins for 10. months schooling
Rand. Jefferson £10.0.0.
Pd. do. for Revd. Mr. Johnson entrance money for
do. in Grammar school £8.0.0.[191]

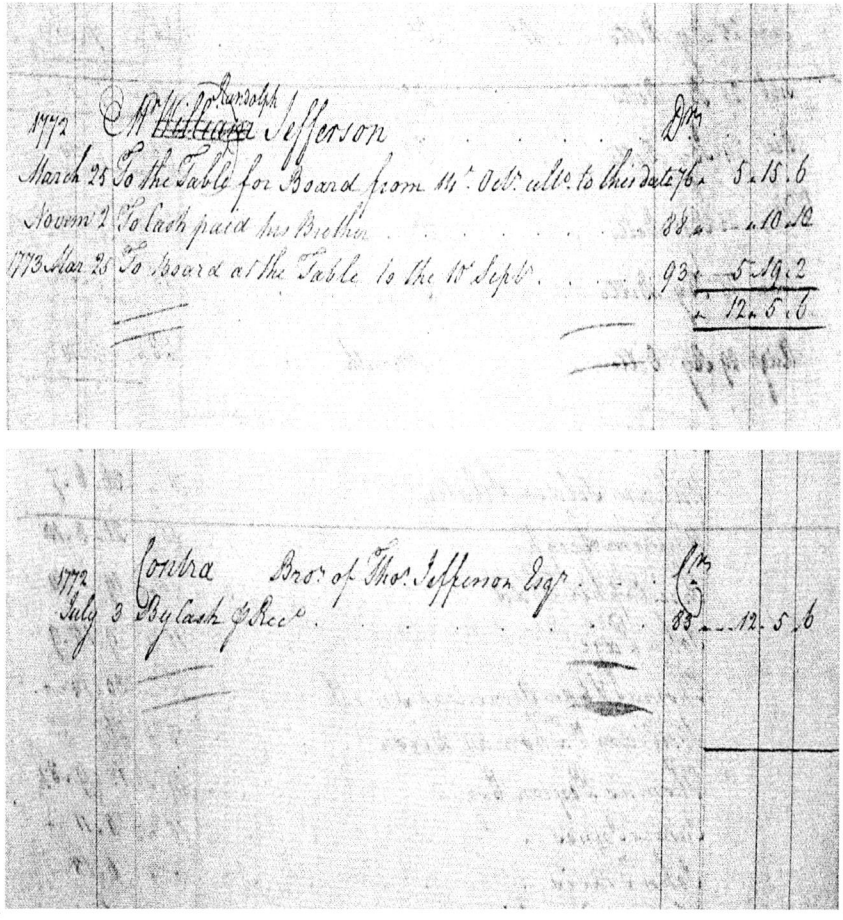

Thomas Jefferson, as Randolph Jefferson's informal guardian, settled his brother's accounts at the College of William and Mary. The Bursar first miswrote Randolph's name as "William," then inserted the correction. (OFFICE OF THE BURSAR RECORDS, SPECIAL COLLECTIONS RESEARCH CENTER, SWEM LIBRARY, COLLEGE OF WILLIAM AND MARY)

The Bursar's Books for William and Mary clearly show that Randolph was charged for and paid for board for two terms, which totaled £12.5.6. An additional, unspecified charge of £6.10.0 was posted in the daily Journal under Randolph's name on July 3, 1772.[59]

Ledger
1772:   Randolph Jefferson
        March 25   To the Table for Board
                      from 14th Oct – to this date   5.15.6
                      November 2 To Cash paid his Brother   0.10.10[60]

1773   March 25   To board at the Table
                      to the 10 Sept. 5.19.2   12.5.6[61]
1772   Bro. of Thos. Jefferson Esq.
        July 3   By Cash Recd   12.5.6[62]

Journal
1772   July 3   To Randolph Jefferson   6.10.0[63]

It is likely that £6/10/0 was the standard charge for board per term. In 1777, three Braxton boys, Carter, Corbin, and George (son of Carter Braxton, Esq.) were debited with £6/10/0 to their balance on July 22, 1777. No explanation is given for the charge. However, Daniel and Theo. Fitzhugh also owed £6/10/0 for "Board charged for six months on December 26, 1777."[64] Randolph's reduced charges for both the spring and fall terms are likely because he arrived "late" (4 October 1771) and left "early" (10 September 1772).

Bursar Miller's dates are additionally confusing, implying that Randolph's tuition expenses extend into 1773. This may have led Bernard Mayo, author of *Thomas Jefferson and His Unknown Brother*, to conclude that Randolph attended classes until early 1773.[65] However, there is no evidence in the Jefferson accounts to indicate that Randolph returned to William and Mary after September 1772.

The irregularities in Randolph's account might be explained further by the adoption of a new collection policy in 1771. Shortly following Randolph's enrollment at William and Mary, Bursar Robert Miller announced in the *Virginia Gazette* that future students must pay their six months board in advance.

A Salutary Order of the Visitors and Governors, directing every Boarder, on his Admission into the College, to pay for six Months Board in Advance, is evaded and rendered nugatory if the like Sum be not paid per Advance by every Boarder at the End of each Half Year. All, therefore, who are in Arrears for Board, are required to discharged them by said 10th of May. On failure, the Boarders shall be dismissed from the College until the 10th of May next, or are recovered by Suit.[66]

This new policy may explain the debit to Randolph for £6.10.0, dated July 3, 1772. It constitutes his "bill" in anticipation of the fall term and is an indication that Randolph intended to stay in Williamsburg for 1772-1773. While there is no direct explanation as to why Randolph left school, it is possible that illness or accident interfered with his plans, causing a premature return to Albemarle.

While in Williamsburg, Randolph was attended by Dr. John M. Galt, whose bill eventually totaled £2.9.4.[67] The nature of Randolph's illness is not disclosed in Thomas Jefferson's accounts, but is simply described as "Pd. Doct. Galt for medicine &ct. to R. Jeff while at college." Additionally, the payment to Dr. Galt was delayed until May of 1777, obscuring precisely when Randolph received treatment.[68]

Dr. John Minson Galt (1744-1808) was a Virginian who had attended the College of William and Mary, then served his apprenticeship in medicine, which was followed by a year at St. Thomas' Hospital in London. There, he trained under London's eminent Hugh Smith and attended lectures on midwifery by Colin McKenzie. Galt continued his studies in Edinburgh and Paris between 1765 and 1767. He worked as a surgeon for the Hudson Bay Company, then settled in Williamsburg where he opened his apothecary. On February 2, 1769, Galt announced his return to Williamsburg in the *Virginia Gazette*.

The subscriber, who is just arrived from London, purposes settling in Williamsburg, where he intends practicing as a SURGEON,

APOTHECARY, and MAN-MIDWIFE; and hopes, from the application he has made in these branches, to be able to give satisfaction. Those who will please to favour him with their employ may depend upon the strictest attendance.
JOHN MINSON GALT[69]

On September 21, 1769, Galt advertised the importation of "a fresh and complete Assortment of DRUGS and MEDICINES, chymical and galenical" which he offered to be sold "at a very low advance for ready money."[70] Concurrent with Randolph Jefferson's arrival, Galt ran an advertisement on October 24, 1771, which announced, "Just IMPORTED in *the* Chatham, *Captain* Anderson *and to be* SOLD *reasonable for ready Money, a large Assortment of* DRUGS *and* MEDICINES."[71]

Patent medicines were promoted even by highly educated doctors, and Galt was no exception. In April of 1772, while Randolph was still in Williamsburg, Dr. Galt advertised the wonders of Dr. Keyser's Famous Pills, which he sold at his shop near the Capitol. Testimonials from four Dukes and a Lord attested to their miraculous powers. Keyser's Pills cured "Rheumatism, Asthma, Apoplexy, Palsy, Giddiness of the Head, Dropsey, the Sciatica . . . Diseases of the Eyes . . . and afforded vast relief in the Gout." They were also recommended for a complaint called "the Flour Albus" and combated "Every Appearance of a Certain Fashionable Distemper." Purchases larger than a dozen received a discount. Interestingly, despite the fact that Dr. Galt's advertisements insisted on "ready money," he waited at least four and one half years for his payment for treating Randolph Jefferson.

Because Dr. Galt's services as a pharmacist, physician, and surgeon were so global, it is impossible to deduce Randolph Jefferson's complaint, which could have been anything from a bad tooth to a broken leg. It is interesting, however, that the younger Galt, rather than the elder Dr. William Pasteur who had established his first Williamsburg shop in 1759, attended Randolph, especially since both the Jeffersons and Pasteur had a connection with the George Gilmer family.[72]

Another delayed payment in the Jefferson accounts further suggests that Randolph may have come home suddenly. When Randolph arrived at Shadwell in September of 1772, his personal property apparently did not accompany him, suggesting that between September of 1772 and January of 1773, Randolph may have expected to return to Williamsburg, but ultimately did not.[73]

On January 26, 1773, Thomas Jefferson noted: "I am to pay Wm. Page 10/ for bringing things from Wmsburgh. for myself. Also 10/ for Randolph Jefferson for bringing his bed and trunk up frm. Wmsburgh." William Page did not get the 20 shillings until February 20th.[74]

From Thomas' notes it is impossible to tell precisely when William Page brought Randolph's personal belongings back to Shadwell; it may have been considerably earlier than January. During January and February of 1773, Thomas was "on the road" when he made the notation about Page's transportation of both his own and Randolph's things from Williamsburg. This may, in part, account for the late payment to Mr. Page.[75]

Interestingly, in the records maintained by William and Mary, Randolph Jefferson's enrollment status is that of an alumnus of the College. In 1941, a provisional list of William and Mary alumni was published. It includes students who attended the Grammar School, and Randolph does not appear among them. Rather, he is listed as an alumnus of the College, alphabetically preceding his brother, Thomas.[76] It is possible that Thomas Jefferson agreed with this perception and thought of his brother as being "at College" rather than "at School," which is reflected in his accounting of Randolph's stay. On October 13, 1772, when Thomas paid Winkfield for his services to Randolph, he wrote "Pd. Winkey at College for R. Jefferson." He also refers to "my brother's board at college." Rev. Thomas Gwatkin's contribution to Randolph's education necessarily meant that his experience there transcended the Grammar School and may have determined Thomas' view of his brother's status.[77]

While many of Thomas Jefferson's biographers have concluded that Randolph Jefferson was ultimately incapable of rigorous academics, he may have been fundamentally disinterested in the world of the mind. Perhaps, like their father, Peter Jefferson, Randolph was essentially an active man attracted to the outdoors. The absence of records concerning Randolph Jefferson, both personal and public, makes it extremely difficult to draw any precise conclusions about what he gained intellectually from his formal education.[78] Whatever Randolph's orientation toward academics, it can be concluded that Thomas Jefferson, who directed his younger brother's education and controlled his funds until he turned twenty-one, felt that Randolph could and would benefit from higher education and society in Williamsburg.

Despite what may have been a premature departure from William and Mary, Randolph spent nearly a year in an atmosphere very different from the Virginia Piedmont and was undoubtedly altered by the experience. At William and Mary, he was exposed to Virginia's most eminent men and lived among the young gentlemen who would become the "power elite" of the next generation. While it is doubtful that Randolph's presence had a significant influence on them, this time in Williamsburg likely afforded him the typical variety of experiences that make living away from home so valuable. His outlook was indeed enlarged, and he may have successfully gained "a more universal Acquaintance."

While at William and Mary, Randolph was immersed in new ideas and learned new skills in the classroom. He encountered the bustle of Williamsburg, complete with everything from the intrigue of Colonial government to fresh imports of all kinds. Boarding at college brought with it the opportunity to form new connections and friends. He met, talked, and studied with numerous men who had been educated in England, Scotland, and France. Rev. Thomas Gwatkin's fondness for his students easily may have extended to young Randolph Jefferson, and Dr. Galt's considerable skill possibly alleviated a serious physical crisis.

Ultimately, the cumulative effects of this "year abroad" for Randolph Jefferson will not be measured in "arts and letters" or even in the political realm. Still, the impact likely resulted in an overall maturation and a general increase in competency, which *will* be reflected in Randolph's management of Snowdon, his contribution as a citizen of Buckingham County, and the rearing of his children.

CHAPTER FIVE

Randolph Jefferson, Patriot

The ground of liberty is to be gained by inches....
Thomas Jefferson to Charles Clay, 1790

Randolph Jefferson's transition to full-time proprietor at his Buckingham County plantation, which he called "Snowden," rather than "Snowdon" as his father had, was a gradual one.[1] Between his return to Albemarle from Williamsburg in 1772 and his coming of age in October of 1776, he became acquainted both with his farm and the nearby plantation owned by John Nicholas, who had served as one of Peter Jefferson's executors.

For the years 1773 and 1774, Randolph appears twice on Buckingham County's tax list. He is recorded at Snowden, along with two other white males, John Bowles and Stephen Slatery. There, Randolph is responsible for 13 tythes in 1773 and 14 tythes in 1774. Bowles and Slatery are likely overseers for the farm and/or associated with the ferry at Snowden. The additional tythables are slaves over the age of sixteen.[2]

The second entry places Randolph with a group of men under John Nicholas' name at his plantation, Seven Islands. Nicholas was responsible for 38 tythes in 1773 and 34 tythes in 1774. Along with Nicholas the following white males were taxed: John Johnston, Randolph Jefferson, Patrick Campbell, Daniel McCallum, Samuel Taylor, Jr., Walker Perkins, Jess Bootwright (sic), and Peter Campbell.[3]

With the exception of Walker Perkins, the nephew of Snowden neighbor Hardin Perkins, the men listed with Randolph Jefferson at Seven Islands are not known to be later associated with Snowden. Walker was the son of Mary Walker and Constantine "Constant" Perkins, the older brother of Hardin Perkins, Sr. Born in Goochland County, about 1755, Walker was Randolph Jefferson's peer in age. Following his time at Seven Islands, Walker returned to Goochland County, where he married Judith Hughes on June 5, 1778. He died as a relatively young man in May of 1781.[4]

During the 1770s, Nicholas operated a school at Seven Islands, and John Johnston was the schoolmaster at that time. Johnston's primary job was to instill classical academics, preparing the boys for college; however, these private plantation schools also "finished" their pupils, preparing them for genteel society.[5] There are no known bills paid to Nicholas for Randolph's board or schooling. Beginning in 1773, Randolph apparently continued his education there, as a guest of his father's executor. He could have been a conventional student, continuing with Latin or Greek, although it seems more likely he was there to learn plantation management.

Randolph lacked the advantage of observing his own father when it came to farming and spending time at Seven Islands offered him the opportunity to learn from Nicholas' operations. As a surrogate uncle or mentor, John Nicholas provided a sterling example. He was a Justice of the Peace in Buckingham, an early Vestryman and Church Warden in the county, and he reliably served for decades as Albemarle's County Clerk. An executor of Peter Jefferson's estate and son-in-law to Joshua Fry, Nicholas was practically a member of the Jefferson family.[6]

John Nicholas (1725-1795) inherited his 2,600-acre James River plantation from his father, George Nicholas (1695-1734), and Seven Islands had been his home since about 1748/49. He and his wife, Martha Fry, reared seven children there. Located downriver from Snowden, just west of where Slate River enters the James near present-day Arvonia, the plantation was named for the seven islands which dot the river at that point. All in all, it was very comparable to Snowden and would have served as an excellent example for Randolph to "study."[7]

Though slim evidence, the 1773 and 1774 tythables suggest a possible scenario for Randolph's transitional years from Shadwell to Snowden. When Randolph did not or could not continue his studies in Williamsburg, he returned home, possibly to recuperate from an illness in September of 1772. Then, in 1773, he went to Seven Islands to continue his interrupted education. While in Buckingham, he spent part of his time at Snowden and part with John Nicholas, Sr., with whom he may have developed a close relationship. In 1790, Thomas and Randolph Jefferson would meet at Seven Islands to close their father's estate.

Although Randolph would not fully come into his inheritance until October of 1776, by June of 1773, Snowden was already known as "Mr. Randolph Jefferson's plantation." The 2,000 acre farm sitting at Buckingham County's Horseshoe Bend had been continually worked since Peter Jefferson's death in 1757, and now Randolph Jefferson was ascending as, at least, the plantation's part-time proprietor. On June 17, 1773, an advertisement was placed in the *Virginia Gazette* by Snowden's overseer. His name is not disclosed, and there is currently no clue to his identity. The fact that he, rather than Randolph, placed the advertisement might indicate that Randolph was in residence elsewhere. The advertisement read as follows:

> Taken up, at Mr. *Randolph Jefferson's* Plantation, a small bay MARE, appears to be very old, and is branded on the rear Buttock R; also a dark bay MARE COLT with a large Star in her Forehead, and a Snip on her Nose, but neither docked nor branded. Posted, and appraised to six Pounds. The Owner may have them by applying to the Overseer on the Plantation.[8]

The degree to which Randolph divided his residence between Albemarle and Buckingham during these transitional years is unknown. While he anticipated his inheritance, there was much to draw him home to Shadwell, including his sisters, Anna Scott and Elizabeth, and their mother, Jane. Brother Thomas had established himself and his bride at Monticello in comparatively rude quarters, while he planned and built and rebuilt his ultimately impressive mansion house.

By the spring of 1773, seventeen-year-old Randolph was stepping out in society, entering the adult social world if not the "marriage market." Thomas' account books for their father's estate show that Robert Nicholson tailored £2/1/7 ½-worth of apparel for Randolph. About the same time, a gown and "curls" were purchased for Anna Scott and Randolph may have found himself escorting his twin sister, who was simultaneously coming of age, to the neighborhood balls.[9] The spending did not stop with new outfits. Randolph and Anna Scott ran up quite a bill with Carter and Trent, using funds still held in the Peter Jefferson estate. Brother Thomas eventually paid the agent for Carter & Trent for sundries furnished to Randolph and sister, Nancy, totaling £23.3.8. Thomas also paid for additional custom-ordered work done for Randolph's wardrobe, including £4.12 to tailor Edward Butler and 27 shillings to Eleanor "Nel" Shepard, who was one of two local women who knitted stockings for the Shadwell household.[10]

During this transitional time for Randolph, the Jefferson brothers experienced the sudden death of their sister, Elizabeth, which took place during the extreme weather of February-March of 1774. Beginning with an earthquake on February 21st and the flooding of the Rivanna River, Albemarle was visited with destruction and death came to Shadwell. Randolph's exact whereabouts during the earthquake are unknown; however, if he was not already at Shadwell, he quickly may have been summoned.[11]

On the afternoon of February 21, 1774, Thomas recorded in his memorandum book that he was at Monticello when the earthquake struck at 2:11 p.m. He recorded aftershocks at 2:45 p.m., "as violent but not so long," and another on the following day.[12] This was the first recorded earthquake in Virginia and was felt throughout the state. Church bells rang as far away as Winston-Salem, North Carolina. The estimated magnitude was 4.5 on the Richter Scale.[13]

On March 1st, Thomas wrote, "my sister Elizabeth was found last Thursday being Feb. 24th." Thomas had made no entries in his book between February 25th and February 28th. During those days, he was dealing with his family, Elizabeth's death, and the rising waters of the Rivanna. As the elder son, Thomas was responsible for the financial well-being of his mother, his sister, Elizabeth, and his minor, unmarried siblings. However, his was an over-arching responsibility, not a day-to-day overseeing of the household. At the time of Elizabeth's death, her mother, her sister, Anna Scott, and brother, Randolph, were most intimately responsible for her welfare. Randolph, as the primary masculine presence at Shadwell, may have been "in charge" during the earthquake until Thomas arrived from Monticello.[14]

Elizabeth's funeral was held on March 7th, amidst flood waters.[15] It is unclear how long Elizabeth had been missing. In fact, virtually all the details of her death are unknown. The earthquake and heavy rains likely created confusion, even panic, which ultimately led to her death. Years later, family friend, Wilson Miles Cary, wrote to Sarah N. Randolph, "I have always understood that she was very feeble minded if not an idiot – & that she and her maid were drowned together while attempting to cross the Rivanna in a skiff."[16]

Elizabeth, who was next to Thomas in birth order, was twenty-nine years old at the time of her death. Her mental deficiencies and inability to judge the danger of the high waters may have led to her demise. She also may have panicked in the strange weather and earthquake. Shadwell servant, Little Sall, died at the same time.[17]

∼

The next significant Jefferson family milestone was the death of Jane (Randolph) Jefferson on March 30, 1776. Despite the fact that Jane had left bequests of slaves to both Randolph and his twin, Anna Scott, this property would never become theirs. Before her death, Jane had "mortgaged" her slaves and transferred them to Thomas to cover her debts to him.[18] Her will had never been rewritten, and the simple document reveals her original wishes to provide for her unmarried children, one of whom had preceded her in death and two of whom were not yet legal adults:

> I Jane Jefferson of the County of Albemarle do make this my Last will & testament, in manner & form following, I give to my daughter Anna Scott my 2 negroes Lucinda old Sals daughter an[d] Belin I – daughter Sharlott, I give to my son Randolph, my two negroes Simon old Sals son and Sirus Little Sals son I give to my daughter Elizabeth all my wearing apparel, with one good bed an[d] furniture, every

thing else I give to my Executor, to be Equally divided among all, perhaps there may be Some debts I am not apprized of, and Lastly I do Constitute my Son Thomas Jefferson my Sole Executor, and Legatee of whatever Else I have power to possess of

Jane Jefferson[19]

When Jane Jefferson's remaining property was sold, twenty-four neighbors and relatives, including her sons, Thomas and Randolph, purchased property from her estate. Total proceeds were recorded on February 19, 1777; purchases were not itemized. Under Randolph's account, Thomas wrote only, "Articles bought at the sale of my mother's estate . . . £4/18/6." As a result, there is no way of knowing exactly what Randolph desired that had belonged to his mother. He might even have purchased something of his father's which had survived the house fire at Shadwell in 1770.

Sister Lucy's husband, Charles Lilburne Lewis, and Anna Scott Jefferson's future husband, Hastings Marks, both acquired items at the sale. Charles Lewis' significant purchase indicates that Lucy enjoyed some of her mother's furnishings at her own home, Monteagle, while Thomas took home the lions' share of Jane Jefferson's property.[20]

Randolph Jefferson	£ 4/18/6
Chas Lilburne Lewis	£25/17/6
Hastings Marks	£11/0/0
Thomas Jefferson	£51/3/70

When Jane Jefferson died in the spring of 1776, Randolph and Anna Scott were still dependent on their father's estate and would not turn twenty-one until October. Shadwell was scarred and very changed from their childhood, shrouded with recent grief. Peter Jefferson's dwelling house was burned, their sisters Jane and Elizabeth deceased, and all the older siblings married. Soon, Randolph would be master of Snowden, leaving Anna Scott adrift for many years, relying on the hospitality of her brothers and sisters.

Randolph Jefferson Comes of Age

Finally, on October 1, 1776, Randolph Jefferson was "his own man" and was presumably well-settled in Buckingham County. His mother's household inventory indicates that Randolph already may have removed his personal furnishings from Shadwell prior to her death, leaving items that accommodated his mother and Anna Scott, as well as the remainder of Elizabeth's things. Jane's inventory lists three bedsteads with bedding, including mattresses, sheets, and blankets. Three chamber pots and two feather beds and a "Virginia Tick [Bed] Bolster and Pillow."[21]

Just as the Shadwell Randolph left behind was very altered from Peter Jefferson's vision, so too was the promise of the Fluvanna lands and his "Snowdon." Randolph Jefferson's eventual "house on the bluff" looked down on a much quieter Scott's Ferry than his father had projected from the 1750s.[22] Having lost the county seat to Charlottesville, the village at Scott's Ferry was much contracted, sleepily waiting for its resurrection. As Scottsville historian, Virginia Moore, put it, "The group of businesses which had grown up around the Old Albemarle Courthouse had, by now, more or less withered away. What use, here, for a smith, saddler, wheelwright, merchant, innkeeper? To survive they needed customers."[23]

Despite Randolph's technical independence, Thomas did not immediately close Randolph's account in their father's estate and, over the next few years, he continued to handle a portion of his brother's fortune, paying a variety of his outstanding bills. During this period, Thomas' memorandum books reveal that the brothers had regular exchanges. These dealings suggest that they now related more or less as peers – both "squires" of their respective plantations – Thomas at the ever-evolving Monticello and Randolph at the newly-claimed Snowden.

Quickly establishing that new life and persona, in January of 1777, Randolph bought a writing table from his brother for £6.[24] While this table graced a main room at Snowden, more importantly, its acquisition creates a picture of Randolph Jefferson's life that typically has been denied him. A writing table indicates he did some writing, or at the very least, some bookkeeping. Though his surviving letters are far from elegant, the simple fact that a single man, in his early twenties, desired a writing table is a somewhat elegant thing. This is especially true for a man frequently characterized as a "muddy-boots" farmer.

A month later, in February of 1777, Randolph purchased a horse from Thomas for the handsome price of £35. The horse was called Ryno, named after a character in James Macpherson's *Fingal: An Ancient Epic Poem*.[25] Apparently, it was stately and beautiful as was Ryno, son of Fingal.

In turn, the following August, Thomas bought a horse named Oroonoko from Randolph, who charged his brother £36. Oroonoko was a variety of tobacco grown in Virginia. It was also the protagonist of *Oroonoko: or, the Royal Slave*, a short novel by Mrs. Aphra Behn, which Thomas Southerne had turned into a Restoration-era tragedy.[26] If Randolph named the horse, he may have been thinking of the tobacco, the slave prince, or both. Many Virginians were indeed enslaved to tobacco. In either case, the name displays some originality, even irony.

Randolph had grown up in a culture attuned to fine, blooded horses (i.e. thoroughbreds), as well as working animals. At the time of Peter Jefferson's death, there were an unusually high number of horses at Shadwell. Of the twenty-two horses itemized in his inventory, roughly half of the entries noted their names. In 1757, Diamond, a sorrel mare, was the prize, valued at £10.1.0. The second most valuable horse, a four-year-old dark bay valued at £10, was not named in the inventory while Jenny Morris, a bay mare who may have seen better days, was valued at only £1. Roans, grays, and bays with the names Fidler, Daubin, Skeltoon, Harry, Jewel, Cupid, Broom, Blaze, Tepsie, and Cherry all graced the fields at Shadwell, surrounded by colts and mares whose names (if they had them) did not appear on the inventory. No stallion is mentioned.[27]

While "an old chair and harness," worth 45 shillings, were included among the miscellaneous tools and tack in Peter Jefferson's inventory, there was no valuable carriage, gig, or other obvious pleasure cart appraised. There was, however, a new "C. Harness four Horse," presumably a harness for a four-horse carriage, valued at 38 shillings. Were the roads so undeveloped in Albemarle that maintaining fancy wheeled transportation was impractical? Later in life, both Jefferson boys will eventually own and pay taxes on pleasure carriages, Randolph's gig being a very modest vehicle compared to Thomas' impressive landau.[28]

Horse racing was a favorite gentlemanly sport in Tidewater Virginia, and Peter Jefferson had grown up immersed in that culture.[29] His father, Thomas Jefferson II, who willed Peter two horses (Norman and Squirell), owned at least one race horse, a winner named Bony.[30] Racing did not appeal to Peter's son, Thomas. His former slave, Isaac, remembered, "Mr. Jefferson had nothing to do with horse-racing or cock-fighting: bought two race-horses: but not in their racing day: bought em arter done runnin."[31] Randolph Jefferson's relationship with racing is unknown; however, in 1797, he stabled a stallion at Snowden, providing an important stud service to the neighborhood to populate the area with quality stock.[32]

Between August 22, 1776, and January 19, 1778, Randolph utilized his brother's blacksmith, resulting in a bill of £2.14.5½. The beautiful Ryno may have been shod there. Since Monticello was a day's ride from Snowden, it seems probable that these were incidental repairs, or perhaps a significant emergency order, when Randolph and/or one of his servants were already in the vicinity. It would have been highly impractical to take Snowden's routine needs to the smith at Monticello.

The entries in the memorandum books for February of 1778 indicate that Thomas Jefferson's smith shop was busy and profitable. His son-in-law, Thomas Mann Randolph, who lived nearby, had work done there as well. Between March 8, 1776, and December 31, 1777, Jefferson's two smiths were credited with £178.9.1 1/4.[33] In *Free Some Day: The African-American Families of Monticello*, Lucia Stanton describes the blacksmith operations at Monticello during the 1770s. The oldest son of the enslaved couple, George and Ursula, was Jefferson's first smith:

> The oldest son, George, learned the blacksmithing trade from Francis Bishop, a white man hired in 1774 to operate Jefferson's first blacksmith shop and to train some of his slaves. After Bishop's departure in 1776, George and another slave, Barnaby, ran the shop until the arrival in 1781 of the hard-drinking British deserter William Orr, whose alcoholic exploits were vividly remembered by George's brother Isaac.... Surviving blacksmithing accounts indicate that, aside from the usual task of shoeing horses, George was primarily engaged

in the repair of agricultural tools and the making of parts for guns and vehicles. Articles that George made in his shop included chisels and plane irons, axes and scythes, bridle bits, and spoons.³⁴

Randolph Jefferson would have known George and the high quality of his work, not hesitating to take advantage of George's shop if he was in the neighborhood.

As 1777 progressed, although correspondence between the brothers does not survive from this period, it becomes clear through Thomas' memos that he and Randolph had ongoing exchanges, apparently representing each other in business and periodically selling things to each other. Thomas twice sold Randolph marrow houghs. On April 20, 1777, he collected £140 from Carter Braxton for Randolph's tobacco crop; it took over a month for the funds to be delivered to Randolph. Later, Randolph paid £6 to an upholsterer in Staunton for some work done for Thomas. In turn, Thomas paid Dr. Galt £2/9/4 for Randolph. In October of 1777, Thomas settled accounts with Dr. Gilmer for both Mrs. Jane Jefferson and his brother. He also collected 30 shillings from Dr. Gilmer for blankets provided by Randolph.³⁵

Then, finally, on September 5, 1779, the brothers closed out Thomas' accounting of Randolph's funds in their father's estate:

1779 Sep 5.
This account was this day settled between us and allowed to be just, the Debtor side amounting to four hundred and ninety one pounds nineteen shillings and eight pence halfpenny, the Credit side amounting to four hundred and thirty one pounds eighteen shillings exclusive of the value of a negro fellow Phill bought by T. Jefferson of R. Jefferson when the latter came of age, at such a price as Mr Nicholas Lewis should value him to, which valuation we omitted to have made at the time, and is therefor still to be done.

TH: Jefferson
Randolph Jefferson³⁶

Randolph Jefferson Joins the Rebellion

Concurrent with Randolph Jefferson's coming of age, the American Colonies ignited into war with England. Every American school child knows that Thomas Jefferson played an enormous role in the American Revolution. His brother's contribution, by contrast, was a very local one, but not entirely without significance. Celebrating his twenty-first birthday in October of 1776, Randolph was old enough not only to witness the rapidly changing climate of colonial politics, but also participate in them.

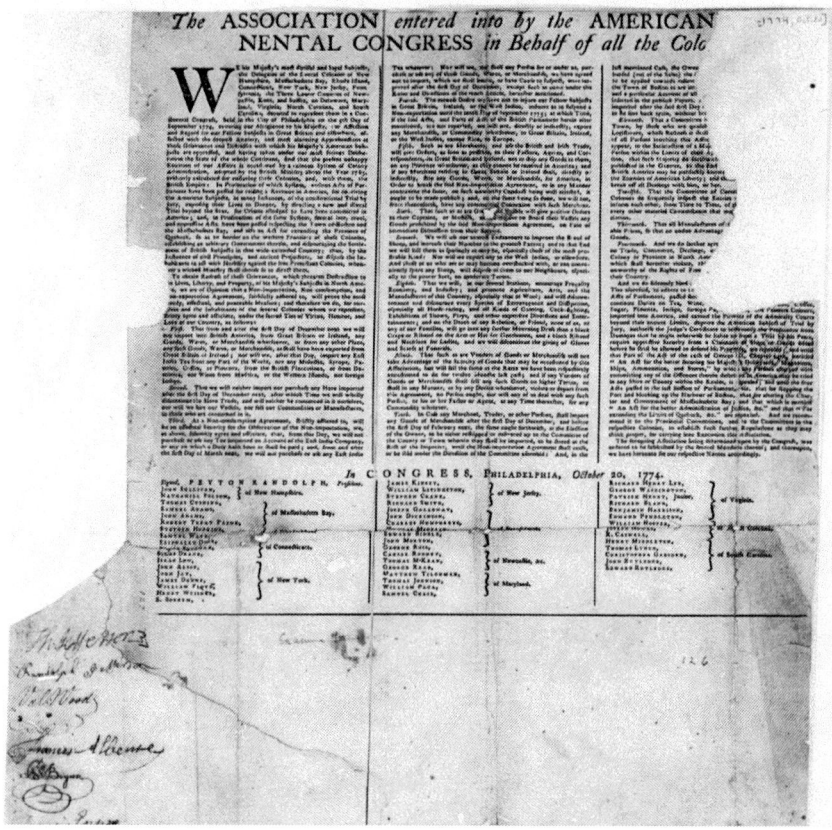

Randolph Jefferson's signature follows his brother's on this Broadside of "The Association entered into by the American Continental Congress in Behalf of all the Colonies," 20 October 1774.
(COURTESY OF LIBRARY OF CONGRESS)

OATH OF ALLEGIANCE IN ALBEMARLE

A handwritten transcript of the so-called "Albemarle Declaration of Independence," dated June 21, 1779, was found in George Gilmer's Commonplace Book. Gilmer's signature is the first listed along with many familiar Albemarle names, including Thomas Jefferson, Randolph Jefferson, and their kinsmen, Charles Lewis, Charles Lilburne Lewis, and Isham Lewis.

We whos(e) names are hereunto subscribed do swear that we renounce and refuse all Allegiance to George the third King of Great Britain, his heirs & successors and that I will be faithfull and bear True Allegiance to the commonwealth of Virginia as a free and independent state, and that I will not at any [time] do or cause to be done, any matter or thing that will be prejudicial or injurious to the freedom and independence thereof as declared by congress and also that we will discover and make known to some one justice of the peace for the said state all treasons or traitorous conspiracies which we now or hereafter shall know to be formed against this or any of the united states of America So help me God

[George Gilmer, *Commonplace Book*, 1775-1820, Mss5:5 G4213:1, Virginia Historical Society.]

Little is known concerning Randolph Jefferson's political orientation. However, an early clue concerning his political leanings is his signature on a somewhat mysterious document. His is one of six signatures on a printed copy of the "Continental Association of 20 October 1774." The Articles implemented a trade boycott with Great Britain. The original document was signed in Philadelphia by Peyton Randolph, President, and the various members of the Continental Congress. In the lower margin of the printed broadside are the handwritten names: Th: Jefferson, Randolph Jefferson, Val: Wood, Francis Alberti, A. Bryan, and Francis Eppes.[37]

All of these men have a connection to Thomas Jefferson. Valentine Wood was one of Albemarle's first Gentleman Justices, serving alongside Peter Jefferson. Francis Alberti was the violin master who instructed both Jefferson brothers. A. Bryan is likely Anderson Bryan, who made a survey for Thomas Jefferson's property, Montalto, in 1774. Francis Eppes, Thomas Jefferson's brother-in-law, was married to Elizabeth Wayles. The precise intent of those signing the broadside is not known. Randolph Jefferson's signature placed directly under his brother's indicates he was an early supporter of dissent.

Though Randolph resided in Buckingham County, during these transitional years of the burgeoning nation he still looked to Albemarle for his political comrades, joining his brother and cousins in their rebellious fervor. Whether he was influenced by Thomas' ideology or had his own brand of political and revolutionary ideas, he certainly signed petitions.

Randolph is listed under Capt. Wingfield's Company on Albemarle's "1776 Militia Return," certified by Thomas Jefferson, County Lieutenant. Randolph is described with the term "Clear," rather than "Service." There is no other known document associating him with the Albemarle Militia, though, after the war, Randolph will serve many years in the Buckingham Militia.[38]

In the spring of 1778, Randolph's attention turned to the Continental Army, and he joined a corps of light horse led by the soldier-statesman, Thomas Nelson, Jr., who followed Thomas Jefferson as Governor of Virginia and fought in the Yorktown campaign. The previous year, in August of 1777, Nelson had been appointed Brigadier General and Commander-in-Chief in the Commonwealth of Virginia.[39] In late 1777, Gen. George Washington felt the need for increased cavalry. Rather than create another regiment of Continental dragoons, Congress passed a resolution on March 2, 1778, calling for "Young Gentleman of Property and spirit of the several states to constitute a troop or troops of light cavalry."[40] Thomas Nelson, partially at his own expense, responded to the call by raising, outfitting, and training such a unit.

In May of 1778, the Virginia Assembly authorized the formation of a regiment. Enlistment lasted for one year, unless Congress released the men sooner. Brigadier General Thomas Nelson was appointed commander and ordered to organize troops.[41] By May 1st, Nelson's brothers, Hugh and Robert, were already in action and placed a notice in the *Virginia Gazette* stating that they were joining the General's Corps of Light Horse and encouraging other gentleman to join them.

> Williamsburg, May 1 [1778]
> HUGH NELSON and Robert Nelson, Esquires, have entered themselves as volunteers in the corps of light dragoons now raising by their brother the General [Thomas Nelson]; and Lewis Burwell, Esq; of Gloucester, after setting on foot a subscription in his county to equip young gentlemen for that service, to which he himself paid 100[?], turned out likewise as a volunteer. When gentlemen of the first rank and fortune set so noble an example, we can hardly entertain a doubt of General NELSON's being soon able to march to join our illustrious Commander in chief with a complete and well appointed regiment of dragoons.[42]

It may have been this very announcement that induced Randolph Jefferson to volunteer. A single man with an overseer to watch his farm, Randolph answered the call. There is no indication that Thomas Jefferson personally approved or disapproved of Randolph's decision; however, Thomas' accounts for Randolph show two important purchases, indicating the young gentleman's preparation.[43] On June 9, 1778, a bill was paid for a "horseman's belt," which cost 30 shillings.[44] Also, nearly a year after the fact, on June 23, 1779, Thomas Jefferson settled an account with William Armistead for a "suit of Uniform clothes" for Randolph. The cost was £3.10.9.[45] Armistead served as the state

Commissary of the Virginia Public Store and as the Commissioner of the Continental loan office in Virginia until June of 1780.[46]

The preparation for uniforms began on April 28th, when the Council directed the Commissary of Stores at Williamsburg to furnish material to each man sufficient to make a cloak, a coat, and a waistcoat. The Council likewise directed the Commissary to provide enough linen for two shirts and two pairs of stockings. Military historian, Charles H. Cureton, describes the uniform.

Randolph Jefferson wore this private's uniform when he served in Nelson's Corp of Light Horse. [BY PERMISSION OF THE ARTIST, LT. COL. CHARLES H. CURETON, USMCR (RET.) AND THE COMPANY OF MILITARY HISTORIANS]

The uniform consisted of blue-faced-blue coats with white lining, buff waistcoats, and leather breeches. Buttons were of gilt brass. Each ensemble also included a cloak of blue cloth with a blue cape (collar). Horsemen's caps were not worn, but the Public Store provided black feather plumes for officers' hats. The weapons carried by the volunteers seem to have been pistols and sabers.[47]

On May 15th, Washington requested that Nelson move his troops to the Main Army, using the most direct route through Baltimore. Precisely where Randolph Jefferson joined the march northward is unknown. The troop stopped to equip in Port Royal, in northeastern Virginia on the Rappahannock River, and he was likely with them by then. On June 5th, any remaining interested young gentlemen were asked to rendezvous at Port Royal. The surviving copy of the *Virginia Gazette* is damaged, but most of the announcement is readable:

> Those gentleman who in - the corps of light horse are desired to rendezvous – – [Port Ro]yal on the 15th instant, or as soon after that day – – horses and accoutrements will be found by the – – for those who are not able to furnish themselves. Clothing will be sent to the place of rendezvous, which will be sold to them at the most reasonable advance, by the publick storekeeper, agreeably to an act of the last General Assembly.
>
> Thomas Nelson, jun.[48]

On June 12, 1778, the pages of the *Virginia Gazette* enthusiastically announced: "We hear that large subscriptions have been made in the different counties of this state for raising and equipping light horsemen to serve under General NELSON. In the small county of King George, the ladies contributed near one hundred pounds – a laudable example."[49]

A letter from Governor Patrick Henry, dated July 14, 1778, authorized £2,500 for the purpose of raising, equipping, and marching the troop of Virginians to Philadelphia, as Congress had requested. Thomas Nelson's costs were far in excess of this, and he furnished "many of the privates with horses from his own estates."[50]

According to George Dabney, a principal manager on one of Nelson's properties, Dabney "was informed by some of the troop, who were gentlemen of respectability, that during the march, general Nelson did not draw on the public for a single ration, but that it was his custom to send an express to the different public houses to have provision made for his troop, which he paid at his own expense." Dabney estimated the expedition cost Nelson over $10,000.[51]

On July 22nd, Nelson and his men headed north to Baltimore. A "return" dated July 29, 1778, indicates that Nelson had with him "103 rank and file, divided into four troops." Ironically, the eager volunteers got no further than Baltimore. By late August, General Washington no longer felt he needed additional cavalry, and Congress released Nelson's unit from Continental service on August 30th. Presumably, Nelson's Corps of Light Horse and Randolph Jefferson returned home to Virginia.[52] Randolph certainly got a four-month adventure, but did he return home satisfied with his efforts toward independence? There is no indication that he attempted to enlist again.

Years later, on May 4, 1823, Green Clay of Madison County, Kentucky, wrote to Thomas Jefferson, remembering his initial foray into war. "I am the Brother of Parson Charles Clay once of Albemarle C°. I went out with a Troop of light horse to the northward in company with your brother Randolph in 1778 from your house."[53] Clay may have referred to Monticello when he wrote to Thomas Jefferson that the young men rode from "your house." Monticello was habitable and expanding in the summer of 1778. However, Jefferson and Clay more likely left from Shadwell, which was very convenient to where the Three Notch'd Road intersected the Old Fredericksburg Road, both major thoroughfares leading out of Charlottesville.[54]

The surviving muster list from Baltimore includes both Randolph Jefferson and Green Clay, along with many other gentlemanly Dragoons present on July 29, 1778. Some of the men, such as George Prince, had extensive war records. For others, like Jefferson, this may have been their primary military experience during the war.[55]

Green Clay (1757-1828) was the son of Charles and Martha Clay of Powhatan County, Virginia. Only nineteen years old when he rode off with Randolph Jefferson to join Nelson's Corp of Light Horse, Clay later fought in the War of 1812, achieving the rank of General. A first cousin to Henry Clay, Green surveyed land in Kentucky, owned several distilleries, a tavern, and numerous ferries along the Kentucky River. At his death, he was purportedly one of the wealthiest men in the state, with extensive land and slave holdings. (COURTESY OF KENTUCKY DEPARTMENT OF PARKS, WHITE HALL STATE HISTORIC SITE)

The Buckingham Home Front

Due to the loss of Buckingham County Militia records for the war years, Randolph Jefferson's role at the county level during the war years is obscured.[56] Neither he nor his widow lived long enough to claim a pension, if he deserved one for service. His name has not been found among the commanding officers mentioned in the surviving pension applications of others, and he was not a commissioned officer in the Buckingham Militia until after the war. This leaves Randolph's public claims as the largest single window into his home front activities.

During the war, citizens were given certificates itemizing their property, impressed or loaned, as well as for services rendered to the Continental Army or local Militias. At the conclusion of the war, between 1781 and 1783, citizens brought their certificates to special sessions held in the county courts. "Court Booklets" were compiled in which the authenticated certificates were listed and then sent to Richmond for settlement. There, two commissioners authorized payment and warrants were issued by the auditor of public accounts.[57]

As Brent Tarter observed in his introduction to *Virginia Revolutionary Publick Claims*, "The appearance of a name in the Public Service Claims is no positive proof that [a] person voluntarily contributed to the American military effort and the absence of a name from the Public Service Claims is no sure proof of loyalism or even of inactivity."[58] That said, the extent to which Randolph Jefferson relinquished and/or loaned his property suggests cooperation on his part. There is no reason to believe that he did not wholeheartedly support the war effort. Indeed, Randolph offered the use of his farm, his crops, his livestock, and his labor force, the total of which ultimately constituted a fairly substantial claim.

In the introduction to the 1981 edition of *Thomas Jefferson and his Unknown Brother*, Randolph Jefferson's contribution to the war effort is summarized: "During the British invasion of 1781, his plantation Snowden furnished provisions for Virginia troops, pasturage for cavalry horses, and Negro laborers, who helped remove military stores from Scott's Ferry."[59] This barely begins to tell the story of his involvement in the war, especially between 1779 and 1781. In fact, he was one of the largest supporters in Buckingham County, and, when placed in that context, Randolph's participation at home comes to life.

The "Old Courthouse Road" (a.k.a. "Ferry Road") ran right through the middle of the Jefferson plantation, making it the major north-south passage for troops and their horses through Buckingham County to the James River. Additionally, and no less important, was the plantation's proximity to the river and Scott's Ferry. The vantage from the bluffs of Snowden was potentially of strategic importance, should the war front ever come to central Virginia.

As a landowner and significant slaveholder, Randolph Jefferson was exempt from the draft. It was imperative that plantations such as his be maintained, producing crops and food not only for the community at Snowden, but also for the Continental Army. Though he may have stayed close to home as the war progressed, Jefferson and his farm were touched both directly and indirectly by a series of events.

In 1778, it was Thomas Jefferson who suggested that nearly 4,000 prisoners of war be housed near Charlottesville. Many Albemarle and Buckingham gentlemen acted as guards, including Randolph Jefferson's friend Hudson Martin, his cousin, Charles Lewis, and several of his Buckingham County neighbors. ("ENCAMPMENT OF THE CONVENTION ARMY AT CHARLOTTE VILLE IN VIRGINIA AFTER THEY HAD SURRENDERED TO THE AMERICANS", N.D. EMMET COLLECTION, MIRIAM AND IRA D. WALLACH DIVISION OF ART, PRINTS AND PHOTOGRAPHS, THE NEW YORK PUBLIC LIBRARY, ASTOR, LENOX AND TILDEN FOUNDATIONS)

During the war years, three particular situations especially involved the Buckingham Militia and the county at large. First, in early 1779, there was the establishment of the prisoner of war barracks outside of Charlottesville. There, Buckingham Militiamen would join others from Albemarle and surrounding counties to guard the prisoners. Then, in the battle at Guilford Court House in the late winter-early spring of 1781, Buckingham citizens and the Buckingham Militia came to the aid of Gen. Nathanael Greene in North Carolina. Lastly, during the following summer, when Banastre Tarleton and his troops invaded central Virginia, the war came to the Horseshoe Bend and Snowden. Each of these united the geographically separated Jefferson brothers in a consolidated effort for American independence.[60]

In the fall of 1778, it was Thomas Jefferson who proposed that British and Hessian prisoners being held in Boston be moved to Charlottesville. At the time, the area was remote in regards to the war, and Thomas argued that the presence of these some 4,000 men would not be dangerous. A potential economic boon to the area, he optimistically estimated that they would pour some $30,000 a week into the Albemarle economy in buying local goods. Among the prisoners were skilled workers and craftsmen whose labor was valuable. On September 15, 1778, John Harvie, agreed that Charlottesville would be ready to receive the prisoners during the fall or winter. Ultimately, Harvie offered some of his land on which to build the Barracks. The effect of the prisoners on the immediate surroundings was, as Thomas Jefferson had imagined, significant; however, events did not proceed exactly as he had envisioned. Lightly guarded, many escaped and their numbers dwindled closer to 3,000.[61]

As a result, military presence increased and several Companies from Buckingham County Militia served as guards; among them were Randolph Jefferson's neighbors, Capt. Hardin Perkins and Lt. Anthony Murray. Albemarle County Militia also provided guards, which included Randolph's Albemarle-based friends and relations.

Col. Charles Lewis of North Garden who was married to Thomas and Randolph's first cousin, Mary Randolph Lewis, served as Commander of the Guards.[62] Under him was Albemarle resident, Hudson Martin, a Lieutenant in the 9th Virginia Regiment. Born on July 3, 1752 (OS), Hudson was about Randolph Jefferson's age and the husband of Jane Lewis, daughter of Col.

Nicholas Lewis. In July of 1781, he would act as bondsman for Randolph's marriage to Anne Lewis. It is probable that Jefferson and Martin had long been associated, perhaps primarily through the Lewis family. After the war, he served as Deputy Clerk of Albemarle for several years and later was appointed Magistrate.[63]

The peculiar village of British and Hessians at Ivy Creek which developed at the Barracks must have held fascination for the entire area; it certainly did for Thomas Jefferson, who appreciated the infusion of officers and European gentleman. Their sophisticated, Continental taste in music, art, food, and wine made for lively conversation.[64] Since the prisoner-of-war camp was a day's ride from Snowden, Randolph Jefferson may have had only occasional direct contact with the Barracks. Still his friends and family who were stationed there likely shared their experiences.

By the autumn of 1780, enemy activity increased and Thomas Jefferson, now Governor of Virginia, grew concerned that during a potential invasion of Virginia, prisoners might be liberated, unleashing a sizable force in the Piedmont; so the British prisoners were moved away from Albemarle to Fort Frederick, Maryland, in October of 1780. The German troops followed a few months later.[65] According to the letters of Thomas Anburey, he and other prisoners left at that time. On November 20, 1780, he wrote from Winchester, Virginia, "About six weeks ago we marched from Charlottesville barracks, Congress being apprehensive that Cornwallis in overrunning the Carolinas might by forced marches retake the prisoners."[66]

As the war deepened, the need for supplies grew. Randolph and his neighbors gave ample and ongoing support as evidenced when the claims and certificates collected during the war were produced at Buckingham Courthouse in 1783. From food stuffs, to horses, to wagons, to guns, the citizens of Buckingham faithfully supported their troops.[67]

In May of 1780, the Virginia General Assembly authorized local commissioners to impress much-needed supplies for the Continental Army. Property owners like Randolph Jefferson and his neighbors at the Horseshoe Bend were given certificates appraising their beef stock. Certificates written on November 22, 1780, indicate that Randolph's neighbors, Thomas Ballow and Anthony Murray, "were sworn to appraise the farm," meaning Snowden. Signed by Commissioner John Bates, Randolph's certificate reads: "Beef Cattel (sic) Agreeable to an Act of Assembly Entitled an Act. for giving further [?] Powers to the Governor & Council & for other purposes Appraised to £11.09.8." Likewise, Thomas Ballow's stock was appraised by Hardin Perkins and Anthony Murray, while Ballow, along with Stephen Perkins, appraised Hardin Perkins' cattle. Ultimately, Randolph supplied 924 lbs. of beef to the Army, Ballow relinquished over 400 lbs of "grass fed beef," and Anthony Murray provided an additional 500 lbs.[68]

∼

Within just two months, following Benedict Arnold's invasion up the James River, Governor Thomas Jefferson sent out a broadside requesting more aid addressed "To the First Magistrate of Each County:"

RICHMOND (In Council) January 20, 1781.
The invasion of our country by the enemy at the close of the late session of Assembly, their pushing immediately to this place, the dispersion of the publick papers, which for the purpose of saving them necessarily took place, and the injury done at the printing office, have been so many causes operating unfortunately to the delay of transmitting you the important act for supplying the army clothes, provisions, and waggons, which I now enclose you.... While we have so many foes in our bowels and environing us on every side, he is but a bad citizen who can entertain a doubt whether the law will justify him in saving his country, or who will scruple to risk himself in support of the spirit of a law where unavoidable accidents have prevented a literal compliance with it.

... No man can say this will be an injury to him, because the times were affixed to compel an early compliance, the delay of which, some days, must rather be matter of indulgence.

PERSONS will be appointed to receive the waggons and appendages to be furnished by your county, who shall give you notice of the place of delivery in due time.

I have the honour to be, with great respect, sir, Your obedient, and most humble servant,

TH: Jefferson[69]

The following month, Buckingham was bustling with war activity. Numerous men in the county gave supplies in support of Gen. Nathanael Greene's army, and ultimately the Buckingham Militia rode off to join Greene in North Carolina. Precisely who left Buckingham is unknown due to the loss of records; however, it is possible that Randolph Jefferson was among them.[70]

During February and March, prisoners moved through Buckingham. Traveling on the Old Courthouse Road, these prisoners would have marched right through the center of Snowden, passing in front of Randolph Jefferson's door.[71] On February 18, 1781, Governor Thomas Jefferson wrote to the County Lieutenants of Augusta and Shenandoah Counties concerning prisoners taken at the Battle of Cowpens. In April, orders went out announcing: "After Fatted Beefs are Consum'd a Call may take Place to Collect Bacon."[72]

In February, Randolph promptly had provided 17 lbs. of bacon and ½ bushel of meal for British prisoners and guards marching through Buckingham. His certificate, which valued the provisions at 14 shillings, was signed by Nathaniel Hunt, who was escorting the prisoners from Gen. Greene's headquarters to the Albemarle Barracks. Snowden neighbor, Capt. Hardin Perkins, provided three bushels of meal and a generous 50 lbs. of bacon for Continental use, valued at £2.6.6.[73]

While prisoners moved through central Virginia, Gen. Greene eventually regrouped to fight Cornwallis in North Carolina at the Battle of Guilford Court House on March 15, 1781. Greene had arrived there on February 9th, eager to fight Cornwallis, but in need of reinforcements. On February 10th, he wrote to Patrick Henry requesting troops: "If it is possible for you to call forth fifteen hundred Volunteers & march them immediately to my assistance, the British Army will be exposed to a very critical and dangerous situation."[74] In Buckingham County, many men and their families provided guns, food, and other supplies to the county Militia which prepared to go to Greene's aid. Several pension applications later stated that the men marched from Buckingham on February 1st, heading for Gen. Greene's camp in North Carolina.[75]

Had Randolph Jefferson lived long enough to apply for a pension, today, a great deal more might be known about his war service. While it is not known if Randolph rode to Guilford Court House with his friends and neighbors, he did leave Buckingham County during this period and in late April was in the vicinity of Richmond when he borrowed £450 from his brother. At that point in the war, £450 represented only modest purchasing power. In April, Thomas Jefferson's many expenditures included £18 for a mockingbird, £150 for a map, £397/10 for mending a chariot, £4800 for two mules, and £675 for 50 lbs. of sugar.[76]

The loan was recorded in Thomas' account books on April 27, 1781.[77] Thomas also noted that this transaction took place at Westham. Situated on the north side of the James River in Henrico County, Westham was about six miles upstream from Richmond. This was the area of Beverley Randolph's Westham plantation, the proposed location of the aborted "Beverley Town," in which Peter Jefferson had purchased the ferry lot. The subsequent village of Westham was eventually home to an iron foundry, public arsenal, and weapons factory during the Revolution. In early January of 1781, British troops led by Benedict Arnold had sacked the site, destroying the iron foundry.[78]

Thomas had served as Governor of Virginia since 1779 and, early in his term, had moved the state capital from Williamsburg to Richmond. The considerable distance between Snowden and Richmond physically separated the Jefferson brothers, limiting their contact. How the brothers communicated between 1779 and 1781 and how often the brothers saw each other is unknown. No correspondence between them survives from this period. Thomas' notation of the loan at Westham is the first documented contact between Randolph and Thomas since he assumed office as Governor.

On April 24, 1781, Thomas Jefferson received intelligence that Gen. Benedict Arnold and Gen. William Phillips were possibly headed toward the capital, approaching from Westover, about thirty miles to the east. Petersburg, another important store of ammunition and weapons, fell that day. The next few days were fraught with concern about another impending invasion of Richmond.[79]

During the third and fourth weeks of April, letters to Thomas Jefferson were flowing in from the likes of the Marquis de Lafayette and flowing out to men like militia leader, Garrett Van Meter, who was struggling with mutineers rioting in Hampshire County. As Governor, Thomas Jefferson's hands were more than full. The pressure on Richmond continued to increase, and on May 10th, when the Virginia Assembly met, they called the May 24th meeting of the legislators to assemble not in Richmond, but considerably further west in Charlottesville. Governor Jefferson, himself, would soon flee the city.[80]

Why, then, in the heat of war and the immediate threat of the British attack, would Thomas Jefferson venture on April 27th to Westham to loan his brother £450? Why would Randolph risk his own safety and, if he indeed came from Snowden, ride approximately sixty miles to receive it?

There is no surviving written communication illuminating their plans. How did the men arrange this meeting? Had Randolph recently spent time in Richmond or somewhere in the vicinity? If so, what business brought him there? That day, did Randolph carry away more than his £450 from Westham, doing some service for his brother, Thomas?

In early 1781, Randolph Jefferson was twenty-five years old, a single man, responsible for no wife or children. In February, when the men of Buckingham rode off to join Gen. Greene, he may or may not have gone voluntarily with them. Records housed at the Library of Virginia reveal that Randolph was given a commission at the rank of Lieutenant in the Buckingham Militia in 1787, but they do not indicate when he first joined. If Randolph rode off in February, taking supplies to North Carolina, he may have remained for the Battle of Guilford Court House on March 15, 1781. Following the battle, joining the retreat to central Virginia, he could have gone to Richmond, where, in April, he visited his brother and secured a loan.[81]

There is another possible scenario in which Randolph came to Westham not from Buckingham but from Albemarle County, delivering a letter to his brother and concurrently receiving a loan. On April 25, 1781, Reuben Lindsay of the Albemarle Militia wrote the following letter to Governor Jefferson. In part, it concerned Maj. Charles Lilburne Lewis, who was married to Lucy Jefferson:

Albemarle 25th. April 1781

Sir

Your favor of the 16th. Instant came to hand covering the Inclosed Commission and one for Major Charles L. Lewis, the receipt of which I most thankfully Acknowledge – but being conscious of my inability to execute an Office of that importance at Present, must be leave to return the Commission. That fill'd up for Major Lewis I have delivered him, which he told me he would Qualify too the first Opportunity.

I am persuaded the Executive was not fully inform'd respecting the recommendation of Officers at our Last Court – particularly respecting myself – for I believe the Court would have recommended me to the Commission of County Lieutenant, had I not told them I would not serve, and beg'd they would think of some one else, and did myself propose Capt. John Marks, who was recommended by a great Majority of the Members then Present.

I shall continue to act under my former Commission until Other Officers are Appointed and do Qualify, which I wish should happen as soon as possible.

I am Sir, with much respect Your Very Hble Servt.,

RN. Lindsay[82]

Col. Reuben Lindsay lived north of Monticello, near the Albemarle-Orange County line at Springfield Farm. His wife was the daughter of Dr. Thomas Walker, one of Peter Jefferson's executors. Lindsay's name appears regularly in Thomas Jefferson's memoranda between 1772 and 1823, indicating their relationship before and after the war. In retirement, Thomas Jefferson stayed at Col. Lindsay's home on several occasions, making note that he tipped the household staff. While nothing has yet surfaced to directly connect Col. Lindsay and Randolph Jefferson, they undoubtedly were acquainted.[83]

Perhaps Randolph was in Albemarle County on April 25th when Lindsay wrote the letter, and offered to deliver it to his brother; then he rode directly to Westham, which is precisely a two day ride away from the Piedmont. Having done so, it would have been a convenient time to borrow funds from his brother.

The War Comes to Snowden

By May, the Continental Army was in the immediate vicinity of Snowden. On May 28th, Gen. Steuben wrote to Governor Jefferson from Albemarle Old Courthouse concerning the "present situation of the Virginia Line."[84] Deeply discouraged, he also wrote to Gen. Greene, "In a word, my dear General, I despair of ever seeing a [Virginia] Line exist. Every[one] seems to oppose it."[85] Despite Steuben's despair, the citizens of Buckingham County came to the aid of the army; these included Randolph Jefferson, who provided "3 bushels oats for 6 teams," sometime in May. His certificate was signed by Hezekiah Gray, A.C.[86] Randolph's much more significant contribution was soon to follow.

At this point in the war, the Marquis de Lafayette and his troops had brought an infusion of skill and energy that must have thrilled and invigorated central Virginia. The "boy General" from France was born September 6, 1757, and was almost exactly two years younger than Randolph Jefferson, a potent combination of a peer and a hero figure.

On May 29th, Thomas Jefferson wrote from Charlottesville to Lafayette, "I sincerely and anxiously wish you may be enabled to prevent Lord Cornwallis from engaging you till you shall be sufficiently reinforced to be able to engage him on your own terms."[87]

Lafayette responded from his headquarters that same day, asking that Col. Anthony Walton White be "impowered (sic) to employ persons to impress two hundred Horses fit for the Dragoon Service, on the South side of James River and in any other part of the State your Excellency may please to direct."[88]

It had finally happened. The war had come to the Horseshoe Bend, Albemarle Old Courthouse, and Snowden. By June, Lt. General Charles Cornwallis and the British troops reached central Virginia and, "suddenly Charlottesville, Monticello and the area around Albemarle Old Courthouse felt the hot breath of war," as Scottsville historian, Virginia Moore, wrote dramatically.[89]

In brief, Thomas Jefferson was just finishing his term as Governor of Virginia as British Col. Banastre Tarleton's raid was approaching Monticello. One of Tarleton's goals was to capture Governor Jefferson, as well as other members of the Virginia legislature. Thanks to Jack Jouett's famous ride to Monticello, Jefferson was warned of the enemy's approach. He packed up his family, sending them to his Bedford County farm, Poplar Forest. Wasting little time, Jefferson followed separately.[90]

Thomas Jefferson's flight was well-documented at the time and by Jefferson himself. As the years passed, however, the Randolph Jefferson family preserved a peculiar legend concerning Thomas' journey from the Little Mountain to Bedford County. Virginia Moore recorded the family story in *Scottsville on the James*:

> There is a tradition that Jefferson hid out in a cave under a bluff below Scott's Landing, about a mile downstream from his brother Randolph's plantation-house; the place is pointed out.
>
> Fortunately there exists some "hard evidence" as to Jefferson's exact movements: the testimony of George W. Foland, a descendant of Peter V. Foland, who as the husband of Jefferson's great-great-niece, had access to family knowledge.
>
> He says that Thomas Jefferson rode south along blind mountain paths. Striking Rockfish River, he followed it to a ford near Mount Alto, crossed, and stopped at a house where he told his story and asked for shelter. The owner thought this road, the only good one in the area, unsafe – Tarleton might gallop up. Conducted two miles upstream, ex-Governor Jefferson (his term had just expired) spent two weeks at the home of Mr. Thomas Farrar. Then pressed on to Scott's Landing, which he knew intimately, having for years, passed through it to visit Snowden, now his brother's. It was probably Randolph Jefferson who suggested the cliff-cave below Scott's Landing as a hideout. Provisions could be sent as long as there remained any danger of a Tarleton or Simcoe raid. Perhaps also such luxuries as pillow, quilt, towel. Already Scott's Landing had heard that Cornwallis was using Jefferson's Goochland seat Elk Hill below Point of Fork as British headquarters, news not calculated to make a cave more comfortable. When at last word came that the foe was riding towards Tidewater, Jefferson hurried to Elk Hills's scene of devastation. Horses, cattle, sheep, hogs had been slaughtered, slaves stolen, barns burned.[91]

This etching of Banastre Tarleton, circulated by the George Washington Bicentennial Committee (c. 1924), represents the portrait painted by Sir. Joshua Reynolds in 1782.

This tale contains anything but "hard evidence." A closer look puts it in the category of family myths. Its roots in reality, if any, are yet to be discovered.

When Thomas Jefferson left Richmond for Monticello, he believed his term as Governor had expired on June 2nd, a point that would be hotly contested later. However, under the assumption that his duty to the Commonwealth of Virginia had been passed to Gen. Thomas Nelson, Jefferson removed his family to safety at Poplar Forest and eventually remained there until mid-summer.

Editorial notes in *Jefferson's Memorandum Books* offer this succinct version:

> Warned of the approach of the British by militia captain John Jouett, Jr., TJ and most of the legislators narrowly escaped capture. Part of Tarleton's force under Captain McLeod arrived at Monticello minutes after TJ's departure to the south, "thro' the woods along the mountain." TJ overtook his family, sent off earlier, at John Coles' Enniscorthy, where they dined. The next day they reached Geddes, the home of Hugh Rose on the Tye River, about seven miles northeast of present Amherst in Amherst County. On 7 June, probably after learning from Mr. Thomas that the British had left his home unmolested after a brief occupation, TJ returned alone to Monticello....[92]

In fact, Thomas Jefferson wrote of those days in his "Diary of Arnold's Invasion and Notes on Subsequent Events in 1781," stating the following:"

> I now sent off my family, to secure them from danger and was myself still at Monticello, making arrangements for my own departure, when Lt. Cristoph. Hudson arrived there at half speed, and informed me the enemy were ascending the hill of Monticello. I departed immediately, and knowing that I should be pursued if I followed the public road, in which too my family would be found, I took my course thro' the woods along the mountains and overtook my family at Colo. Cole's, dined there, and carried them on to Rockfish after dinner, and the next day to Colo. Hugh Rose's in Amherst, I left them there on the 7th. and returned to Monticello. Tarleton had retired after 18. hours stay in Charlottesville. I then rejoined my family at Colo. Rose's and

proceeded with them to Poplar Forest in Bedford 80. miles S.W. from Monticello. On a ride into the farm about the end of the month, I was thrown from my horse, and disabled from riding on horseback some months. I returned to Monticello July 26. . . . [93]

In another variation, "The 1816 Version of the Diary and Notes of 1781," Thomas Jefferson commented, "The nonsense which has been uttered on the coup de main of Tarleton on Charlottesville is really so ridiculous that it is almost ridiculous seriously to notice it."[94]

Following a simple retelling of his passage to Poplar Forest, Jefferson added, "Here, too, I must note another instance of the want of that correctness in writing history, without which it becomes romance."[95] No two week-stay at Thomas Farrar's. No secluded cave dwelling at Snowden for Thomas Jefferson. The poetic Virginia Moore, as it turned out, had traded History for Romance.

~

Over the years, this myth of Snowden's role in Thomas Jefferson's "Escape from Tarleton" has obscured the real contribution made by Randolph Jefferson and the community of the Horseshoe Bend at what turned out to be a pivotal point in America's Revolution. During the months of June and July of 1781, the war came to Randolph's door and he indeed stepped up to support the cause of revolution at this very juncture. In fact, not only Randolph Jefferson, but also the fields of Snowden herself and her community of African Americans supported the Continental Army, particularly the 3rd Regiment of the Light Dragoons.

A significant amount of Randolph Jefferson's aid during the war was offered in May, June, and July of 1781. Once Cornwallis and Tarleton were able to penetrate central Virginia, one of their primary targets would be the military store at Albemarle Old Courthouse. With the likes of Westham and Petersburg already destroyed by the enemy, the munitions newly moved to central Virginia were extremely important. Robert R. Howison, in his *History of Virginia*, published in 1848, related the situation:

> At Albemarle Old Courthouse, the Virginians had collected a large quantity of valuable military stores. To destroy these now

In 1834, artist Joseph-Désiré Court remembered Gilbert Motier, Marquis de Lafayette, as he appeared as a Lieutenant General, c. 1792.

became an important object to Cornwallis: to protect them an equally important one to Lafayette . . . .

[T]he Marquis moved cautiously from Culpeper through Orange and the upper part of Louisa, to Boswell's Tavern, near the Albemarle line. Cornwallis marked his movement . . . There was a rough road, long disused, leading from a few miles below Boswell's, to a point of Mechunk Creek; forthwith Lafayette set to work his pioneers and axemen; the road was opened, the army passed along it, and the next morning, to the utter astonishment of Cornwallis, his adversary was encamped in an impregnable position on the Creek, and just between the British army and the stores of Albemarle Courthouse! His English lordship was once more baffled and . . . he changed his front, and marched slowly towards the eastern coast.

. . . Now at last Cornwallis was on the retreat, and Lafayette was the pursuer.[96]

On June 6th, Capt. William Moseley, Buckingham County Militia, wrote Randolph a certificate for "5 days service of Negro fellow in removing Military Store to fork of Tye River." His services were valued at 15 shillings. Just as Randolph Jefferson was donating his property in Buckingham, Col. William Cabell wrote in his diary concerning Thomas Jefferson's property at Elk Hill:

> *June 7*: Cornwallis with his army joined Simcoe and Tarleton at [Thomas] Jefferson's and Ross's plantation near Point of Fork, and remained there until the *13th*, sending out his light troops from time to time to destroy warehouses, etc.[97]

Two days later, Lt. Hugh Rose, the Lieutenant of Nelson County, wrote to Col. Cabell:

> Tis not certainly known that the enemy are nearer than Point of Fork [Columbia]; but yet the supplies at Albemarle Old Court House [near Scottsville] and Irving's store are an object and may be, indeed certainly are, in danger. For the security of these stores our forces must be employed. No jealous Whig will refuse to turn out on such an occasion.[98]

The citizens of Buckingham, and particularly the men of the Horseshoe Bend with James River frontage directly across from the old courthouse, rose to the occasion. On June 1, 1781, Anthony Dibrell, who had acted earlier as a guard at the Albemarle Barracks and had fought at Guilford Courthouse, joined others who were ordered to remove military stores from Scott's Ferry to New London in Bedford County. Allen Scruggs and John Watkins were paid for five days service for removing the stores. Each man received £2.10 for his labor. Ware Oglesby was paid £1.4.6 for four days of "poling canoes" to the Tye River.[99]

Randolph Jefferson did not join them, but he volunteered 20 bushels of oats, 2 barrels of corn, and 38 lbs. of bacon for the guards, as well as a wagon for removing the military stores to New London in Bedford County. These

Charles Mackubin Lefferts (1873-1923) devoted much of his life to the study of uniforms worn during the American Revolution. This watercolor depicts a member of the 3rd Continental Light Dragoons, the mounted regiment which camped at Snowden, during the summer of 1781.
(COLLECTION OF THE NEW-YORK HISTORICAL SOCIETY)

provisions were later valued at £3.18.6. Randolph also loaned the services of one of his slaves to help with the trip to the fork of the Tye River. Upon the slave's return to Snowden on June 6th, William Moseley, C.M., wrote out a certificate to Randolph for the value of 15 shillings. Capt. Thomas Ballow, a neighbor, also contributed to the effort, volunteering 15 ½ lbs. of bacon and ½ bushel of meal for the use of the "canoe men."[100]

The removal of the military store in early June was just the beginning of military presence at Snowden. The 3rd Regiment of Light Dragoons would camp there for over a month and a half. Randolph eventually claimed the use of his pastures for seventy-three head of cattle for forty-seven days. Feed included seventy-six feet of tops and 600 bundles of fodder which fed seventy-three beeves for twenty days and an additional eight head of cattle for thirty days. This claim alone totaled £6.11.0.[101]

The Dragoons still were utilizing pasture at Snowden as late as July 16th.[102] Randolph's overseer, Stephen Perkins, provided seven bushels of oats at 1/6 and pasturage for eighteen horses, at a value of 15 shillings. Capt. Presley Thornton of the Continental Army wrote a certificate for the goods on July 16, 1781. His certificate stands out, as most of those written to the men of Horseshoe Bend were provided by the Buckingham Militia.[103] Thornton's presence at Snowden is a tangible reminder that during the summer of 1781, Randolph Jefferson enjoyed not only troops on his plantation, but also the company of officers.

The events of the summer of 1781, particularly the successful protection of the military stores at Scott's Ferry, proved to be a definitive moment in the war. From that turning point on, it was a march towards Yorktown and victory for the Continental Army. Along with other patriotic citizens of Buckingham County, Randolph Jefferson and the fields of Snowden rose to the needs of the day.

Though the true and vivid events in June of 1781 eventually were overshadowed among Randolph Jefferson's descendants by the family tradition of Thomas Jefferson and the cave at Snowden, some stories from these days may also have survived as personal legends in the Randolph Jefferson family. Thirty years later, Randolph's only daughter would name one of her sons Lafayette Nevil.

∽

On April 5, 1782, Buckingham Court held its special session for public claims. Randolph may not have attended, sending overseer Stephen Perkins as his assignee.[104] Buckingham Sheriff, William Cannon, also represented Randolph and at least thirty other Buckingham citizens, including Jefferson's friends and neighbors John Nicholas, Samuel Jordan, and Anthony Murray.

Buckingham's Clerk headed the entries noting: "At a court held for Buckingham County 5 April 1782. Pursuant to an act of Assembly entitled An Act for Adjusting claims for property impressed or taken for Public Service. The Court proceeded to adjust and receive the claims of the different claimants produced to them in specie as hereafter mentioned."[105] Claims presented for Randolph Jefferson included:

7 bushels of oats at 1/6, pasturage for 18 horses furnished
by Stephen Perkins (Overseer) for the use of 3rd Regt. Of
Light Dragoons as per Certificate of the 16th July 1781
given by Presly Thornton Capt.   0-15-0

5 days service of Negro fellow in removing Military Store
to fork of Tye River as per Certificate given by Wm. Moseley C.M.
Store June 6th 1781.   0-15-0

3 bushels oats for 6 teams pr. cert
Hezekiah Gray A.C. May 1781   0-4-6

20 bushels oats, 2 barrels corn, 38# bacon for guard & waggon
removing military stores to New London in Bedford pr.
Certificate by Wm. Moseley C.M. Stores June 1781.   3-18-6

17# bacon, ½ bushel meal for British Prisoners & guards pr.
Certificate of Feb. 1781 by Nathaniel Hunt.   0-14-0

                                                        28-11-10[106]

There is an additional entry for "4 days pasturages" and another for "35 beeves," valued at £2.6.8. Randolph also produced the two certificates signed by Buckingham County Commissioner, John Bates, dated November 20 and November 22, 1780.[107]

After the claims at Buckingham Courthouse were consolidated, they were evaluated by the Commissioner in Richmond. In the Commissioner's Books, entries are organized under various headings: "Prisoners of War," "Army," "Militia," and even one specifically for beef furnished for the "United States." The Commissioner's Books indicate Randolph Jefferson's actual payment authorized in Richmond; he was awarded the following:

August 1, 1783	3 warrants for pasturage 1 for £5/11/7, 2 for £10 each	25/11/7
August 1, 1783	Prisoner of War: 1 warrant bacon furnished the Guard of PW assignee Steven Perkins	18/9
August 1, 1783	Army: 1 warrant for oats assignee Steven Perkins	5/2/6
August 24, 1783	Randolph Jefferson assignee William Cannon[108]	11/11/4

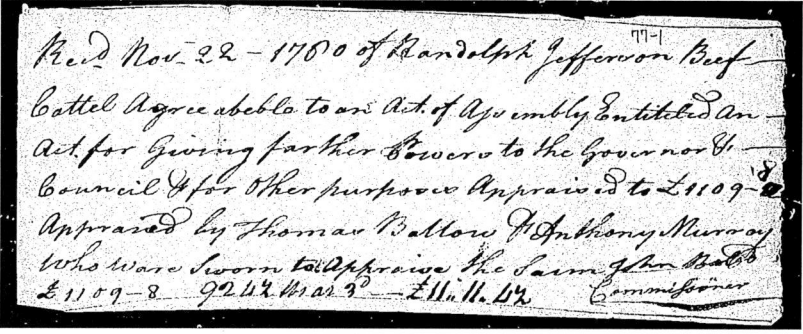

This certificate, signed by Buckingham County Commissioner, John Bates, recorded the appraisal of Randolph Jefferson's beef cattle. His long-term neighbors, Thomas Ballow and Anthony Murray, acted as appraisers. (COURTESY OF THE LIBRARY OF VIRGINIA)

Following the war, the surviving exchanges between the Jefferson brothers dwindled to almost nothing.[109] In a lone reference on September 23, 1783, Thomas Jefferson gave a slave named Tom money for ferriage to Snowden. The reason for Tom's trip remains unknown. That year, Thomas Jefferson's career took him to the national and international stage, first to Congress and then to France as Commissioner and Minister, beginning a long separation between the Jefferson brothers.

The war years and the challenges facing the emerging nation inspired further service from Randolph Jefferson. As the decade progressed, he served in the Buckingham Militia. In a "Recommendation of Militia Officers" addressed to the Governor of Virginia, dated May 4, 1787, Randolph was suggested as "Lieutenant of the Company commanded by Anthony Murry." Anthony Murray was Randolph's neighbor and the son of Richard Murray, one-time proprietor of Snowden's ferry house tavern.[110]

Seven years later, in Buckingham Court held May 13, 1794, Randolph was recommended as the "sixth Captain of the first Battalion." William Perkins was recommended to the Governor as Lieutenant Colonel and Samuel Allen, Gent. was recommended as Major of the first Battalion. In April of 1795, Randolph continued in his position as Captain. Then, on June 13, 1796, the following request went to the Governor: "Hardin Perkins is recommented to his Excellency the Governor as a proper person to fill the office of Capt. In the room of Randolph Jefferson who has resigned."[111] Hardin Perkins, Jr., Randolph's friend and neighbor, was a very appropriate replacement as the Captain of Company 6. After at least nine years of service in the Militia, Randolph's reason for not continuing remains a mystery. He was content to enjoy the title of Captain the remainder of his life.

POLITICS IN BUCKINGHAM

A piece of Buckingham ephemera, a surviving polling list, reveals Randolph Jefferson's political leanings, at least in 1788. On April 14, 1788, at a poll held at Buckingham County Court House, Jefferson voted for delegates Joseph Cabell and Thomas Anderson to represent the county in the next Virginia Assembly. His neighbor and fellow militiaman, Anthony Murray, preceded Randolph at the polls. Murray voted for Hickerson Barksdale and Thomas Anderson. The other proposed delegate was John Cabell. During the Revolution, Capt. Thomas Anderson led a Company in the Buckingham Militia. He marched his men to Albemarle to act as guards at the Barracks, and at least some of his men were engaged in "boating down the James River for the use of the army and continued until Cornwallis was captured at Little York." Col. Joseph Cabell (1732-1798) long represented Buckingham, Albemarle, Amherst, and Fluvanna Counties in the Virginia Assembly. The Cabell name would continue to surround the greater Jefferson family for much of the next three decades. Eventually, the families would intermarry when Randolph Jefferson's great-granddaughter, Lou Mundy, married P. Carrington Cabell in 1873.

[Jeanne Stinson, *Buckingham County Virginia Extant Poll Lists 1788, 1840, 1841, 1848* (Athens, GA: Iberian Publishing Co., 1994), 13.]

Chapter Six

Randolph Jefferson, Planter

> *Cultivators of the earth are the most valuable citizens.*
> Thomas Jefferson to John Jay, 1785

In July of 1781, as the war moved east and away from Snowden's front door, Randolph Jefferson committed to marriage. He was twenty-five years old when his bachelor days at Snowden came to an end. It was, as it turned out, a significant coincidence of events – the winding down of the Revolutionary War in Virginia and his, perhaps, long-anticipated wedding.

At twenty-five, who was Randolph Jefferson? He had studied at the College of William and Mary in Williamsburg and possibly received additional training at Seven Islands. He had briefly served under Gen. Thomas Nelson and supported the Revolutionary cause with his fellows in Buckingham County. He was soon to be, if he was not already, a stalwart member of the Buckingham County Militia. Now well established on his patrimony, the squire of 2,000 acres of prime Virginia farmland, Randolph was also responsible for thirty slaves.

Randolph Jefferson, Bridegroom

Randolph Jefferson's marriage to his first cousin, Anne Lewis, took place in Albemarle County about July 30, 1781.[1] Anne was the daughter of Charles Lewis, Jr. and Mary Randolph, Randolph Jefferson's aunt. Neither Anne's birth date nor death date is known.[2] Her marriage bond identifies her as Anne Lewis,

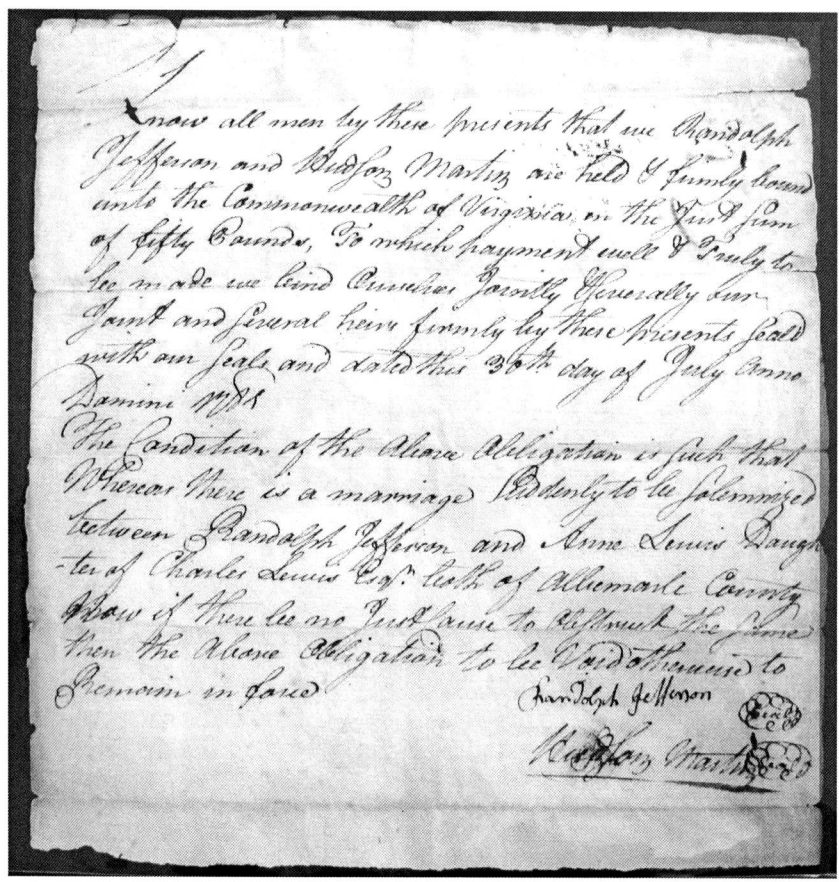

Marriage Bond for Randolph Jefferson and Anne Lewis. (ALBEMARLE COUNTY MARRIAGE BONDS)

although many other documents reveal that she was known as Nancy, especially within the Randolph Jefferson family.

The couple was likely married at Nancy's home, Buck Island, located in southern Albemarle County. Both of her parents were alive to see her wed. The bondsman was Hudson Martin, who was married to a Lewis cousin.[3] He may have been a particular friend of Randolph's or the Lewis family's or both.

Randolph and Nancy were a classic first-cousin marriage of the day, and their marriage may long have been assumed between the families. It also may have been delayed by the war, or Nancy may have been a bit younger than

Randolph, postponing the marriage. Their "common culture" was deep, and, in addition to being products of rooted Virginia gentry, Randolph and Nancy shared a distant Welsh connection.[4]

The Charles Lewises made their home at Buck Island, where they reared eight children. The mere six to seven miles between Buck Island and Shadwell placed the Lewis-Randolph and Jefferson-Randolph cousins in close proximity. Randolph and Nancy had known each other all their lives, and, while Randolph lived at Buck Island from 1762-1765, they potentially grew even closer. They may have studied together. As Randolph did not remember his father, it is also possible that Charles Lewis, who was in his early forties when Randolph lived at Buck Island, served as a father figure for the boy.

When Randolph and Nancy wed, they followed in the footsteps of his sister, Lucy Jefferson, and her brother, Charles Lilburne Lewis, who had married on September 12, 1769. Charles Lilburne was the eldest of Charles and Mary Lewis' three sons.[5] Lucy was Randolph's next older sister. They, too, had known each other all of their lives. Lucy was just shy of seventeen when she married her first cousin, probably at Shadwell, as weddings at the bride's home were the custom of the day. If so, it may have been one of its last celebrations.[6] The Jefferson home there burned to the ground five months later.

Nancy Lewis would have a longer wait for her wedding. Over the next few years, Randolph went away to William and Mary, took possession of Snowden, removed to Buckingham County, and participated in the Revolutionary War. As a result, it would be over a decade before there was another Jefferson-Lewis marriage.

There is no known surviving correspondence between Randolph and Lucy, though in the autumn of 1807, when Lucy (Jefferson) Lewis was about to remove to Kentucky to join "the bulk of her children," she wrote a heartfelt *adieu* to her elder brother, Thomas. Lucy includes Randolph in her sentiments: "I feal much hurt at leaving two Brothears, for evar, and not seeing eathar, I wish you all the happiness that can possibly be expressed by an affectionate Sustar."[7] If she also wrote to Randolph, a letter is not known to exist. However, the mention of him, on equal footing with Thomas, certainly confirms ongoing affection among the Jefferson siblings.

The strength of the relationship between the Randolph Jefferson and the Charles Lewis families should not be underestimated. In the past, Randolph has been viewed almost exclusively from "the balcony" at Monticello, where he appears small and distant, even insignificant. Dramatically overshadowed by his older brother, Randolph lived in relative obscurity in Buckingham County. But when another perch is taken, that of the Lewises of Buck Island and Monteagle, Randolph emerges as an equal among other Jeffersons.

On September 17, 1810, three of Randolph's nieces – Martha C., Lucy B. and Ann M. Lewis – wrote to their uncle, Thomas Jefferson, concerning the death of their mother, his sister, Lucy. Isolated from their Virginia home, residing in Livingston County, Kentucky, the sisters wrote an appeal for love and support from their Virginia relations. They began the letter by imparting the news of their mother's death, calling Lucy (Jefferson) Lewis "the best mother and sister."[8] Later in the letter, they made a plea for an intercession by their Virginia family into the complicated situation with their brother-in-law, Craven Peyton, who had married their oldest sister, Jane Lewis.

The letter is first addressed to the relation with the highest rank and most power, the elder statesman and "clan chief," their uncle, Thomas. Other men in the family are named in the following order:

1. "yourself" (Thomas Jefferson, their mother's elder brother)
2. Uncle Randolph (their mother's younger brother)
3. Mr. Randolph (Thomas Mann Randolph, b. 1768, cousin Martha Jefferson's husband and a cousin in his own right)
4. Mr. P. Carr (Peter Carr, b. 1770, eldest son of Dabney Carr and their aunt Martha Jefferson)
5. Mr. D. Carr (Dabney Carr, b. 1789, youngest son of Dabney Carr and their aunt Martha Jefferson)
6. and all our dear connections[9]

Historically, and in many cultures existing today, this kind of codified respect for males and/or elders in a family group would be instantly understood. In part, the Lewis sisters' approach reflects this kind of formal respect. It also indicates a hierarchy of power and affection from the Lewis family's point of view. From the point of view of Martha, Lucy and Ann Lewis, Randolph Jefferson was second in command.

While the Jeffersons of Snowden were deeply involved with the Charles Lilburne Lewises of Monteagle, the Jeffersons of Monticello were equally cool towards these cousins. Despite the fact that Thomas Jefferson had served as Charles Lewis, Jr.'s executor, he (and his family) kept a definite distance from the next generation. Conveniently, during most of the years the Lewises resided at Monteagle, Thomas Jefferson was living away from Monticello. Once Jefferson retired, many of the Lewises had already left for the west. Later, when scandal touched the Lewis family in Kentucky, its impact on the Jeffersons at Monticello was faint.

Thomas did not involve his daughters with the Lewises. When he left for Paris, he did not leave Polly (a.k.a. Maria) with his sister, Lucy, who lived nearby. Instead, he chose his sister-in-law, Elizabeth (Wayles) Eppes, to care for her. Nor did he engage these Lewis men, despite their proximity, to assist in the management of his farms when he was away from Monticello for long stretches of time.

Additionally, various aspects of the Lewis "culture" clashed with Thomas Jefferson's views. These Lewises were vocal Presbyterians, and Jefferson felt Calvinism wrought a certain element of "fanaticism" in Virginia.[10] He also openly disapproved of the culture of marrying early in life, to which many Virginia gentry and the Lewises subscribed. Thomas Jefferson saw that it cut short education, especially for the women, who were later responsible for the education of their children. Lucy was not yet seventeen when she married Charles Lilburne Lewis, and, by all evidence, her education and experience were indeed limited.[11]

In the beginning, Thomas appeared fairly neutral, if not positive, concerning his brother-in-law, Charles Lilburne Lewis. In 1781, Thomas wrote a letter recommending him to the rank of Colonel in the Albemarle Militia.[12] In 1792, Thomas stated, "C. L. Lewis my brother-in-law is becoming one of our wealthiest people."[13] During the 1790s, however, he watched the Lewis fortune decline in the hands of Charles Lilburne and eventually compared the family's circumstances to a "shipwreck." They borrowed money from the Jeffersons, the Carrs, and other in-laws. Their business practices became suspect; their debts mounted. They even took advantage of their son-in-law, Craven Peyton.

Unsurprisingly, in the end, Thomas cooled toward his cousins, sharing little but cordiality and genetic makeup with the Charles Lilburne Lewises.[14]

To further complicate their relations, Thomas Jefferson actively disliked several members of the extended Lewis family – Lucy's in-laws and his own cousins. Business dealings, especially in and around the town of Milton where the Lewis family was heavily invested, had soured Thomas towards specific members of the family. Charles Lilburne Lewis' sister, Elizabeth, had married the enterprising Bennett Henderson, with whom she had twelve children. Bennett was a Magistrate in Albemarle and a key developer in the town of Milton, where he erected a large flour mill and tobacco warehouse. His death in 1793 left Elizabeth a widow with many children to rear. Over the course of increasingly complex dealings with Elizabeth concerning her husband's estate and business matters, Thomas Jefferson came to dislike her and her family intensely, to the point of believing the Hendersons were not to be trusted.[15]

Likewise, Thomas distanced himself from Edward Moore, the husband of Charles' sister, Mildred. Like her sister, Elizabeth, Mildred also "married well," but according to Rev. Edgar Woods, Moore declined into insanity:

> Edward [Moore] occupied a position of considerable prominence, but unfortunate habits seem to have ruined both him and his estate. He was a magistrate, and in the decade of 1790 represented the county in the House of Delegates. His plantation of five hundred acres, which he bought from John Harvie ... was sold under a deed of trust.... Overwhelmed with debt, stripped of his property, and declared insane in 1807, he was by order of Court placed in the Asylum, where he died the next year....[16]

The degree to which the Randolph Jeffersons associated with the Moores and Hendersons is currently unknown; however, their son, Isham Randolph Jefferson, eventually married Mary Ann Henderson, a granddaughter of Bennett and Elizabeth (Lewis) Henderson.

Despite Thomas Jefferson's coolness to the Lewis clan, he sustained positive feelings for his sister, Lucy. Over the years, he wrote cordially, even warmly, to her. Their geographical closeness naturally resulted in contact. They lived near enough to each other for Thomas to "ride down" to see her, paying a casual call that did not require formal socializing. In 1790, Thomas entertained Lucy and his brother-in-law for at least one dinner on the mountain top. Addressing Charles Lilburne Lewis, Thomas wrote, "Some of our neighbors are to come and dine with us tomorrow; We shall be very happy if you and my sister will be of the party."[17]

In letters to Lucy and her husband, Thomas Jefferson closed warmly, "Present my friendly respect to Colº Lewis and my salutations to the family, and be assured yourself of my constant and tender affections."[18] Lucy, in turn, expressed affection for Thomas. On May 26, 1806, she wrote the following from her home at Monteagle to Monticello: "The receipt of your letter to day, gave me much real pleasure, but your presents hear much more, if your health would of permitted it, which I hope [it] will, before you leave us, as to mine I enjoy more perfect health, thin I have done for many years past...." She closes saying, "Mr. Lewis begs you will accept his worm respects. And be leav me to be with real esteem. Lucy Lewis."[19]

Despite Thomas Jefferson's lack of involvement with his Lewis kin, there is no evidence that he disliked his Lewis sister-in-law. Only three Jefferson-to-Jefferson letters survive written prior to Nancy Jefferson's death, to give a glimpse at their relationship. Thomas had lived abroad from 1784 to early 1789, missing many details of Randolph's marriage, though he was aware of his namesake, Thomas Jefferson, Jr., who was born before Thomas left for France.

When Thomas wrote from Paris on January 11, 1789, he extended an invitation to Randolph and Nancy to join him at Monticello, adding, "Remember me affectionately to my sister." In another letter written from Thomas to Randolph on February 28, 1790, which discusses, among other things, the family business of Elizabeth Jefferson's estate, Thomas closes with the especially warm remarks, "Give my love to my sister and the little ones. I am, Dear brother, yours affectionately, Th: Jefferson." Lastly, in the autumn of 1792, Thomas closed a letter to Randolph saying, "My love to my sister." As always, Thomas was cordial, if not revealing.[20]

The reasons for this distance between the Charles Lilburne Lewis family and the Thomas Jefferson family were no doubt complex. Charles Lilburne and his siblings were not only Lewises; they were also Randolphs.

Little is known about Thomas Jefferson's relationship with his mother, Jane Randolph, and many biographers have speculated that it was conflicted. This may or may not explain his coolness toward her sister's family. The erratic, even insane, streak in the Randolph bloodline may have disturbed Thomas deeply, and possibly he sensed indications of emotional and psychological problems in the Lewis-Randolph offspring. Eventually, instability and madness in the group did indeed erupt violently.

The children of Charles Lilburne and Lucy (Jefferson) Lewis were, after all, double Randolphs, and there may have been various congenital abnormalities among them.[21] Whatever may have repulsed Thomas Jefferson, however, did not serve as a similar warning for Randolph Jefferson, and his children paid a price for it. The further inter-breeding between his children and the Lewises indeed may explain the incidence of mental deficiencies in some of Randolph's grandchildren and great-grandchildren.

The distance maintained by the elder Jefferson decidedly left room for his younger brother to make the Lewis family his own, and that is precisely what happened. Commensurate education, levels of erudition (or lack thereof), general interests -- all may have played a part in the congeniality between the younger Jeffersons and the Lewises. Additionally, closeness in age between Lucy and Randolph Jefferson no doubt played a major factor. Unlike Thomas, they had both spent part of their youth connected to Buck Island, drawing these particular Jeffersons closer to the Lewises.

While there is no direct evidence of the nature of the relationship between Nancy and her brother, Charles Lilburne, indirect evidence – the multiple intermarriages among their offspring – indicates they were compatible and close as siblings. Anna Scott "Nancy" Jefferson would become engaged to (but ultimately did not marry) her first cousin, Charles Lewis. Thomas Jefferson, Jr. would wed his first cousin, Mary Randolph "Polly" Lewis. Younger members of the Randolph Jefferson family would marry the next generation of Lewises. Isham Randolph Jefferson married first Mary Ann Henderson, granddaughter of Bennett Henderson and Elizabeth Lewis, and second Margaret Gwatkins Peyton, granddaughter of Lucy Jefferson and Charles Lilburne Lewis. Peter Field Jefferson's bride was Jane Woodson Lewis, another granddaughter of Lucy and Charles Lilburne.

There was no doubt much coming and going between Snowden, the old home at Buck Island, and the Charles Lilburne Lewis home at Monteagle. If there was regular correspondence between the Jeffersons and the Lewises, it has not survived, leaving little documentary evidence of family exchanges. One especially poignant example connecting the Lewis and Jefferson families is documented in an exchange of letters between Isham Lewis (born c.1788), son of Charles Lilburne and Lucy (Jefferson) Lewis, and his uncle, Thomas Jefferson.

In the spring of 1809, Isham was visiting Snowden where he enjoyed Randolph Jefferson's hospitality, as well as the company of his many cousins and his sister, Polly, who was now Mrs. Thomas Jefferson, Jr. Five years earlier, Isham had inherited 230 acres of relatively poor land on the Three Notch'd Road. He did not hold it long and had joined the Lewis migration west. When he returned to Virginia on business in the fall of 1808, Isham remained in Albemarle.

On April 27, 1809, he wrote to Thomas Jefferson indicating that he was in want of useful employment and blaming his father for his "neglect to bring me up in any useful pursuit."[22] Thomas replied promptly on May 1, 1809, offering no concrete job, but inviting Isham to stay briefly at Monticello where he would teach him basic surveying techniques.[23] Isham accepted, leaving Snowden for Monticello. Following successful instruction, Thomas wrote two positive letters of introduction for Isham and gave him 50 shillings toward his journey to Tennessee. Ultimately, Isham returned to Kentucky where, in 1811, he and his brother, Lilburne, committed the gruesome murder of the Lewis-owned slave, George.[24] Reverberations of this family tragedy were no doubt felt at both Snowden and Monticello, but no details have survived.

~

When Anne "Nancy" Lewis arrived at Snowden she brought with her personal property, which included a slave named Lucy, daughter of Phoebe.[25] Nancy also brought with her the Lewis sensibility, which would shape her children and strongly influence their life choices. Her religious beliefs were likely no small part of this legacy.

The Lewises were devout Presbyterians across several generations, making them "dissenters" from the Anglican Church in Colonial Virginia. Robert Lewis, the Welsh-born lawyer and patriarch of the family, had arrived in

There is no known image of Anne (Lewis) Jefferson. Given her social rank, she would have followed the fashionable dress of the day. Towards the end of Anne (Lewis) Jefferson's life, Abigail Adams (1744-1818) was America's First Lady (1789-1797). Mrs. Adams, pictured here, attempted to influence fashion away from the revealing Napoleonic-style of the day, presenting a more conservative image. (EMMET COLLECTION, MIRIAM AND IRA D. WALLACH DIVISION OF ART, PRINTS AND PHOTOGRAPHS, THE NEW YORK PUBLIC LIBRARY, ASTOR, LENOX AND TILDEN FOUNDATIONS)

Virginia in 1645. He and early generations of the family were subject to the domination of The Church of England in the American colonies. However, by the Revolutionary era his great-great-grandson, Charles Lewis, Jr., and his son, Charles Lilburne Lewis, were strong advocates of religious freedom. On October 22, 1776, Charles Lewis and Charles Lilburne Lewis signed a petition joining dissenters in Albemarle, Amherst, and Buckingham counties, complaining of their required support of the established church. In fact, the first signature on this petition was that of Charles Lewis, Jr. It was closely followed by the names of his sons, Charles Lilburne and Isham, as well as three of his sons-in-law.[26] Dissenters like the Lewises made fierce revolutionaries, fighting for their religious freedom from the Anglican Church as well as political freedom from England. No wonder Charles Lilburne and his brother, Isham, promptly volunteered for military service when the fighting began.[27]

As there were few Presbyterian churches in the Virginia Piedmont, the Lewis family held services in their home. Presbyterian ministers visited Buck Island, and later the Charles Lilburne Lewis home at Monteagle, serving the Lewises and their like-minded neighbors. Among them were local Reverends Samuel Black, Samuel Davies, John Todd, John Brown, John Martin, and Henry Patillo.[28] The Lewis' private library included a large Bible, a prayer book, a hymn book, and several volumes of sermons to carry them through times when ministers were not available.[29]

While a youthful Randolph Jefferson lived at Buck Island, he may have been influenced by these family services. If he did not come to follow the Presbyterian doctrine, he certainly must have tolerated it and, in turn, the Lewis family must have tolerated whatever beliefs Randolph held. Anne Lewis no doubt arrived at Snowden professing the Presbyterian faith and it may have been strong enough to influence the leanings of the Randolph Jefferson family.

During the decades the Jeffersons lived at Snowden there was no Presbyterian church in the neighborhood. Cove Presbyterian Church in Albemarle County, founded in 1769 and located at Covesville, south of Charlottesville, was no more convenient to Snowden than it was to Buck Island. Within Buckingham County, Trinity Presbyterian at New Canton and Maysville Presbyterian at Buckingham Courthouse, both post-date the Jeffersons' era at Snowden, as does the Presbyterian Church of Scottsville, which was not founded until October

18, 1827. If the Jeffersons continued the Lewis family orientation, their services likely remained private.

When Randolph Jefferson took possession of Snowden it lay in Tillotson Parish, which covered all of Buckingham County. Rev. William Peaseley (1714 - c.1786) served there as the Anglican, then Episcopal, rector for almost two decades. Well-established in his role when Randolph Jefferson took possession of Snowden, Peaseley was born in Dublin, Ireland and educated there at Trinity College. He arrived in Buckingham in about 1760, bringing with him a somewhat checkered past, which included an accusation of adultery during the 1750s while he was rector in St. Helena's, Beaufort County, South Carolina. Peaseley moved to Buckingham with his second wife, Lucy Sanders of Lunenburg County, Virginia. They had at least three children, lived on 300-350 acres of land, and owned two slaves.[30]

Rev. Peaseley was responsible for the parish's four churches: Buckingham, Goodwins, Buck and Doe, and Maynards. The mother church was "old Buckingham," located near Gold Hill in the northeastern section of the county. Buck and Doe Church was named for a stream near Willis Mountain in the southeastern section of the county. Goodwins Church was situated about eleven miles north of Buckingham Courthouse, near today's Highway 20, roughly twelve miles south of Snowden.[31]

Randolph Jefferson's first experience of church-going in Buckingham County may have been during 1773-1774, when he spent time at John Nicholas' Seven Islands. Buckingham Church was most convenient to Seven Islands, and as a Vestryman and Church Warden, Nicholas may have insisted that Randolph attend church with the Nicholas family. However, by the time of Randolph's marriage in 1781, the tight grip the Church of England once held on the colonies had loosened and Nancy's presence at Snowden may easily have turned the Jefferson household into a Presbyterian one.

Over the years, it is possible that Randolph Jefferson reflected the style of his older brother, who committed to no particular church and regarded his religious views as a private matter. In February of 1777, he joined Thomas and a few other men in Albemarle pledging support to Rev. Charles Clay in the formation of a Calvinistical Reform congregation. Thomas Jefferson drafted a document, "Subscription to Support a Clergyman in Charlottesville," engaging

Rev. Clay to hold divine services and preach a sermon for a self-selected group, meeting monthly at Albemarle Courthouse.[32] Clay would be paid annually on December 25th by this voluntary membership. It was Thomas Jefferson's answer to the Act of Exempting Dissenters of 1776, which read: "Whether a general assessment should not be established by law, on every one, to the support of the pastor of his choice; or whether all should be left to voluntary contributions."[33]

Rev. Charles Clay was well-known to the Jefferson brothers. He read the funeral service for their brother-in-law, Dabney Carr, who died at Shadwell in 1773; for their sister, Elizabeth Jefferson (d. 1774); and for their mother, Jane Jefferson (d. 1776).[34] His brother, Green Clay, was a friend of Randolph Jefferson.

Thomas was the strongest supporter among the original nineteen subscribers, pledging £6. Col. Nicholas Lewis was second, pledging £3.10, and Randolph Jefferson was third, pledging £2.10. The majority of the men pledged ten, fifteen, or twenty shillings, indicating that Randolph's commitment was not a casual one.

Services were to be held monthly in the Charlottesville courthouse, initially on the 4th Saturday, then on the 4th Sunday. The first meeting would be Saturday, March 1, 1777. If membership increased, the number of meetings might do likewise. By the spring of 1777, Randolph was already settled at Snowden and would have to make a day's ride to attend services with his Albemarle compatriots. Could this commitment to Rev. Clay's congregation indicate that he was not living full-time in Buckingham County, or, perhaps, that Randolph had regular business in Albemarle County?

The men supported Rev. Clay through 1778 and into 1779. Thomas not only met his pledge of £6, he also paid Rev. Clay for some of the other men, including his neighbor, Philip Mazzei, and Thomas Garth. On March 9, 1778, Thomas paid 50 shillings towards Randolph's subscription to Rev. Clay. In April of 1778, subscriptions were paid for a second year and, on August 15, 1779, Clay was paid £30 "in consideration of parochial services."[35] It is not known how much longer the group remained together or if Randolph Jefferson subscribed beyond the spring of 1778. He may have found himself settled and busy at Snowden, infrequently making the trip to Charlottesville.

Early in his marriage, Randolph Jefferson signed a petition against Virginia's proposed assessment to support the Christian ministry. Presented to the Virginia House of Delegates, Randolph's signature stands out among nearly 500 signatories, bolder than many. He may have been as much against increased taxes as he was against state support of the clergy.[36]

Randolph may never have aligned with a particular Christian denomination; however, sometime before 1803, at least some of his children were influenced by local Baptist preaching, perhaps specifically by the sermons of Rev. Martin Dawson. Ordained about 1774, he preached for many years in southern Albemarle County on Totier Creek, situated near Porter's Precinct, at what was commonly known as Dawson's Meeting House.[37] Dawson officiated at the weddings of Randolph Jefferson's only daughter, Anna Scott, who married Zachariah Nevil in 1803, and Thomas Jefferson, Jr., who married his cousin, Polly Lewis, in 1808.[38]

Another of Randolph's younger sons was, at the very least, sympathetic to the Baptists. In 1851, Robert Lewis Jefferson gave an acre of his land at Porter's Precinct to Baptist Trustees, John H. Nicholas and John N. Moorman. A house, built for a church, already stood on the land.[39] This may have been Dawson's Meeting House, still standing from the last century. Eventually, Robert Lewis Jefferson's only son, Elbridge Gerry Jefferson, would become an ordained Baptist minister and help establish Gooseberry Baptist Church in Buckingham County.[40]

Dwelling at Snowden

Even though forty years had passed since Peter Jefferson established himself in Albemarle County, there was still no town to speak of near Snowden, and Randolph Jefferson found himself in a familiar situation. The local population was very small, as it had been near Shadwell when Peter and Jane Jefferson chose to settle there rather than back east in Goochland. The significant difference between Shadwell in the 1730s and Snowden in the 1770s was its proximity to the James River. Randolph had little need to go out into the world, for the river, flowing right past his door, provided a steady stream of commerce and travelers.

This photograph of the James River's Horseshoe Bend was taken by William E. Burgess, c. late 19th century. Burgess made his home in Scottsville and created thousands of picture postcard images during his career. The land visible on the south side of the river, across from the town of Scottsville, constitutes the northern tip of Snowden. (COURTESY OF SCOTTSVILLE MUSEUM)

The world came to him. Additionally, the road running south from the ferry landing cut right through Snowden, so the Jeffersons enjoyed foot, horse, and wheeled traffic as well.

When Randolph first took over his estate, he stepped into a neighborhood that had been established by his father and his father's peers. Peter Jefferson had been a figure of wide experience and associations, not only in the Piedmont, but throughout the James River culture to Richmond and on to Williamsburg, where he served as a Burgess representing Albemarle County. In short, Randolph was the son of a well-known and influential man. In 1776, when he established himself at Snowden, his older brother was showing signs of similar, if not greater, ambition. Randolph, however, was apparently comfortable with a "local life," his family and his neighbors forming the core of his relationships. In 1808, he would choose two immediate neighbors – Hardin Perkins and Robert Craig – to serve as executors of his will, along with his brother Thomas.

Randolph's neighborhood was remarkably stable over the years. A survey made before Peter Jefferson's death reveals Snowden's adjacent neighbors at that time.

> Peter Jefferson; 2,050 ac; no date. "This is an Inclusive Patent of 2,050 Acres of Land Lying on the South side the Fluvanna River opposite to Totier Creek in Albemarle County, 400 acres of which purchased of John Ladd, 100 acres purchased of Noble Ladd, and the other part New Land. Ferry Road runs through the center of the tract. Joining Colo. John Bolling (300 acre parcel), Wm Battersby (200 acre parcel), Thomas Ballowe (400 acre parcel), Wm Battersby (119 acre parcel), Patterson corner, Perkins. Shows George's Creek, Marr's Road, Bibbins Branch, Perkins Path, and Rocky Ridge.[41]

By 1776, when Randolph came into his inheritance, Snowden's earliest neighbors, the Ladds, William Battersby, and Richard Murray, had already died or left the area; however, the descendants of some of these men assumed their fathers' lands adjacent to Snowden. Randolph would become well acquainted with the next generation; among them were Richard Murray's son, Anthony Murray, and the sons of Hardin Perkins, Price Perkins and Hardin Perkins, Jr.[42] Like them, Randolph Jefferson was a stable planter, living among equally established men.

There is no description of the dwelling Randolph Jefferson provided for his bride in the summer of 1781. His bachelor quarters at Snowden, 1776-1781, might have been a rude cabin or what had formerly been an overseer's or a tenant's house. He may have started out at the ferry house or in another structure which dated back to the Ladd family. Bachelors, especially young ones, did not tend to build elaborate homes until a wife and growing family entered their lives. When Thomas Jefferson brought his bride, Martha, home to Monticello, there was no mansion house. They started married life in a 20x20-foot, one-room brick "outbuilding."[43]

Nor does a description survive of any later Jefferson-built or occupied dwelling at Snowden. However, in 1815, the year of Randolph's death, his dwelling house was valued at $900. A bit lower than his neighbor's, Hardin Perkins, Jr., whose house was valued at $1,000, but considerably higher than neighbor Thomas Ballow's house, which was worth less than $500.[44] That year in Albemarle County, Thomas Jefferson's Monticello was valued at $7,500 and Farmington, which Jefferson designed, was valued at $9,000.[45] Placed in this context, it is clear that Randolph had a good, solid house at Snowden, but not a grand one. Still, his total tax obligation was a hefty $20.83½.[46]

Likewise, there is no inventory for Randolph Jefferson's personal property, except a few luxury items revealed on the 1815 personal property tax record. That year, many household goods were subject to tax. They included cut glass, silver tea pots, various types of furniture, and even ice houses for private use. Randolph was assessed on his silver watch, four pieces of Mahogany furniture (including calico curtains), and a chest of drawers made of wood other than Mahogany. The "calico curtains" refer to the imported decorative fabrics enjoyed by the Jeffersons.[47]

The Mahogany furnishings, which may have included the writing desk which Randolph purchased from his brother, and fine fabrics indicate that Snowden's dwelling house was handsomely and solidly finished. These speak of quality necessities in Randolph and Nancy's life.

The items on which Randolph was *not* taxed are equally useful in shaping a picture of life at Snowden. He paid no tax on a bookcase, nor a clock, silver-plated items, or cut glass. Additionally, Randolph was not taxed under "pictures, prints, engraving, mirrors or looking glass."[48] If Randolph ever sat for a portrait, it did not hang in his home in 1815. A few others in Buckingham were taxed for "portraits in oil" that year, though only one is recorded on Randolph's particular page. Likewise, no one on his page in the 1815 record was taxed for mirrors or looking glasses. Was there a lack of vanity in Buckingham County, or did this particular tax collector overlook the new category?

Growing up at Shadwell, a fairly "showy" place for its day, Randolph easily could have acquired a taste for expensive living. There, Peter and Jane Jefferson served their guests with a silver coffee set and imported china. Peter Jefferson certainly liked personal decoration, which included silver spurs and a silver hilt on his sword. Likewise, Randolph's cousins, the Lewises, spent their wealth on tangible things. When his cousin, Charles Lilburne Lewis, built his house, Monteagle, he thought it worth insuring against fire and Randolph's brother, Thomas, was one of the most conspicuous consumers in the area.[49]

Given this background and family milieu, it comes as a bit of a surprise that the Randolph Jeffersons did not own a few more luxury items. If they did, and did not report them among their personal property, it would not be the first time a silver tea set had been hidden from the tax man. However, it is entirely possible that Randolph never acquired such "fancy" things and he had his own ideas about extravagant spending, which may have been connected to his distaste for accumulating debt. While he may have equivocated on other ideas, Randolph apparently held fast to his frugality. His one known extravagance was his silver watch, which brought him both comfort and anxiety.

Randolph had the funds to purchase highly refined items and certainly had the collateral to go into debt for them. In addition to avoiding creditors, Randolph may have been of a utilitarian mindset. If a stoneware teapot served his purpose, he did not require one made of silver. Still, the Jeffersons definitely enjoyed imported goods. In the 1790s, Randolph maintained an account with the Glasgow firm of Donald, Scott & Co. Like his mother, sister, Anna Scott, and brother, Thomas, Randolph purchased significant merchandise from the importer. In 1797, he settled his accounts with the firm which totaled £80.13.10½.[50]

As simple and solid as Randolph's standard of living may sound, when compared to Buckingham's norm, he clearly stands out as one of an affluent few. Among the thirty households named on his page, Randolph pays the second highest personal property tax. Only two other silver watches are listed, along with eight clocks of varying kinds.

His neighbor and friend, Hardin Perkins, Jr., paid tax on a clock and a chest of drawers made of wood other than Mahogany, despite his more valuable dwelling house. Like Randolph, Hardin did not pay tax on an oil painting, silver, or a looking glass, yet his estate inventory, made just a few years later in 1821, included eight silver spoons and a looking glass. Were they not part of Hardin's property in 1815? Did the tax man overlook these, or did Hardin fail to disclose them?

The only other known reference to Snowden interiors is a peculiar and possibly erroneous identification of surviving furniture which once belonged to Randolph Jefferson. On May 15, 1929, a letter written from R.B. Chaffin & Co, Real Estate, Auctioneers and Brokers, located at Main and 12th Street, Richmond, Virginia claimed to have the following for sale: "We have . . .

In 1815, the dwelling house of Hardin Perkins was valued at $1,000. Known today as the Hermitage, this sketch by Margaret Allen Pennington depicts the elaborated house as it stands today. (HISTORIC BUCKINGHAM)

'Sheraton' sideboard formerly owned by Thomas Jefferson, having been a part of his Monticello furniture, antique sofas, tables and chairs, a part of the Randolph Jefferson (younger brother of Thomas Jefferson) [furniture] at 'Snowden.' Price $50,000."[51]

Pastimes in Buckingham County

Dining with friends, visiting family, and entertaining at home constantly occupied the Virginia gentry, and the Randolph Jeffersons were no exception. The Jefferson-to-Jefferson letters confirm that Randolph's twin, Anna Scott (Jefferson) Marks, was a frequent guest at Snowden, especially after the death of her husband. Thomas Jefferson called, but apparently not often. Members of the Lewis and Jefferson families stayed for extended periods. Randolph and his second wife, Mitchie, spent lengthy visits with her family, especially at Woodlawn with her mother, Susannah Pryor, and the Pryors no doubt visited Snowden.

Among the planter class, family members doubled as friends and business associates. Correspondence reveals that Thomas Mann Randolph, niece Martha Jefferson's husband, acted at least once as Randolph's agent.[52] Two cousins named John Jefferson spent time at Snowden.[53] Craven Peyton, who married Randolph's niece, Jane Lewis, was also in Randolph's sphere. Peyton was an affluent and important business man at the town of Milton, near Buck Island, and served as an agent for Thomas Jefferson.[54] In later years, Thomas "Jefferson" Randolph, Martha's son, who married Jane Hollins Nicholas, and his brother-in-law, young Wilson Cary Nicholas, were regular guests at Snowden.[55]

During Thomas Jefferson's retirement, Randolph visited his brother and vice versa, though the twenty miles between Snowden and Monticello was a full day's travel by horseback or by stage, *if* the roads were travelable. The plantations were not so close that one could "drop in" for the afternoon or even go up and back for the day. Additionally, the James River had to be crossed. Sometimes it was frozen, sometimes a rushing torrent. Still, letters could be, and often were, answered the next day, carried from one plantation to the other by the Jefferson slaves or, in Randolph's case, delivered by one of his sons or one of his Pryor brothers-in-law.

During this period, planters such as Randolph Jefferson were busy men and typically traveled widely within their local realm, which extended to at least the neighboring counties. For Randolph those included Albemarle, Fluvanna, Cumberland, Prince Edward, and Nelson Counties.[56]

No diary or memorandum book survives to illuminate Randolph Jefferson's daily life; however, diaries such as those preserved in the Bolling family reveal these active gentlemen farmers both at and beyond their plantations. Sisters and other family members were frequently visiting; men were busy calling on each other. Crops and weather greatly occupied them. Fires were a constant threat, and many houses burned to ashes. Hunting and fishing provided sport and food. Cock fights and horse races offered diversion. For those living further east, trips to Petersburg and Richmond relieved the tedium of country life. Snowden's isolation, however, made trips to sizable towns less likely, giving the Jeffersons even more reason to rely on friends and family for entertainment.

The Bolling family intermarried with both the Jeffersons and the Randolphs. Randolph Jefferson's sister, Mary, married Maj. John "Old Indian" Bolling. Many entries in the Bolling diaries refer to the area of Virginia where Peter Jefferson was born, reared, and lived – Osbornes, Fine Creek, and various Randolph-owned properties. Maj. Thomas Bolling (1735-1804), Maj. John Bolling's brother, kept a diary from 1787-1804, which has survived.[57] Occasionally, he wrote about something of larger importance, such as when a theater burnt in Richmond in January of 1798, or of local interest, such as the excitement of a ball at Osbornes on December 20, 1798. However, mostly he noted the comings and goings at home, highlighting who dined with whom.

Thomas Bolling's son, Col. William Bolling (1777-1845), was an even more enthusiastic and colorful diarist.[58] On February 3, 1794, the youthful William wrote, "Went to a dance at Mr. Edward Coxe's where we had a very agreeable company and a very merry dance. We danced till 1 o'clock in the morning. Mrs. Christian Friend was my partner."[59] Things picked up again the next morning: "We began to dance before breakfast and danced until abt. 12 o'clock (at noon)

Col. William Bolling (1777-1845), diarist, painted by William James Hubbard (1807-1862). (Muscarelle Museum of Art at the College of William & Mary in Virginia; Gift of Mrs. Robert Malcolm Littlejohn)

when we broke and returned home."⁶⁰ Entertainment being scarce in the country, dances and musical evenings were frequent and may have been among Randolph's favorite pastimes, where he could indulge in his fiddling.

Weddings were important local events and the men looked forward to them as much as the women did. From 1793-1798, William Bolling kept a list of the marriages of acquaintances, a total of fifty-seven weddings. If the young gentleman only attended a fraction of these, he enjoyed a lot of wedding parties. He may even have seen Randolph Jefferson at one or two of the celebrations. The list included several Randolph family weddings and Maria Jefferson's marriage to John W. Eppes, held at Monticello in 1797.⁶¹ William describes their mutual cousin's wedding in February of 1794, which turned into a four-day affair:

> February 20 - Went to Field Archer's wedding. He married Cousin Martha Bolling and Cousin Edward Bolling was married on the 17th to Miss Dolly Payne of Goochland and he brought his wife down to the wedding and also Miss Patsy Payne, a fine girl. The wedding lasted till the 20th. I got acquainted with Miss Nancy Manlove here, a most agreeable girl.⁶²

In November of 1795, William found another wedding an exhausting affair:

> November 3 - Went to a Frolick at Col. Goode's given in celebration of the marriage of his daughter, Polly, to John Spotswood, Jr. & of his son, Frank to Miss Martha Hughes (Powhatan). There was as large a company as I ever saw at a private house. Miss Polly was my partner. I got but little sleep as my bed was composed of my greatcoat & part of a sofa. A little before day, for they played cards the whole night in the room where I sleep, they opened the faro bank. There were three fiddlers which were divided into two rooms, one room not being sufficient to contain all the company.⁶³

Col. William Bolling's home at Bolling Hall was in Goochland County, and he eventually married Jefferson cousin, Mary Randolph. A more mature

Bolling Hall, c. late 19th Century. (LANCASTER, V.45.47.22.02 - "BOLLING HALL," GOOCHLAND COUNTY, C. 1875-1900, ROBERT A. LANCASTER, JR. COLLECTION, VALENTINE RICHMOND HISTORY CENTER)

Bolling Hall, c. 1970. (COURTESY OF VIRGINIA DEPARTMENT OF HISTORIC RESOURCES)

William opened his 1809 diary on January 1st, mentioning an array of day-to-day local activities, the type that would have occupied many men of his station, including Randolph Jefferson. He noted the talk of the continued embargo, with its threats of war and insurrections; the annual public hiring of Negroes; snow, which confined him to the house, so he worked on his blacksmith's accounts; hunting wild hogs; putting up ice in the icehouse; attending a barbecue and winning at marbles; shooting a fox; hunting for partridge and hares; hearing an excellent sermon; setting his fish trap; losing his pointer, who swam to Mt. Pleasant and did not come back, etc. On the 15th, Bolling added, "Spent the day as usual when alone. Had a dish of fish for dinner. Got my pointer, Rondo, the day after he was lost."[64]

In February of 1809, William Bolling qualified as a Magistrate for Goochland County, and his diary is riddled with over 200 mentions of court business at the Goochland Courthouse.[65] In all counties, Court Day was an important event. Buckingham held its court monthly, while Albemarle County and others held theirs only quarterly. Apparently, Randolph Jefferson was never asked to serve in Buckingham as a Gentleman Justice (a.k.a. Magistrate), or if he was asked, he declined. His neighbors, brothers Hardin Perkins Sr. and Price Perkins, were both Justices. His neighbor, Anthony Murray, had his name put forward, but he never qualified.[66]

As a result, for Randolph Jefferson, attending Court Days at Buckingham Courthouse was not a duty. Like the majority of farmers in Buckingham County, he could enjoy commerce and social exchange when Court was in session. Auctions were held for land and for slaves. Valuable horses were on display; spontaneous races occurred. Itinerant peddlers displayed their wares, turning the courthouse square into a bustling market. Entertainers also took advantage of the crowds, as did politicians who delivered extemporaneous speeches. Gambling and heavy drinking added to the sport. Religious meetings attempted to balance what could become prolonged public drunkenness, particularly during elections.[67]

The courthouse had long been accessible from Snowden. A road connecting the plantation as far south as the Slate River had existed since the 1740s. An Albemarle road order, dated July 9, 1747, granted the petition of William Diuguid and others for a road to be opened from "Slate River near Mr. Marrs's to the Court House."⁶⁸ Diuguid, Noble Ladd, and John Ladd were ordered to clear the road, with Noble Ladd as overseer. Noble Ladd and Gideon Marr were ordered to mark it out. This road, sometimes called Marrs' Road, Old Courthouse Road, and the Ferry Road, bisected Snowden and the Horseshoe Bend. Still, the twenty-three mile distance from Snowden to Buckingham Courthouse likely deterred frequent trips. Randolph often would have been faced with the uninviting prospect of inclement weather or muddy spring roads detracting from the ride.⁶⁹

Despite the fact that it came alive monthly, a magnet drawing folk from the vast countryside, during Randolph's lifetime the area around Buckingham Courthouse never successfully developed into a town. Buckingham's first courthouse, which sat to the west of the present site, was built sometime after November of 1761 when the Governor's Council ordered that "the court house be located at Richard Taylor's on the land of Samuel Glover."⁷⁰ The modest wooden courthouse sat at the center of an unincorporated collection of enterprises, likely similar to those that grew up at Scott's Ferry when it was the location of Albemarle's courthouse.⁷¹ At the very least, the courthouse square quickly included a jail, clerk's office, and lawyers' offices. Always nearby was at least one tavern, or ordinary, which provided "entertainment" – food and lodging. Among the tavern keepers were John Cox, Bolling Branch, also Deputy Clerk at the courthouse, and Guerrant Staples. The Raleigh, just to the west, was kept by Daniel Guerrant.⁷²

For twenty years, development was quiet. Then, in 1782, efforts finally were made to incorporate a town at Buckingham Courthouse. In May, the Virginia General Assembly agreed that the proposed town of Greensville would be of "great utility" to the county. Tavern keeper, John Cox, put up ninety-six acres where the town would be laid out. Lots in Greensville languished undeveloped and, in 1790, the citizens of Buckingham requested of the Assembly that the act establishing the town be repealed. It wasn't until after Randolph Jefferson's death in 1815 that the town of Maysville was officially created at Buckingham Courthouse and a new effort was made to establish a town in the center of the county.⁷³

Nor did Randolph live to see Buckingham's new courthouse building designed by his brother, Thomas. In 1821, Charles Yancey, who was among the commissioners appointed to select plans for a new courthouse in Maysville, approached Thomas Jefferson for a design, which was accepted on July 4, 1822. Constructed between 1822 and 1824, it was destroyed by fire in 1869."[74]

Chapter Seven

Matters of Money

The want of money cramps every effort.
Thomas Jefferson to George Washington, 1780

Loans between the Jefferson brothers were executed without the embarrassment of hiring lawyers or appearing in the courts. Over the years, Thomas' memos reveal the ease with which the brothers loaned each other funds.

Randolph was firmly established at Snowden when there was more than the usual exchange between the Jefferson brothers, during February and March of 1778, when Thomas borrowed £25.8 from Randolph. It is worth noting that Thomas quickly paid back his debt to Randolph, which was not always his habit.

Feb 25 Borrowed of Randolph Jefferson £25-8.
Mar 9 Pd. Mr. Clay for Rand. Jefferson 50/.
Mar 12 Pd. Randolph Jefferson £25-8.
Mar 14 Gave Bob to pay ferrg. to Snowden 2/.[1]

Thomas' memorandum books reveal a series of loans to Randolph; however, because Thomas was in the habit of noting monies out, but not always monies repaid him, it is impossible to know how, when, or if these obligations were settled between the brothers.

On February 1, 1777, Thomas lent Randolph £4.10; on April 11, 1777, he gave Randolph 11 shillings; on August 9, 1779, he lent him another £33. In April of 1781, Randolph borrowed £450 from his brother. In the 1790s, Randolph's longstanding bill with Donald, Scott & Co. of Glasgow for £80.13.10½ was no longer his concern when Thomas assumed the debt in exchange for two of Snowden's slaves, Ben and Cary.[2] In August of 1813, when Randolph was "compelled" to bring $40.00 to court, Thomas covered the emergency.[3]

Friends like Philip Mazzei, Thomas Jefferson's neighbor, also provided loans to Randolph Jefferson. Filippo "Philip" Mazzei (1730-1816) was a Florentine merchant, surgeon, and horticulturist who arrived in Virginia in 1773, establishing himself at Colle, adjacent to Monticello.[4] Mazzei became a close friend of both Thomas Jefferson and the American Revolution. In 1778, he risked much when he sailed to Italy to raise funds for the American cause.

In the autumn of that year, following his ride with the Cavalry, Randolph borrowed £146.4.6 from Mazzei. On January 29th, Thomas gave his bond to Mazzei as security for Randolph's loan. Payment was due on November 4, 1779, with interest from November 4, 1778.[5] Mazzei, who was old enough to be Randolph's father, may have taken a paternal interest in the young man. At the very least, they were close enough that Randolph felt comfortable asking him for a loan.

Sometimes debts among extended family members put Thomas in the uncomfortable position of collecting funds. Upon his return from Paris, he was desirous of closing his father's estate, as well as those of his two deceased sisters, Jane and Elizabeth. On January 25, 1790, Thomas wrote to Dr. Thomas Walker, one of Peter Jefferson's executors, describing the agreement that he, Randolph, and fellow executor, John Nicholas, had reached. Among the accounting details, a cryptic sentence concerning Randolph stands out: "I knew that my brother's circumstances required indulgence, and was unwilling that advances made by you for the whole family should be delayed after it was signified that further delay would be inconvenient."[6]

A letter written to Jefferson's brother-in-law, John Bolling, gives a further clue to Randolph's "circumstances:"

Richmond Mar 6, 1790.
Dear Sir
I intended to have the happiness of seeing you and my sister, and sat (sic) out for that purpose the day before yesterday, but the day was so bitter cold that I was obliged to return back after getting to Manchester. I was anxious to settle the inclosed account with you, because that is all which is wanting to close the two administrations of my sisters estates. As I set out for New York the day after tomorrow, and shall remain there, I must get you to inform me by letter whether the account be right, and if not right, to correct it. A settlement is all I wish, as the balance may wait your convenience. You will receive demands for the other distributees. Those from Mr. Marks and my brother are I believe much needed. There is an execution out against the latter. Present my best affections to my sister & be assured of the esteem of Dr. Sir Your sincere friend & servt.,

TH: JEFFERSON

P.S. My address will be To Thomas Jefferson, Secretary of state, New York. Lodge your letter in the Richmond post office and it will come safe.[7]

The week prior, Thomas had written to Randolph delineating the amount due him from their sister Elizabeth's estate. The summary showed that Randolph could expect to collect a total of £36.7.2 from the following sources:

John Bolling (brother-in-law, husband of Mary Jefferson)   7-17-2
Charles Lilburne Lewis
   (brother-in-law, husband of Lucy Jefferson)   8-11-0
Nicholas Lewis (TJ's share, friend and neighbor of TJ)   19-19-0 [8]

When or if Randolph collected the funds is unknown, as is the nature of the writ of execution out against him. Apparently, Randolph had lost a judgment (likely in the Buckingham Court) and settlement had been awarded to the plaintiff. The Buckingham Sheriff may have already impounded property equal to its value.

Apart from this debt, which appears to be a typical, and temporary, cash flow issue, Randolph's life was stable. In 1789, he was married, with several children. His overseer was Thomas Hall. He owned fourteen slaves above age sixteen and an additional taxable slave aged between twelve and sixteen. There were undoubtedly more slave children under the age of twelve living at Snowden; however, they were not taxable that year, leaving their number unknown. Randolph also paid taxes on eight horses and on his 2,000 acres at Snowden.[9] One notable difference at Snowden that year was the presence of a guest, kinsman John Jefferson.

On January 7, 1790, John Jefferson wrote to Thomas Jefferson concerning his own financial embarrassment.[10] The letter was written at Snowden and carried to Monticello by Randolph. In it, John Jefferson claimed that he was in want of funds to pay his attorneys to settle a suit in which he was owed significant funds to cover those costs:

> . . . I did in May Cumberland Court 1784 recover a Judgment against [Col. Richard James] for about £400, he replevied it and then Superseded the Judgment, which yet is Undertermin'd, and now is sent to Prince Edward District, and I have another Suit against [him] in Our Court for a Considerable larger Sum than the above, but Cant raise Money to pay my Lawyers, for I have Nothing left me but Lands and Cant Sell any of them, and have two other Suits for 6 Negroes I am unjustly kept out of, and for want of about Ten pound for the Attorneys . . . I have not a Horse in the World to do the Necessary riding and without I can Make a Friend of you for the above two favours . . . . Your Brother will deliver this Letter togather (sic) with a Copy of Colo. James(') Bond . . . .
>
> I am Sir With the Greatest Esteem Your Distress'd But Mo. Obt. Friend &c.
>
> Jno. Jefferson[11]

John Jefferson may have first appealed to Randolph, finding him unable or unwilling to provide him with £10. It is clear from Thomas' correspondence that Randolph was short on funds. Thomas kindly, but formally, answered his kinsman on February 14, 1790, passing the somewhat sticky business to Col. Nicholas Lewis:

> Monticello Feb. 14. 1790
> Dear Sir
> I have duly received your letter of the 7th. of January by my brother, and by him returned the copy of the bond you inclosed. You mention that for the want of 10. or £15 to pay your lawyers you are unable to prosecute to effect actions commenced for the recovery of your rights. I wish my stay in the country would have permitted me to charge myself with the satisfaction of your lawyers. But be so good as to inform them that I leave instructions with Colo. Nicholas Lewis to answer their demands, authorized by you, to this amount, which I am in hopes will suffice. – Wishing you a successful issue from your embarrasments (sic), and speedy and substantial justice I am with great esteem Sir your affectionate friend & servt.,
> TH: JEFFERSON[12]

In 1796-1797, Thomas assumed several debts for Randolph in exchange for Randolph's enslaved boys, Ben and Cary. The largest obligation was Randolph's account with Donald, Scott & Co. of Glasgow for £80.13.10½. Additionally, Randolph owed a man named Anderson £9.6.0 for an "order." He also was indebted to David Anderson, assignee of Christopher Hudson, for £11.10.[13]

The debts to Anderson and Hudson were local matters, and although the exact nature of these obligations is unknown, the interweaving of money, friends, and family is a familiar story. Both Christopher Hudson and his assignee, David Anderson, were well known to the Jefferson brothers. In 1769-1770, Thomas Jefferson had represented Charles Hudson, Christopher's father, in a Chancery case in which Christopher was one of the minor children.[14] Years later, during the Revolutionary War, Capt. Christopher Hudson rode to Monticello to warn Thomas Jefferson that the British were coming only to find that Jack Jouett had beat him there. According to Hudson, he discovered Jefferson still at Monticello, "perfectly tranquil, and undisturbed." Hudson insisted Jefferson leave immediately, which he did. Within ten minutes, the British had surrounded the house.¹⁵ The Hudsons were also connected to the town of Milton, intermarrying with the Lewis and Henderson families. In 1813, Randolph's son, Isham Randolph Jefferson, would marry Mary Anne Henderson, daughter of John Henderson and Ann Barber Hudson, Christopher Hudson's granddaughter.¹⁶

David Anderson, assignee, may have been representing Brown, Rives & Co., a vast company which had a store in Milton.¹⁷ Until the war of 1812, Milton was the county's chief commercial center.¹⁸ If so, he was the son of Richard Anderson, Christopher Hudson's brother-in-law.¹⁹ Richard Anderson's brother, Nathaniel, had married Sarah Carr, sister of Dabney Carr, Thomas and Randolph's brother-in-law, making all these Andersons in-laws of Thomas Jefferson's oldest friend, thus revealing Albemarle for the small, small world that it was.

~

Like most Virginia gentry, Randolph Jefferson took advantage of the county court system for settling disputes and debts, occasionally finding himself at the Buckingham Courthouse on serious business, where he appeared as plaintiff, defendant, and witness.²⁰ He took at least one complaint to the Buckingham Court where he argued for damages valued at £32 owed him by his cousin, John

Prince Edward County Courthouse. (HISTORY OF PRINCE EDWARD COUNTY, VIRGINIA, BY HERBERT CLARENCE BRADSHAW, 1848)

Jefferson. Randolph won this case in 1806 with the aid of Buckingham attorney, Archibald Austin.[21]

Due to the loss of Buckingham's records, the frequency with which Randolph Jefferson found himself in court over matters of money forever will remain unknown, although two cases, which were appealed to the Prince Edward District court in 1802 and 1804, impressively reveal the potentially serious and threatening consequences of even short-term debt.

Given that both the era and the Jefferson family expressed a penchant for "the good life," it is remarkable that Randolph Jefferson chose to live within his means, reflecting the general values of his neighborhood in northern Buckingham rather than the aristocratic ambitions of many of the Virginia gentry.[22] Whether a planter led a high or modest lifestyle, debt was a constant problem for Virginians; many were ruined by it, and it is significant that Randolph Jefferson was not crippled by a system which offered seemingly endless credit. This does not mean he never carried debt or bought on credit. Randolph's was a virtually cashless economy, in which credit played a central role.

Local merchants typically sold planters goods on credit and did not expect to get paid until crops went to market. If there were several lean years in a row, debt could mount. In addition to running "tabs," planters sometimes borrowed money from merchants because they were among the few "cash rich" citizens in Virginia. Eventually, creditors would press for funds. In 1802, this is precisely what happened to Randolph Jefferson when local merchant, William Moon, Jr., took him to court.[23]

Sometime before March 17, 1802, William Moon, Jr. complained to the Buckingham Court concerning Randolph's obligation to pay him £30. Indications are that the bond had come due two years earlier and that Randolph was considerably delinquent, which was not a rare or extraordinary thing in the Virginia credit economy. While most of the details of the Buckingham Court case which ensued are lost, initially the case was judged in the plaintiff's favor.[24]

On March 17, 1802, Bolling Branch, Deputy Clerk of Buckingham, addressed a memorandum to the Sheriff of Buckingham summarizing the state of Moon's case. Rolfe Eldridge, Clerk of Buckingham, was also alerted. Jefferson and Benjamin Morris owed £60 for the judgment in Moon's favor, plus court costs of $6.19.[25] The £60 represented a £30 debt, with lawful interest beginning on February 10, 1800.[26] Sheriff David Coupland was expected to collect the monies before the second Monday in May, when the Buckingham Justices would render the funds to said William Moon, Jr.

Deputy Sheriff Ambrose Ford served Randolph with this information. He was unable to produce the £60 and Ford impounded one of Randolph's slaves, a woman named Jane, whose sale was set for April 28th. The resulting funds would be used to pay Moon when Court reconvened on Monday, May 10th. In order to keep Jane at Snowden until the date of the sale, Randolph and a man named Nathan Wash presented a bond, guaranteeing the sum of £67.7.6 and promising to deliver Jane at the courthouse.[27] The bond, dated April 1, 1802, reads:

Randolph Jefferson's slave, Jane, may have been a field laborer or a household slave. ("A Typical Mammy," 1897, SOCIAL LIFE IN OLD VIRGINIA BEFORE THE WAR, BY THOMAS NELSON PAGE, ILLUSTRATED BY GENEVIEVE COWLES AND MAUDE COWELS)

Know all men by these presents, that we Randolph Jefferson and Nathan Wash are held and firmly bound unto William Moon junr in the just and full sum of sixty seven pounds seven shillings and six pence to be paid unto the said William Moon junr his certain attorney, his heirs, Executors, administrators or assignees; to which payment well and truely to be made, we bind ourselves jointly and severally and each of our joint and several heirs, Executors and administrators firmly by these presents Sealed with our seals, and dated this first day of April one thousand eight hundred and two. The condition of the above obligation is such that whereas William Moon jr. hath sued out of the county court of Buckingham a writ of fieri facias against the goods and chattles of the above bound Randolph Jefferson and Benjamin Morris upon a Judgment obtained in the said court; which Writ with the legal costs attending the same amounts to the sum of thirty three pounds thirteen shillings and nine pence and directed to the Sheriff of Buckingham, and whereas Ambrose Ford deputy for David Coupland Sheriff of the said county of Buckingham by virtue of the said Writ hath taken the following property belonging to the said Randolph Jefferson to satisfy the same; to wit: a negro woman by the name of Jane. And the said Randolph Jefferson being desirous of keeping the same in his possession until the day of sale thereof hath tendered the above bond Nathan Wash as security for the forthcoming and delivery thereof on the day and at the place of sale. Now if the above bound Randolph Jefferson and Nathan Wash or either of them do and shall deliver the aforesaid property to the said Sheriff or either of his deputies at Buckingham courthouse on the twenty eighth day of April instant then and there to be sold to satisfy the said William Moon jnr. Execution, then the above obligation to be void, else tho remain in full force and virtue.

Randolph Jefferson (Seal)
Nathan Wash (seal)

Signed in the presence of,
Amb. Ford.[28]

Sketch artist Lewis Miller witnessed a slave auction in Christiansburg, Montgomery County, Virginia. In it, Miss Fillis, her infant child, and Bill were offered for sale. (Lewis Miller, "Sketchbook of Landscapes in the State of Virginia, 1853-1867," Abby Aldrich Rockefeller Folk Art Museum, The Colonial Williamsburg Foundation, Williamsburg, Va., Gift of Dr. and Mrs. Richard M. Kain in memory of George Hay Kain)

There is no indication why Jane was selected to satisfy the debt. Because Randolph wanted her to remain at Snowden until the date of her sale, Nathan Wash provided security that Jane would appear on April 28, 1802, when she would be auctioned on the steps of Buckingham Courthouse. No more is known of Jane's fate. The sale may or may not have taken place, for when Buckingham Court convened in May, the Justices did not award Moon the full amount he demanded. On May 13, 1802, the Buckingham Court judged:

> ... that the plt have execution against the said Jefferson for the sum of sixty seven pounds and seven shillings and six pence the penalty of the said bond.... But this Judgment is to be discharged by the payment of thirty three pounds thirteen shillings and nine pence with interest thereon to be computed after the rate of six p. centum p. annum from the first day of April one thousand eight hundred and two till paid, and the costs. From which Judgement the deff Jefferson prayed an Appeal to the next District Court to be held at Prince Edward courthouse in

September next which is allowed he having given bond and security according to Law.
Appellants costs – $2.69
Appellees do. – $4.20

Rolfe Eldridge, CBC[29]

On September 3, 1802, the District Court heard "Randolph Jefferson, Appellant, against William Moon, Jr., Appellee." Counsel represented Moon, while Randolph was "solemnly called" but "came not." The Justices of Prince Edward found that there was "no error" in the previous judgment, concluding that Randolph still owed Moon £33.13.9, with interest discharged at six percent per year, beginning April 1, 1802, plus costs. The Prince Edward County District Court Execution Book also notes that Randolph owed £0.19.6, plus 8 cents in damages. The court costs for Buckingham were $4.20 (April 1, 1802) and $5.54 (May 13, 1802).[30]

It is unclear why Randolph appealed the reduced judgment, then neither sent counsel to Prince Edward nor attended court. Was the appeal designed to "buy time," so that he might be able to meet the reduced debt without selling Jane?

The threat of having a slave impounded and possibly sold to satisfy a debt was an ever-present reality in Virginia. Whether Jane was ultimately sold or not, this event may have affected Randolph Jefferson deeply and no doubt distressed everyone at Snowden. There is no way of knowing how long she had been living with the Jeffersons, or if she had extended family there. There was a woman named Jane among Randolph's slaves in 1782, but she does not appear to be living at Snowden in 1783.[31]

Precisely how Randolph ultimately raised funds to settle his obligation to Moon is unknown, though he may have decided to sell some of his land rather than part with Jane. During 1802, he sold 123 acres, his taxable land dropping from 2,000 to 1,877 acres.[32] Years later, in the spring of 1815, Randolph informed his brother, Thomas, that he had sold seventy acres of his low grounds at $100/acre to Charles A. Scott. This decision, wrote Randolph, will "take me out of debt with every man that I am involved with and which will enable me to keep all my slaves as long as I live."[33] Clearly, Randolph did not want to sell

Prince Edward County District Court Execution Book

1802 Sept 8   Wm Moon jr v Randolph Jefferson
               (No security of any kind to be taken)   £33.13.9
               damages   £0.19.6 + 8 cts

opposite page:
Apr 1st 1802 to   4.20
May 13th 1802 &   5.54   Bbm (Buckingham)
Sept 3rd 1802

In the space where some of the executions read "Satisfied," the Jefferson v. Moon case is blank. Since the Prince Edward District Court upheld the Buckingham Court's decision, perhaps it was the business of the Buckingham Court to satisfy Moon, not the District Court.

[Prince Edward County, District Court Executions Book 1797-1809, "M," 1802, Vi.]

Prince Edward County District Court Executions Book

10 April 1805   Benj. Bondurant, assignee,
                 v. Randolph Jefferson   £31.1.3

opposite:
1804 June 4   6.69 Bgm (Buckingham)

11 Sept   Benj. Bondurant v. Randolph Jefferson   £37.2.8
               no security of any kind to be taken

opposite page:
1805 Aug 9th   4.42 Bgm (Buckingham)   Satisfied with the money
                                          paid the Plff.
                                          Samuel Jones, D. Sheriff

In light of the court papers, Randolph Jefferson did not pay the £31.1.3 indicated by the April 10, 1805 Judgment. Rather, he was responsible for the ultimate sum of £37.2.8, the Judgment delivered on September 11, 1805. This satisfied the Court and was paid to Benjamin Bondurant by Samuel Jones, Deputy Sheriff Buckingham County.

[Prince Edward County, District Court Executions Book 1797-1809, "B," 1805, Vi.]

his slaves. While the specific reasons for this went unstated, keeping his slaves meant more to him than preserving his land.

~

Within two years, Randolph Jefferson found himself in the Prince Edward District Court for a second time. In June of 1804, he gave his friend and neighbor, Hardin Perkins, a note for £30.1.2, which was no small matter. In August of 1805, Thomas Jefferson would hire a new overseer, J. Holmes Freeman, whom he paid £60 per year.[34]

Nothing about the note itself or the other items remaining in the court papers indicate the nature of Randolph's indebtedness. The date of the note coincides with a hearing in the Buckingham Court, the details to which are now lost.[35] The obligation might have been for a direct loan, though in this virtually cashless agrarian society, planter Perkins was not an obvious source of ready funds. It could have been a settlement for damages; hogs, for example, could do sizeable damage to a neighbor's property, and Jefferson shared a common line with Perkins. Importantly, it was payable on demand and the terms were harsh.

> On Demand I Promise to pay or cause to be paid unto Hardin Perkins of Buckingham County the just sum of Thirty one pounds one Shilling & Two pence for value Rec^d to which payment well and truly to be made I bind myself my Heirs & in the Penal sum of Sixty two pounds Two Shillings & 4 as witness given under my hand this Fourth day of June one Thousand eight Hundred & four
>
> Randolph Jefferson
>
> Test E. Perkins[36]

Inexplicably, two months later, Hardin assigned the bond to their mutual neighbor, Benjamin Bondurant, who owned land just above them on Rock Island Creek and Sharps Creek. On the reverse side of Randolph's note, Hardin wrote:

> "I do hereby Assign the written bond to Benjamin Bondurant this 12th Augst 1804.
> Hardin Perkins
> (witnesses) George Perkins, E Booker[37]

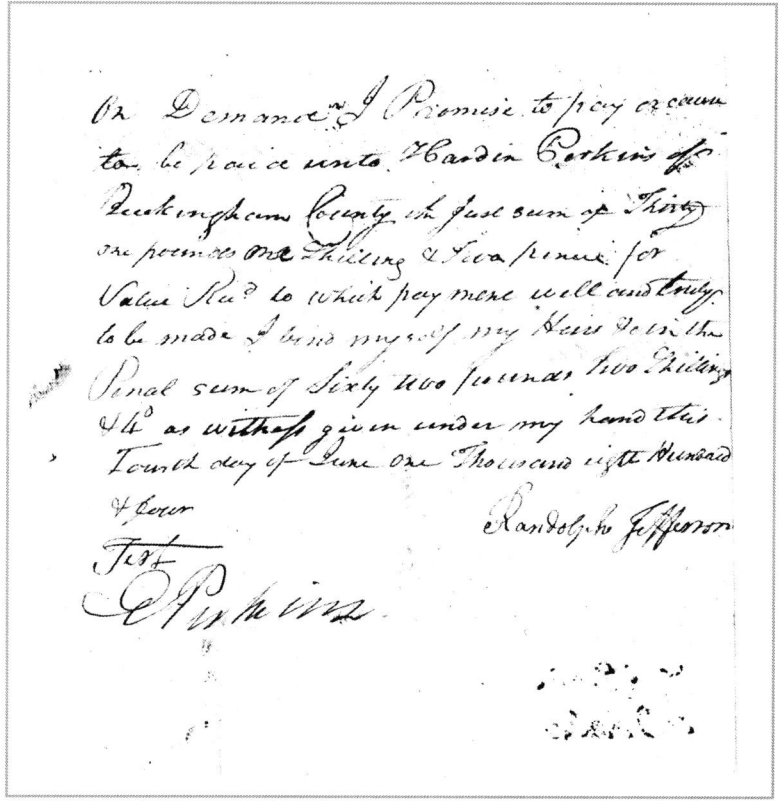

(Prince Edward District Court Records, Courtesy of Library of Virginia)

Did Perkins owe Bondurant funds? Did Bondurant accept the £30.1.2 note from Perkins with plans to further press Jefferson for the penalty, using the county courts? That is exactly what he chose to do. Losing no time, Bondurant immediately executed plans to collect the full sum *plus damages*. On August 14th, a summons was issued stating that the Buckingham Sheriff should take Randolph into custody and keep him safely until the first day of September, when he was to appear in front of the Judges at Prince Edward County District Court. It also announced that Bondurant expected to collect £62.2.4, plus damages of £20.00.

An undated bill of complaint, written by E. Booker, probably Bondurant's counsel, begins: "Buckingham County to wit: Benjamin Bondurant assignee of Hardin Perkins complains of Randolph Jefferson in custody &c. of a plea that

KNOW all men by these presents, that we Randolph Jefferson and Davd. Oglesby are held and firmly bound unto Saml. Allen ——— Gentleman, sheriff of Buckingham ——— county, in the just and full sum of one Hundred and fifty pounds to be paid unto the said Saml. Allen ——— his certain attorney, his heirs, executors, administrators, or assigns; to which payment well and truly to be made, we do bind ourselves, our heirs, executors and administrations, jointly and severally, firmly by these presents. Sealed with our seals, and dated this 16th day of August one thousand eight hundred and four.

THE Condition of the above obligation is such that whereas Benjamin Bondurant asee of Hardin Perkins hath sued out of the District Court of Prince Edward Writ of capias ad respondendum against the body of the above bound Randolph Jefferson for a plea of debt for Sixty two pounds, two Shillings and four pence damage twenty pounds which writ has been duly executed. Now if the above bound Randolph Jefferson — — — — — do and shall well and truly make his personal appearance before the Judges of the next district Court, to be holden for the said district of Prince Edward — then and there to answer the suit of the said Benja Bondurant ——— ——— and do not depart from thence, without leave of the said Court: then the above obligation to be void, or else to remain in full force and virtue.

Signed, Sealed and delivered in the presence of

Saml. Jones

Randolph Jefferson (Seal)
David Ogles by (Seal)

(Prince Edward District Court Records, Courtesy of Library of Virginia) he render to the plaintiff the sum of sixty two pounds, two shillings and four pence...." In the complaint, Bondurant goes on to state that Randolph "unjustly detains" from paying his debt, and although Bondurant often requested Jefferson to pay, Randolph refused and still refuses.[38] When had this protracted refusing taken place? Bondurant had held the note for only two days, and the situation had quickly escalated into a very stiff demand for a £30 note.[39] It is reasonable to assume that Randolph always intended to pay the original debt – whether to Perkins or to Bondurant – but that he resisted being pushed for the original sum, *plus* a crippling penalty, more than doubling his original obligation.

The men involved in this dispute were all well known to each other, close neighbors in a sparsely populated rural setting. They were, however, not all peers. Randolph Jefferson was forty-eight years old, a widower, and the owner of a very valuable farm on the James River. Hardin Perkins, though younger than Randolph, also owned a valuable farm at the Horseshoe Bend, which was graced with a handsome two-story house. In 1796, Hardin had filled Randolph's place as a Captain in the Buckingham Militia. In 1804, he owned ten taxable slaves, six horses, and expected more from his father's yet unsettled estate. His brother, Price Perkins, was a Buckingham Justice.[40]

Benjamin Bondurant was about thirty years old when he initiated his suit against Randolph Jefferson. Married to Sarah East Moseley, they had four small children and would soon be expecting a fifth.[41] At least in terms of land holdings, the Bondurant family was not nearly as affluent as either Perkins or Jefferson. The combined Bondurant farms on Rocky Island Creek and Sharps Creek, which totaled a few hundred acres, were inferior acreage compared to the fertile spreads on the Horseshoe Bend held by Perkins and Jefferson. At the time Bondurant took Jefferson to court, he was paying personal property on himself and one slave. He did not own a horse.[42]

Bondurant's pressure to collect from Jefferson may have been, in part, an opportunity to demand funds from a man he knew was good for the amount. Additionally, behind Randolph stood the U.S. President, which also might have influenced Bondurant's actions.

The August 14th summons demanded bail, its reverse side reading: "Debt due by writing obligatory, & bail required."[43] This Randolph could neither ignore nor delay, and bail did not take long to produce. On August 16th, just two days following the issuance of the summons, David Oglesby provided Randolph Jefferson's bail in the form of his personal security that Randolph would appear in Prince Edward at the next District Court. They co-signed the bond for £150.00. Buckingham Deputy Sheriff, Samuel Jones, witnessed the document.[44]

David Oglesby (1744-1821) lived in Buckingham County and had long known Randolph Jefferson. Eleven years Randolph's senior, Oglesby was already established at the Horseshoe Bend when Randolph settled in at Snowden. In 1773 and 1774, Oglesby resided with Hardin Perkins, Sr. In 1800, Oglesby paid personal property tax on himself, one slave, and one horse. That same year, Oglesby Scruggs & Co. was taxed for a merchant's license in Buckingham's Northern District. If David Oglesby was a partner in this store, he was a potential "cash rich" individual, good for a bond of £150.00.[45]

If Bondurant's case against Jefferson was heard in September or October of 1804, no judgment resulted and here the record falls silent for seven months.[46] Despite all the rush on the part of Benjamin Bondurant to collect his funds, the wheels of justice turned slowly in central Virginia, and the case was not settled for another year.

In April of 1805, Randolph Jefferson countersued Benjamin Bondurant and the Prince Edward District Court heard "Jefferson (v) Bondurant, assignee." A lone item surviving from this step in the lengthy dispute is a "bail piece," written by Buckingham Justice Mathew Branch, which guarantees that Randolph will appear in court.[47]

Dated March 12, 1805, Branch's memorandum states that John Thomas of Buckingham County, representing Randolph Jefferson, had appeared before him claiming that Jefferson "will pay and satisfy the commendation of the court, or render his body to prison, in execution for the same."[48] In his memo, Justice Branch first wrote "Garland Jefferson," then struck through Garland and inserted "Randolph." Branch's slip is a curious one.

John Garland Jefferson (d. 1815), called "Garland," was the son of Elizabeth Garland and George Jefferson, first cousin to Randolph and Thomas Jefferson. Garland lived south of the James River, to the east of Buckingham, in Amelia County, though he had spent time at Monticello where Thomas advised him in the law and Garland benefitted from the extensive library there.[49] Justice

Mathew Branch (who also may have been a distant cousin to all these Jeffersons) apparently knew Garland as well or better than he did Randolph. Once again, the interconnectedness of the Virginia gentry is revealed, a reminder that the Justices who handed down decisions in the county courts frequently faced members of their own extended families.

Randolph Jefferson's representation, John Thomas, was yet another of Randolph's neighbors. He is likely Capt. John Thomas, who served in the Buckingham Militia and had headed a company during the Revolutionary War. His land was situated near Bondurant's farm on Rock Island Creek. Thomas may have represented Jefferson only to Branch or he may have been acting as Randolph's general counsel.[50]

In April, the Prince Edward District Court judged that Randolph Jefferson owed Benjamin Bondurant only £31.1.3, a generously reduced judgment.[51] Randolph appears to have won his appeal and now owed only the original debt, plus some lawful interest and court costs. Yet more months elapsed, and Randolph did not settle the obligation with Bondurant. This time he was not avoiding an unjust demand; rather, it is likely he still lacked the ready cash to meet the debt and did not want to face raising funds by selling personal property (which could mean parting with a slave) or selling valuable land, both of which provided the Jeffersons with income.

This delay unsurprisingly infuriated Benjamin Bondurant and, on August 9, 1805, the Buckingham Court once again heard the matter, listening to Bondurant's ongoing complaint. He demanded the entirety of the debt, which he now claimed totaled £74.5.4.[52]

The Buckingham Court agreed that Jefferson was delinquent in his payment and, as a result of the Court's decision, Sheriffs, Samuel Jones and Samuel Allen, were ordered to impound two of Randolph Jefferson's horses, one bay and one sorrel. August 23, 1805 was set as the date for the Sheriff's sale of the horses, to take place at the dwelling house of Bernard Hatcher in Buckingham County.[53] At this point, Randolph was forced into precisely what he had not wanted -- the impounding of his property.

As in the case of the slave Jane, Randolph desired to keep the horses in his possession until the day of the Sheriff's sale, so he provided security that he would appear, with the animals, at Hatcher's farm on August 23rd. This time the security was Randolph's eldest son, Thomas Jefferson, Jr., who was about twenty-three years old, a bachelor, and still living at Snowden.[54]

The sale may or may not have taken place, though it probably did, for Benjamin Bondurant received £37.2.8. Still not satisfied, on August 24, 1805, the day following the scheduled sale, Bondurant informed Randolph and Thomas Jefferson, Jr., that he yet again was bringing suit in the District Court, claiming that Jefferson owed him £74.5.4 in toto, *"to be discharged by the payment of thirty seven pounds two shillings and eight pence."* Darby Bondurant delivered a copy of Benjamin's notice to Randolph, certified by Daniel Bagby, Buckingham Justice of the Peace.[55]

> Gentlemen
>
> You will be pleased to take notice that on Wednesday the third day of the District Court to be held at Prince Edward Court House I shall move to the Honorable Court by my Attorney for a Judgement against you on your forthcoming Bond, bearing Date the ninth Day of August 1805 for the payment of Seventy four pounds five shilling & four pence, to be discharged by the payment of thirty seven pounds two Shilling and eight pence.
>
> Your most Obt.   Benjamin Bondurant
>   Assignee of Hardin Perkins
>   24th August 1805[56]

In September of 1805, the Prince Edward County District Court, for the final time, heard the case of Benjamin Bondurant, assignee of Hardin Perkins, Plaintiff against Randolph Jefferson and Thomas Jefferson, Jr., Defendants. Randolph was "solemnly called" to court, but "came not" to defend himself. The Justices concluded: "this Judgment is to be discharged by the payment of Thirty seven pounds two Shillings and Eight pence with Interest thereon at six per Centum per Annum from the 9th day of August 1805 till payment and the costs."[57] Court costs were noted at $4.20. The Prince Edward Executions Book duly noted, "Satisfied with the money paid to the Plff. Samu^l Jones, D.S."[58] It seems the higher court had agreed with Randolph Jefferson. Enough was enough. £37.2.8 was an appropriate amount to settle the original note to Hardin Perkins. All that remained was interest beginning August 9th and court costs.

~

Clearly, Randolph Jefferson did not face these financial crises alone. Buckingham was a small place, in terms of population, and Randolph had good and loyal friends. Gentleman Justice Benjamin Morris, Nathan Wash, and David Oglesby stepped up as securities for him. Even his eldest son, Thomas Jefferson, Jr., backed his father. At the time, Thomas owned a horse of his own and, of course, was an heir to Snowden and his father's substantial property. Militia comrade, Capt. John Thomas, was, at the very least, supportive and may have acted as Randolph's counsel.

Randolph's brother, Thomas, who might have come to his aid, was occupied elsewhere serving as President of the United States. Concurrent with this very local argument with Benjamin Bondurant, Thomas was orchestrating the ambitious Lewis and Clark expedition, among many other duties.

During July and August of 1805, however, Thomas Jefferson just happened to take a break from the vicissitudes of Washington, D.C., came home to Monticello, and made a trip to his Bedford County property. There, he laid out a tract of land for his Eppes grandchildren and may have begun plans for the building of a dwelling house, a retreat from Monticello. On August 1, 1805, just prior to Randolph's last round in the Buckingham Court, Thomas stayed at The Raleigh, an ordinary located just west of Buckingham Courthouse and operated by Daniel Guerrant. There is no mention in Thomas' memoranda that he encountered Randolph in Buckingham or that he was there on business concerning his brother. The coincidental timing, however, of Thomas' appearance in Buckingham and Randolph's escalating dispute with Bondurant is notable and Thomas might have advised his brother concerning his ongoing legal matters.[59]

During these last days of Randolph's difficulties with Benjamin Bondurant, Thomas lingered at Monticello. About the time the Prince Edward District Court entered its final judgment for Bondurant v. Jefferson on September 11th, Randolph may have gone to Monticello to spend time with his brother before his return to Washington. A widower, Randolph could travel unencumbered by a wife. A protracted stay at Monticello might have been just the thing to put distance between himself and Bondurant's suit.

Significantly, on September 28th, Thomas noted that he lent Randolph $10.00. Thomas' memorandum does not indicate where the loan took place, though it was likely Monticello, for that day Thomas was at home closing up business. On September 29th, he set out for Washington, D.C., spending his first night on the road at Gordon's tavern in Gordonsville, which sat to the north on the line between Orange and Louisa counties. With $10.00 tucked in his pocket, Randolph presumably rode home to Snowden, perhaps planning a trip to the courthouse to pay final fees, relieved that the long litigation with Bondurant was finally over.[60]

In the end, Benjamin Bondurant received far less than he hoped for and may or may not have been paid with funds raised from the sale of Randolph's horses. Interestingly, Randolph's stable actually increased during this period. In 1804, he paid tax on seven horses, while in 1805 and 1806, he paid tax on nine.[61] Additionally, there were apparently no hard feelings between Randolph and Hardin Perkins, for in 1808, Perkins agreed to be one of the Executors for Jefferson's will.

As to any further debts, short-term or otherwise, there are few other details concerning Randolph Jefferson's financial life. It is notable, though, that he died with the majority of his patrimony intact. Overall, he was a good steward to his inheritance, preserving it for his children. A stark contrast to his ever-expansive and debt-ridden brother, Randolph both raised a large family *and* maintained his fortune.

Chapter Eight

Snowden: A Plantation in Buckingham County

In the lotteries of human life . . . even farming is but gambling.
Thomas Jefferson, 1813

Despite Randolph Jefferson's ascent to Snowden in 1776, the estate of Peter Jefferson did not close finally until January of 1790, when Thomas and Randolph met at Seven Islands, the home of John Nicholas, Sr. Thomas' memos are quiet between January 17th and 22nd. Did he spend some time at Seven Islands? Did he visit Snowden? On January 21st, Thomas spent the night in Fluvanna County at Bremo, the home of John Hartwell Cocke. The next morning, he tipped Cocke's servants and rode home to Monticello.[1]

With the outstanding details to Peter Jefferson's estate finally resolved, Randolph Jefferson's life as a planter was, at last, completely in place. He would live the rest of his days at Snowden, and while Randolph's holdings were somewhat reduced by the end of his life in 1815, he never acquired crippling debt, unlike his brother, nor did he know the complete financial collapse of his Lewis cousins and many Virginians. Apparently conservative in his own spending and satisfied with the expanse of his property, Randolph maintained enterprises at Snowden that did not overwhelm him, his overseers, or his laborers.

The first known change in Randolph's land holdings came in 1796 when he sold his 1/6 percentage in 400 acres (roughly 66 2/3 acres) located on the Hardware River to his brother, Thomas. A separate tract from Snowden, this part of Randolph's inheritance lay across the James in Albemarle County and included part of a limestone quarry. This decision to sell the Hardware tract enabled Randolph to focus his attention and efforts exclusively in Buckingham County. It also proved fortuitous, because the property, which Thomas may never have quarried, became a long-term nuisance. In 1753, the entire 400 acres was purchased by Peter Jefferson and five others for £17.[2] Thomas Jefferson paid Randolph five shillings and "certain articles of household furniture."[3] No inventory of the furniture exists, so its value is unknown. These may have been family pieces from Shadwell.

That year, Thomas ordered a survey of land owned by Randolph. It is not entirely clear which land is being surveyed, though it was likely the Hardware River property, as the work was done by Albemarle's surveyor, John Slaughter. On June 6th, Thomas paid Slaughter "for R. Jefferson for copy works 1/6." There was another payment to Slaughter of $1.00 on June 20th and another in November. Such a survey might be a routine part of transferring a deed; however, on June 4, 1796, Thomas noted that he "sent Bushrod Washington order on Chas. Johnston for £5. fee in defending Rand. Jefferson's land." Johnston, a Richmond merchant, served as Thomas' commission agent. Just why Randolph's land needed "defending" is unstated and a bit mysterious.[4]

Three years later, Randolph Jefferson's ownership of Snowden was being challenged. It may only have concerned the farm's boundaries or, more seriously, Randolph's right to the land actually may have been contested. On June 18, 1799, Buckingham County surveyor, John Patteson, produced a new plat of part of Snowden showing 1,327 contiguous acres belonging to "Capt. Randolph Jefferson."[5] Randolph was not the only planter at the Horseshoe Bend concerned about his boundaries. Over a five year period, a series of surveys recorded for Snowden's neighbors indicates there was considerable activity all around the farm's borders. A warrant for 1,402 acres of land held by a man named Isham Richardson was the catalyst. Buckingham surveyor, John Patteson, was occupied from April of 1797 through December of 1802 establishing lines for Snowden neighbors, Price Perkins, Robert Craig, and William C. Murrey.[6]

Isham Richardson (b. abt. 1725 - 1789), the son of John Richardson, lived in neighboring Cumberland County.[7] The use of the name Isham suggests a possible connection to the Randolph family. Richardson's will was recorded in

Cumberland County on September 10, 1789 and, while it mentions land he owned in southern Buckingham on Willis Creek, it does not refer to land in the vicinity of the Horseshoe Bend. It also names a son, Isham Richardson, Jr., who may be the man holding the warrant.[8] Additionally, in 1789, concurrent with Richardson's death, a man named Martin Richardson initiated a Cumberland Chancery Case against John Jefferson, who may be the same man who lived at Snowden the following year. Where Randolph Jefferson, Snowden, and other lands at the Horseshoe Bend fit into this larger story is yet another mystery.

Patteson's survey was followed by a surprising new land grant executed on May 30, 1800, and signed by James Monroe, then Governor of the Commonwealth of Virginia. The grant matches the survey, describing the 1,327 acres. A note in the grant's margin indicates that it was delivered to Thomas Jefferson in June of 1802. Why it was not delivered directly to Randolph Jefferson is unclear. Whatever the basis of the dispute, Randolph's land was now irrefutably protected by the new grant, and Thomas' efforts on his brother's behalf proved successful.

The Overseers of Snowden

On plantations the size of Snowden, overseers were a common fixture in Buckingham County and central Virginia. The landowner hired a man to manage his slaves and the day-to-day aspects of farm production. During the 1780s and 1790s, Randolph Jefferson employed a series of such men. A good, reliable overseer could contribute a great deal to the success of a farm. His presence also freed the country squire to occupy himself with commerce, politics, the County Militia, or even simply leisure.

Living in a modest dwelling near the "big house" or, depending on the design of the plantation, closer to the fields, an overseer and his family existed apart from Virginia's affluent class; however, important bonds could and did form between overseers and plantation owners. Widows often relied heavily on their managers and when an executor, administrator, or curator was put in charge of a plantation, a trustworthy overseer was essential.

This was a salaried job, either for cash or a percentage of the crops. Contracts were typically year-to-year, as evidenced by the gradual turnover at Snowden.

Land, Negroes and Stock
AT AUCTION.

WILL be offered for Sale, at the late dwelling of John Puller, decd. on the premises, (4 miles above Jefferson) on Thursday the 19th of November next, if fair, if not, the next fair day, the remaining part of his property, consisting of

15 to 20 Likely Negroes,

Stock of every kind, crop of Corn, Rye, Oats, Hay and Fodder, Household and Kitchen Furniture, plantation utensils, &c. &c. Also

1000 or 1200 Acres of Land,

Which will be laid off in convenient lots to suit purchasers. The terms of sale, for the Land will be one third the purchase money in hand, the balance in two equal annual payments, with interest from the date, but if punctually paid the interest to be remitted, with a lien on the same to secure the payments, and such title as is vested in the Executor will be made to them. For the other property bond with approved security to bear interest from the date, will be required for all sums over Ten Dollars; but if the money is punctually paid at the end of twelve months the interest will be remitted, Ten Dollars and under cash. Should any purchaser fail or refuse to comply with these terms, the property will be resold and the first purchaser held accountable for the loss, if any.

Those having claims against the estate of the deceased, are required to exhibit them properly adjusted, on or before the day of Sale, otherwise they may be by law excluded from all benefit of said estate. Those indebted are required to make immediate payment.

THOMAS SPINDLE &
JOHN P. SMITH, } Exrs.

Culpeper County, October 7, 1818.

The estates of many Virginia planters ended up on the auction block and the death of a slaveholder produced great anxiety among his enslaved people. Individuals were sold like livestock; families were separated. Upon Thomas Jefferson's death in 1826, this would be the situation at Monticello. (COURTESY OF LIBRARY OF VIRGINIA)

Relations between owner and overseer might end for a variety of reasons. Single men frequently married and moved on. A man might save enough money to buy a farm of his own. Not infrequently, employers and/or employees became dissatisfied with each other.

Some overseers ran farms for absentee landlords, which was the case for many years at Monticello. There, Thomas Jefferson hired overseers he did not oversee personally; his absence from the farm inevitably created room for abuses. When Hannah died at Snowden at the hands of overseer, Isaac Bates, there was not a Jefferson proprietor in residence. By contrast, once Randolph Jefferson came of age, he maintained his residence at Snowden, living in close contact with the roughly thirty individuals – men, women, and children – who made up his enslaved labor force. Compared to the sprawling Monticello and quarter farms, Snowden was a more intimate setting of 2,000 acres, an environment where a friend's nephew could work for a year or two before settling down on his own acreage or a neighbor's son could learn the basics of plantation management.

Just as the gentry rotated their sons to other plantations for an academic education, so too did they help prepare young men to farm and manage slaves. This relaxed, even friendly, environment was one in which Randolph could maintain close relationships with his overseers, acting as he would toward a nephew or surrogate son. In at least one case, Randolph signed the bond for his overseer's marriage.[9]

The first known overseer at Snowden after Randolph Jefferson came of age was Stephen Perkins, the son of Mary Walker and Constantine "Constant" Perkins, the older brother of Snowden neighbor, Hardin Perkins, Sr. Born in Goochland County on April 22, 1760, Stephen was less than five years younger than his employer. While he does not appear by name with Randolph Jefferson on the initial Buckingham County Personal Property Tax lists, he was likely already working at Snowden when he served as one of the two men who assessed Hardin Perkins' cattle for John Bates in November of 1780. As the war progressed, Stephen was in charge of the fields when the 3rd Light Dragoons camped at Snowden. On July 16, 1781, it was Stephen who provided the troops with seven bushels of oats and pasturage for eighteen horses. He may have worked continually for Jefferson until 1787, when he was replaced by Thomas Ware.[10]

Sketch-artist Lewis Miller's observations of everyday life in Virginia included this simple scene of sawing lumber to make shingles to cover houses. (LEWIS MILLER, "SKETCHBOOK OF LANDSCAPES IN THE STATE OF VIRGINIA, 1853-1867," ABBY ALDRICH ROCKEFELLER FOLK ART MUSEUM, THE COLONIAL WILLIAMSBURG FOUNDATION, WILLIAMSBURG, VA., GIFT OF DR. AND MRS. RICHARD M. KAIN IN MEMORY OF GEORGE HAY KAIN)

Randolph Jefferson's relationship with Thomas Ware transcended that of employee and employer. On October 5, 1787, he signed a marriage bond for Thomas "Weir" & Mildred "Milley" Bryant. Though the record was lost in the Buckingham Courthouse fire, a copy was later presented as evidence for a Revolutionary War pension for Ware's widow.[11]

Between 1784 and 1797, Randolph Jefferson paid personal property tax for several overseers. Not all are named:

1784 - 1786: 2 white males "above 21 years," 2nd man not named
1787:   2 white males, Thomas Ware/Worn [Weir], overseer
1788:   2 white males, Thomas "Ware"
1789 - 1793: 2 white males, Thomas Hall[12]
1794:   2 white males, Jos. B.
1796 - 1797: 2 white males, Joseph Bradshaw

Following Joseph Bradshaw, Randolph's sons would naturally take over the work of the overseer, preparing for their lives as planters. In 1798, Thomas Jefferson, Jr. was not yet twenty-one years old, but was certainly old enough to assume some responsibility for the plantation.[13] At that point, Randolph no longer paid taxes for a hired man.

Crops at Snowden

The primary occupation at Snowden was, of course, agriculture. Randolph Jefferson's land was ideal for growing tobacco, corn, and wheat, as well as a variety of fruit trees. Crops at the farm were discussed in the Jefferson-to-Jefferson correspondence and were also documented in Thomas' memorandum books. Randolph's crops were typical for the area, and there is every reason to believe he was a successful farmer.

Randolph mentions "getting" his wheat crop "down" in October of 1811 and, in July of 1813, he was "busy with wheat." As it turned out, the spring of 1813 led to a summer of drought which ruined crops at Monticello. Crops at Snowden no doubt suffered as well. On August 8, 1813, Thomas wrote to Randolph of the devastating effect the lack of rain had on his crops:

"Stacking Wheat." (COURTESY OF LIBRARY OF CONGRESS)

...We are experiencing the most calamitous year known since 1755. The ground has been wet but once since the 14th. of April. My wheat yielded but a third of an ordinary crop, about treble the seed. Of 230. acres of corn, about 15. acres may make 2. or 3. barrels to the acre; and about 215. acres will not produce a single ear; not half of it will tassil, a great deal not 2. feet high. We usually make about 7. or 800. barrels; we shall certainly not make above 30. I shall be obliged to drive all my stock to Bedford to be wintered, and to buy 400. barrels of corn for bread for my people.[14]

In 1815, as the War of 1812 resolved, Thomas warned Randolph about the timing for selling his crops:

Feb. 17. The news of peace is confirmed since yesterday so that I have little doubt of it. Wheat and tobo. will be immediately at a good price. Corn which was at 27/ at Richmond will tumble down instantly, because their supplies which are always from the great corn county of Rappahanoc will come round by water now freely and immediately.[15]

Over the years, the brothers exchanged seed. In 1801, Thomas paid Randolph's slave, Squire, for 25 1/4 quarts of white clovers seed "@ 1/ £1-17-10 ½." In 1807, Thomas bought more white clover seed (for Monticello) and greensward seed (for Poplar Forest) from Randolph, paying him $20.00.[16]

In 1813, Thomas wrote to Randolph at length concerning crop rotation. This should not necessarily be taken as a reflection on Randolph's skill to operate his "very valuable farm." Thomas frequently offered both solicited and unsolicited advice to various members of his family, and Randolph may or may not have asked his brother for his opinion.[17]

In May of 1813, Thomas visited Snowden and at his leisure observed many things. Having talked with Randolph, Randolph's wife, Mitchie, and Randolph's youngest son, Lilburne, Thomas wrote to his brother on May 25th, offering thoughts on their various conversations:

Reflecting on the manner of managing your very valuable farm, I thought I would suggest the following which appears to me the best, and of which you will consider. To form your lowgrounds into two divisions, one of them to be in wheat, and the other to be half corn and half red clover, shifting them every year. Then to form your highlands into three divisions, one to be in wheat and the other two in red clover, shifting them from year to year. In this way your low ground fields would be in corn but once in 4. years, in wheat every other year, and in clover every fourth year: and your highland in wheat once in every three years, and in clover two years in every three. They would improve wonderfully fast in this way, and increase your produce of wheat and corn every year. If it should be found that the low grounds should in this way become too rich for wheat, instead of putting them every fourth year into clover, you might put them that year into oats. Your annual crop would then be half your low grounds in wheat, a fourth in corn, and a fourth in oats or clover: and one third of your highland in wheat, and two thirds in clover: and so on for ever, and for ever improving. I suggest this for your consideration.[18]

Randolph answered his brother immediately, the next day in fact, saying, "I am extreemly oblige to you for your advice as to managing my farm but am a fraid it will be two great an undertaking for me."[19]

Thomas had given up on tobacco years earlier, after returning to Monticello in 1793.[20] Both his land and his labor situation, however, were quite different from Randolph's at Snowden. The wealth of Snowden lay in its prime tobacco land.[21]

Beginning with the years when Martin Dawson leased the farm from Peter Jefferson's estate until far beyond Randolph's death in 1815, the land remained incredibly fertile for tobacco. In 1860, a spectacular year, Snowden still yielded 31,000 lbs. of the cash crop.[22] Clearly, giving up the cultivation of tobacco at Snowden could have been a serious economic error on Randolph's part had he followed Thomas' advice.

The Orchards of Snowden

Central Virginia was, and still is, known for its suitable environment for a wide variety of fruit trees and grape vines. At Monticello, the peach dominated. This was in part because Thomas Jefferson simply loved peaches, and, in part, because of their versatility. As dessert fruit, dried and chipped, or as brandy, the peach served well. A lovely decorative tree, Thomas Jefferson also used the peach tree as fencing.[23]

It should be of little surprise, then, that shortly after Thomas Jefferson returned from France, Randolph asked his brother for some peach stones to plant at Snowden. In response, in late February of 1790 Thomas gave directions to his servant, George, in "relation of the peach stones," and immediately sent to Snowden "very fine" apricot and plum stones. The elder Jefferson instructed that these were "to be planted immediately and to be cracked before they are planted."[24]

Randolph might have found it difficult to grow peach trees at Snowden. His brother's success at Monticello was due to a specific climate on "the little mountain." Sitting at 867 feet, Thomas Jefferson's elevation was significant enough to produce a micro-climate imitating conditions hundreds of miles south of Albemarle. In the spring and fall, the cold air settled in the low lands and the warm air sat on top, protecting Thomas' peach trees from spring frosts that often nipped his neighbors' plants.[25]

Snowden's land contained both high and low lands, with an appreciable difference in altitude. Its highest point is about 400 feet; however, Snowden had a significant feature that Monticello lacked: it is wrapped by the James River, creating another micro-climate of damp air.

Randolph probably grew fruit for his own family's consumption. In February of 1815, Randolph was still asking his scientific-minded brother for "scions" (a shoot for grafting purposes) of fruit trees. That month Randolph had in mind some "of your good fruit [trees], of apple and cherry," thinking it would not be too late to move them and adding that he would be happy with "any other fruit that you would oblige me with."[26] Apparently, Randolph had learned long ago, you can ask your older brother for peach stones, but he may send you apricot and plum instead.

Randolph must have been disappointed when Thomas wrote back, "I have for some years so entirely neglected my fruit trees that I have nothing in my nursery but refuse stuff, unknown and of no value."[27] It is possible that Randolph specifically asked for apple and cherry starts from his brother's nursery because they had previously proved successful for him. Apples are particularly famous in adjacent Albemarle and Nelson counties, especially the Albemarle Pippin.

Lands around Snowden were certainly conducive to excellent orchards. In 1771, Randolph's immediate neighbor, Thomas Ballow, offered his farm for sale with the following description:

> To be Sold by the Subscriber, in Buckingham County,
> ONE THOUSAND ACRES OF LAND Lying on James River, and joining the Lands of Colonel Robert Bolling, at the Seven Islands. There is a Plantation in the low Ground that will work about eight Hands, with a very fine Apple Orchard, and the Land but little hurt by the late Fresh.[28] About the same Quantity of Woodland on the River, not cleared. For Terms apply to the Subscriber, living on the Premises, who will show the Land to any Person desirous of purchasing the same. THOMAS BALLOW[29]

Orchards in the Virginia Piedmont not only produced fresh fruit, which could be preserved as jams or dried, but also provided the basis for hard ciders and spiked punches. Randolph and Nancy Jefferson grew up in a culture that heartily embraced alcohol. Many Virginians began their day with a "pick-me-up" and ended it with a "put-me-down," finding still other opportunities throughout the day to imbibe.[30]

Alcohol was certainly abundant at Peter Jefferson's Shadwell during the mid-18th century. In September of 1769, Thomas Jefferson took an inventory of the cellar at Shadwell, which then contained over 250 bottles of alcohol, much of which was imported. The stock included: eighty-five bottles of rum, fifteen bottles of Madeira wine, fifty-four bottles of cider, four bottles of Lisbon wine, and beer. Beer was typically made on Virginia plantations, the women in the family supervising the operation. In 1808, Randolph's twin, Anna Scott, assisted her niece, Martha, and Thomas Jefferson's butler, Burwell, in bottling over 200 bottles of wine at Monticello. It is entirely possible that she helped with similar work on a smaller scale at Snowden.[31]

The Kitchen Garden

Just as the Jefferson brothers corresponded about major crops, they also discussed the planting of garden vegetables and exchanged these seeds. When Randolph received seeds from his brother, he was the beneficiary of yet another example of Thomas' vigilance and scientific curiosity. Thomas commented on his own efforts to cultivate the best for his table: "I am curious to select one or two of the best species or variety of every garden vegetable, and to reject all others from the garden to avoid the dangers of mixing or degeneracy."[32]

As winter ended in 1813, Randolph wrote to his brother on February 24th about some seeds he had promised to send. Randolph and his second wife, Mitchie, were visiting her ailing mother, Susannah Pryor, at Woodlawn in central Buckingham County. However, they expected to return to Snowden within a day or two and were thinking ahead to the planting of their summer garden. Randolph also extended the invitation to Thomas to visit, writing, "we are now living at home and would be happy to see you and familey."[33]

Randolph sent the letter to Monticello by Squire, asking Thomas to please "write to me by Squire."[34] The trip from Woodlawn to Monticello was twice as far as the trip from Snowden to Monticello. So, if Squire left from Woodlawn, he likely did not reach Monticello until after the 26th. Apparently he had to wait several days, either for Thomas or for the seed to be prepared.

On March 2nd, Squire, Thomas' response, and the seeds headed back to Buckingham. In a rare expression of deep brotherly affection, Thomas closed by saying he was "disappointed" that Randolph had sent his servant and not come himself for the seeds. Thomas explained his enclosed selection:

> Having been from home the last fall during most of the season for saving seeds, I find on examination that my gardener has made a very scanty provision. Of that however I send enough to put you in stock: to wit Early Frame peas, Ledman's peas, long haricots, red haricots, grey snaps Lima beans, carrots, parsneps, salsafia, spinach, Sprout kale, tomatas. I have sent you none of the following because all

your neighbors can furnish them, and my own stock is short. To wit Lettuce radish, cucumbers, squashes, cabbages, turneps.[35]

Additionally, Martha (Jefferson) Randolph included flower seeds for Mitchie Jefferson. The varieties were not mentioned.

Mitchie was likely in charge of both the flower and the kitchen gardens, and, apparently, after receiving so many seeds in March, she asked her brother-in-law to please send instructions with seeds. He responded by sending her a copy of *The American Gardener, Containing Ample directions for Working a Kitchen Garden...,"* by John Gardiner and David Hepburn.[36]

Again, in February of 1815, Randolph wrote to Thomas requesting some seeds for his garden, specifically cabbage and ice lettuce. Thomas responded quickly on the 16th, with both seed and advice:

> ... I send you some green curled Savory cabbage seed. I have no ice lettuce, but send you what I think better the white loaf lettuce. The ice lettuce does not do well in a dry season. I send you also some sprout kale, the finest winter vegetable we have. Sow it and plant it as cabbage, but let it standout all winter. It will give sprouts from the first of December to April.[37]

If Randolph planted the sprout kale, he would not live until December to enjoy it.

Animal Husbandry and Industry at Snowden

Many plantations the size of Snowden operated a variety of small "industries" to meet the needs of the farm, maintaining a certain degree of self-sufficiency. Later in the 19th century, the statewide agricultural censuses reveal that many farms kept bees for the production of honey, herds of beef cattle, small dairy herds for the production of butter, stud horses for breeding stables, etc. Likewise, many plantations provided services for themselves and their neighbors, exchanging skilled laborers such as carpenters and coopers. Grist mills and blacksmith shops were scattered across the county. The milling of lumber evolved from hand sawing over pits, to water-powered saw mills, to portable steam-powered saws.

For at least one season, Randolph offered a stud service at Snowden. In 1797, he paid a 15 shilling tax for a stud horse.[38] Randolph likely stabled an imported animal, offering it in the neighborhood for a single season. "High blooded" horses from England were frequently made available as described in an 1807 broadside, which announced a stud called Hambleton, bred by the Duke of Grafton, imported by William Lightfoot, Esq. of Charles City County, and to be stabled at Temple Hill, James Kinsolving's plantation in Albemarle, for the "ensuing season." Arrangements varied. In the case of Hambleton, Virginians might pay $10.00 for a single leap, $40.00 to ensure their mare was with foal, or six Guineas to cover mares for the season.[39]

Stud services of imported horses were offered on plantations across Virginia. In 1789, Randolph Jefferson paid for a stud license. This Broadside, dated 24 February 1807, advertises the services of Hamleton, at Temple Hill, owned by James Kinsolving, who purchased Dinah and her children from Thomas Jefferson over a decade earlier. (COURTESY OF VIRGINIA HISTORICAL SOCIETY)

Similar advertisements announced, "The high bred horse, Bremer . . . will stand in my stable in Rockingham County," during the 1796 season. Another read, "Hark !!! The noted horse Thunderbolt . . . will stand the season of 1809."[40] Randolph Jefferson likely produced such a broadside for the season he kept a stud horse, though no copy has survived.

Randolph's Revolutionary Public Claims are evidence that he kept a significant herd of cattle on his farm. He also kept sheep and, at least occasionally, used Thomas' flock to improve his own stock. On June 21, 1813, Randolph wrote to Thomas, "I would be extreemly oblige to you to mark out the ram to your overseer and I will send up this fall for him as soon as the weather gits cool so that he can be brought with safety."[41]

Unfortunately, there is no clue as to whether this was a Merino ram; however, Randolph was surely borrowing the animal for breeding purposes. Thomas Jefferson and others had worked valiantly over the previous few years to introduce the fine Spanish breed to North America. Randolph may indeed have been swept up in the enthusiasm, though many Virginians were not "sold" on the Merino breed, despite the extraordinary quality of the wool. In the south, much of the demand was for coarse, inexpensive materials to clothe enslaved laborers. Eventually, Merino sheep flooded the market, with 20,000 imported from Spain in a two-year period, and prices plummeted. In 1813, Thomas Jefferson wrote, "The Merino fever is so entirely spent . . . our country people will not even accept of them; preferring those breeds giving most wool to what gives the finest."[42]

Sheep at Snowden meant mutton on Randolph Jefferson's table. Some, perhaps, were sold at market. A flock of sheep also suggests the possibility of wool production on the farm, which could be turned into homespun materials. At the turn of the 19th century, the raising of sheep in central Virginia had become nothing short of a political stance, since homespun meant less dependence on England's woolen industry. As Thomas Jefferson put it, "Homespun is become the spirit of the times."[43]

In September of 1811, Thomas earnestly established a small textile industry at Monticello, hiring William Maclure to build spinning and weaving equipment for him. After some trial and error with the design, Jefferson ultimately preferred a 24-spindle spinning jenny and looms with flying shuttles.

Maclure also trained some of the Monticello slaves to spin and weave. The *Thomas Jefferson Encyclopedia* summarizes the operation:

> Jefferson's annual goal was 1,200 yards of cloth woven from purchased cotton and from wool and hemp produced on his farms. He never sought to make fine cloth; his only ambition was coarse cloth for summer and winter allotments for the 130 slaves on the Monticello plantation. . . . Dolly and Mary were the weavers, several women and young girls were spinners, and young boys did the carding. As Jefferson said, "[We are] able to clothe our own people by the labor of a few of the less useful of them."[44]

Thomas Jefferson highly approved of homespun independence and, in 1813, he encouraged Randolph to follow suit by installing a spinning jenny at Snowden. Randolph's slave, Fanny, was trained to operate the machine. Between the summer of 1813, when Fanny went to Monticello and the spinning jenny arrived at Snowden, until Randolph's death in August of 1815, mechanized

A Merino sheep. (COLLECTION OF JOANNE L. YECK)

spinning presumably took root at Snowden. Fanny would have continued to spin and may have taught others who were not needed in the fields.[45]

It is doubtful that Randolph actually planned to manufacture woolen goods and/or homespun material beyond the needs of his servants and his family, and his operation would likely mimic the Monticello industry, on a much, much smaller scale. Compared to Monticello's 130 slaves of varying ages, at this time Snowden had only twelve slaves above age sixteen, which included no excess labor to occupy with weaving.[46]

There are suggestions of two other industries at Snowden. Early in the plantation's history, there was iron-working on the place, including a smith shop. Peter Jefferson's 1757 inventory listed ninety-eight lbs. of "old Iron" stored there. Additionally, a letter from Thomas to Randolph dated May 25, 1813 refers to "the old road to the old smith's shop."[47] While it is unclear just how old the "old smith's shop" was, in 1808 the Randolph Jefferson family business, "Nevil & Jefferson," purchased both German steel and iron at Moon's Stony Point store, indicating they may have reestablished a shop at Snowden.[48]

It is also possible that Snowden ran a sawmill during Randolph Jefferson's lifetime. Not all of Snowden's acreage was cultivated; a percentage was uncleared woodland, which provided lumber for the plantation. On August 14, 1802, Thomas noted: "Sent by Orange to Jesse Moore 9/ for ferrge of plank from Snowden," indicating he had acquired lumber from or via Randolph. Currently, no more is known about the relationship between Randolph Jefferson and the Moores.[49] Again in 1807, Thomas paid Randolph $2.00 for ferriages for Jesse Moore.[50]

The James River Culture: Randolph Jefferson's Neighborhood

Beyond the Horseshoe Bend and Randolph Jefferson's immediate neighborhood, his world naturally extended towards the more developed northern side of the James River and southern Albemarle County. While the town of Scottsville did not incorporate until after Randolph's death, during his lifetime some services remained near Scott's Ferry even after the relocation of Albemarle's courthouse. Very convenient to Snowden, these included the prosperous general store operated by William Moon, Jr., and located at his

Randolph Jefferson to Thomas Jefferson, 8 June 1800. (SMALL SPECIAL COLLECTIONS, UNIVERSITY OF VIRGINIA)

THE STORES AT STONY POINT AND WARREN

In the summer of 1810, Randolph was at William Moon's store at Stony Point when he chanced to meet Thomas Jefferson's waggoner, Jerry, who then carried a quickly scribbled note from Randolph to his brother at Monticello. The letter nicely links William Moon's plantation store with Capt. Patterson's store in Warren. In it, Randolph explains that the axle on his gig is in need of repair, and that he is heading to Warren for iron to fix it. Perhaps, William Moon was out of iron or didn't carry it. Once the gig is repaired, Randolph promises to visit Monticello, accompanied by his second wife, Mitchie. Randolph's wording in this hastily written missive reflects his spoken English, giving insights into his speech pattern. He writes, "I should of bin over before this but have bin very much put to it to git Iron to make me an axiltree to my gigg and have not got any yet. I understand there is some landed at Capt Pattersons grocery at Warren with in a few days past, and I intend going up there tomorrow morning to try and git some for that perpose. . . ."

plantation, Stony Point, about a mile north of the river. Surviving ledgers and journals reveal that Randolph, his sons, and his neighbors in northern Buckingham all frequented Moon's store.[51]

Throughout the 18th century, Snowden's isolation was relieved by the steady activity along the James River and, beginning in the late-1780s, there was a flurry of development, much of it designed to accommodate the inspection of tobacco and its transportation from the western plantations downriver to warehouses in Richmond. Author Virginia Moore characterizes the atmosphere thick with "bateaux fifty to ninety feet long, some carrying sixty to eighty tons, tied up at Scott's Ferry to load and unload wheat, flour, beeswax, venison, butter, tobacco."[52] These shallow-bottomed, pole-driven boats carried both cargo and passengers and were perfectly suited to the James River. If the water was not too low or too high or too frozen, crossing at Snowden was comparatively easy. There were fords for men on horseback and, of course, the ferry landing was on Jefferson's property if he needed drier transport.

As Randolph came of age, the ownership of Scott's Ferry transferred from Daniel Scott to his brother, John. Attempting to establish a tobacco and a flour inspection point on his land, John Scott presented a series of petitions to the Virginia Assembly between 1789 and 1792. Scott's Landing seemed an ideal place to develop a riverside town. Located halfway between Columbia, established in 1788 and located downstream on David Ross' land in Fluvanna County, and Warminster, which had been established the same year upstream on Nicholas Cabell's land in Amherst County, a town at Scott's Landing would fill in the gap along the north bank of the James.[53]

Surprisingly, all of Scott's petitions were rejected, until 1792, when "Scott's warehouse" was finally approved. The statute required that John Scott construct, at his own expense, a substantial tobacco warehouse of "brick or stone, to be covered with slate or tile." The gates were to be made of iron. Salaries for tobacco inspectors were established at $134.00.[54] Oddly, Scott elected not to build, perhaps because stiff competition already had been established by Wilson Cary Nicholas at Warren, just six miles upstream on Ballinger's Creek. Warren's head start long delayed what would become Scottsville. The eventual towns of Warren and Scottsville would compete for many years, with Scottsville winning in the end.[55]

THE SNOWDEN FERRY AND ORDINARY

During Randolph Jefferson's years at Snowden, virtually nothing is known about his relationship to the ferry landing on his property. The ferry was definitely operating in December of 1792 when the Virginia Legislature established rates "from the land of John Scott, over the Fluvanna, to the lands of Randolph Jefferson." At some point, the ferry landing on both sides of the James River moved around the Horseshoe Bend to the east, at least by the development of Scottsville in the 1820s. In 1936, R.E. Hannum, field worker for the Virginia Historical Inventory, stated: "After the court house was moved to Charlottesville the site of the ferry was moved down river to a point just below the present iron highway bridge." As for the Snowden ferry house, it is unclear if an ordinary continued to operate at Snowden beyond the period during which Richard Murray leased the property. Scottsville historian, Virginia Moore, believed that Randolph Jefferson "ran a tavern down below at the end of the ferry." The source of her information has yet to be discovered. No ordinary license has been found in Randolph Jefferson's name, though he may have continued to rent the buildings while other men acquired licenses to operate an ordinary there. A neighbor, John Bryant (a.k.a. Briant), paid for an ordinary license in 1789 and may have been one of Jefferson's tenants. In 1802 and 1803, Randolph's kinsman, John Jefferson, paid for an ordinary license in Buckingham County. His tavern might have been located at the old Snowden ferry house.

Beginning in 1790, Nicholas' Landing and the town of Warren quickly eclipsed what remained at Albemarle Old Courthouse, offering Randolph Jefferson and his family significant services for the next twenty years. Though not as convenient as Scott's Ferry, in country terms it was quite accessible. Crossing the river at Snowden, Randolph would have traveled comfortably on horseback along the James River Road (today, Route 726), west through the Scott property, Valmont, then down through the extensive Nicholas property to Warren.[56]

Warren founder, Wilson Cary Nicholas, had served in the Continental Army, represented Albemarle County in the General Assembly (1784–1789 and 1794–1799), and was a delegate to the Virginia Convention of 1788 that approved the United States Constitution. Later, he was a member of the U. S. Senate (1799-1804), served in the House of Representatives (1807-1809), and was Governor of Virginia (1814-1816). A close friend and political ally of Thomas Jefferson, Nicholas preceded Thomas in death and is buried at Monticello.[57] Randolph Jefferson had a strikingly different relationship with Wilson Cary Nicholas, making use of Warren's numerous services and encountering the younger members of the family in Warren, including Wilson Cary, Jr., and Nicholas' son-in-law, John Patterson.[58]

W. C. Nicholas lived large. His residence, Mount Warren, was an extensive plantation, and he owned over 7,700 acres in Albemarle County.[59] His dwelling house, built about 1780, sat 2.5 miles north of the spot where the town eventually was laid out. In 1805, he insured his mansion house for $5,000. The two-story wooden dwelling, 40 x 40 feet, had an "entry" which was one story high, 20 x 20 feet wide. This was connected to a "wooden wing" which was one story high, 40 x 20 feet wide.[60]

Beginning in the late 1780s, Nicholas founded a number of enterprises on his property at the mouth of Ballinger's Creek where it flowed into the James River. Ultimately, he envisioned a town which he hoped would attract other investors and developers. There, he would accommodate both tobacco and wheat farmers. Nicholas was one of the first Virginians to support the switch from tobacco to wheat in the 1790s. Thomas Jefferson and his son-in-law, Thomas Mann Randolph, were also convinced that wheat crops would prove more profitable than tobacco. Milling and distilling grain was fundamental to the

Warren Tavern. (Fred Briggs Collection, Albemarle Charlottesville Historical Society)

development of towns like Warren, Milton (near Monticello), and, eventually, Scottsville.[61]

The year 1789 was pivotal for the launching of Nicholas-owned operations. He received a permit to build a dam at the mouth of the creek. There he erected an impressive brick flour mill.[62] By the mid-1790s, Randolph Jefferson grew wheat, as well as tobacco, and used both Nicholas' mill and distillery. In December of 1796, Randolph acquired two gallons of whiskey from the distillery, and in June of 1797, sixteen gallons of whiskey were charged to Randolph Jefferson, per his "agent," Mr. William "Murry."[63] He is undoubtedly William Murray, son of Snowden neighbor, Anthony Murray, who was running an errand for Jefferson.[64] During 1799, Randolph used Nicholas' overshot, flour mill with some regularity, as did his neighbors, Hardin and Price Perkins.[65]

In "Architectural Surveys Associated with Early Road Systems," authors Edward Lay and Nathaniel Mason Pawlett sum up Warren's advantageous position in relation to wheat production in the area. Importantly, the building and maintaining of reliable roads were a necessary part of Nicholas' plan for Warren's success.

Engaged in flour milling on a large scale, Nicholas required Valley wheat to feed his mills and distillery there, and a more convenient road connection would ensure this besides serving to further the town as a river port and the Valley's connection with the east. This road was created by using portions of existing roads and cutting some additional sections to rationalize the route in a few places. Styled "the new-cut road to Warren" in a number of early road orders, its eastern terminus finally became Scottsville, and a turnpike company was formed to improve and maintain the road from Staunton in the Valley to the James River.[66]

The Virginia General Assembly also gave Nicholas permission to operate his tobacco warehouse. The wages of the inspectors were established at forty pounds annually.[67] It is unknown where Randolph Jefferson took his tobacco to be inspected, though it seems doubtful that he would have used Nicholas' Warehouse, which was up river from Snowden, especially when the Rivanna Warehouse, established in 1785 on David Ross' land at Point of Fork, was down river from his plantation and therefore closer to its destination in Richmond. Both warehouses, however, were on the north side of the James.[68]

The ferry at Warren was established on December 11, 1789. It ran from Wilson Cary Nicholas' land in the county of Albemarle to the land of John Hardy, on the opposite shore in the county of Buckingham. The regulated price for man or horse was three pence.[69]

By 1795, the town of Warren was finally incorporated on thirty acres of Wilson Cary Nicholas' land.[70] Ultimately, only a few lots were sold. The town included a blacksmith's shop, several houses, a country store, and the ferry landing.[71] "Beyond these enterprises," wrote Rev. Edgar Woods, "Warren never made much progress."[72] Woods may be judging the vitality of the town from the vantage point of 1900, for there are indications that, for many years, a number of services including a bustling tavern were utilized by surrounding planters, including Randolph Jefferson.

During the early part of the 19th century, Warren was the center of commerce closest to Snowden, and there is no question that Randolph was well-acquainted with Warren and its inhabitants. His correspondence especially mentions Capt. Brown and Capt. Patterson, who owned a grocery.[73] Late in life, Randolph became distressed when his second wife, Mitchie, ran up considerable debt at a Warren store.[74]

By 1803, Samuel Shelton, in partnership with William Walker and John Staples, operated the large mill and distillery in Warren. Styled as Samuel Shelton & Co., the distillery was located two miles north of Warren near today's Route 627. Shelton eventually owned numerous acres near Warren, and, in 1810, he bought Boiling Springs from Wilson C. Nicholas, but did not hold the property long, selling it to Nicholas' son-in-law, John Patterson. Patterson, in turn, sold it to the very wealthy Robert Rives for $60,000. In 1820, the impressive mansion house at Boiling Springs was valued at $8,000.[75]

The large stone tavern, which sat on the east side of the Warren road, was a later addition to the town and may have been built in 1804 by Wilson C. Nicholas' brother, George Nicholas.[76] Other reports claim it was built by Jacob Kinney's father and later operated by Jacob Kinney. Randolph Jefferson could have known it under the management of both the Nicholases and the Kinneys, as well as under the proprietorship of the tavern's next owner, William Brown.

According to Rev. Woods, William Brown made the tavern "a prominent figure in its day." Thomas Jefferson, in his retirement, used the ferry at Warren and stayed at Brown's ordinary. An impressive structure, the two-story rectangular building had a basement, seven rooms and two halls, a slate roof, and pine doors. Servant quarters were in the rear. According to architectural historian, Edward Lay, it survived until the 1970s, while the country store and depot lasted a bit longer.[77]

CHAPTER NINE

The Children of Randoph Jefferson

> *To be a good husband and a good father at this moment*
> *you must be also a good citizen.*
> Thomas Jefferson to Elbridge Gerry, 1797

In 1789, Thomas Jefferson wrote from Paris to his brother Randolph that he "almost" envied his "quiet and retirement" at home in Virginia. Randolph was thirty-four years old, an active planter, enjoying the relatively quiet country life which Thomas loved so much. About to embark on a six-month leave, Thomas anticipated the reunion with his siblings, including Randolph's growing family at Snowden.

> . . . This will permit me to pass two months at Monticello during which I hope I shall see you and my sister there. You will there meet an old acquaintance, very small when you knew her, but now of good stature. Polly you hardly remember and scarcely recollects you. Both will be happy to see you and my sister, and to be once more placed among their friends. They will remain in Virginia, and are happy in the idea. Nothing in this country can make amends for what one loses by quitting their own. I suppose you are by this time the father of a numerous family, and that my name sake is big enough to begin the thraldom of education. Remember me affectionately to my sister, joining my daughters therein, who present their affectionate duty to

you also; and accept yourself assurance of the sincere attachment and esteem of Dear brother Your's affectionately,

Th: Jefferson[1]

Randolph's first six children all grew to adulthood at Snowden, yet virtually nothing is known about their youth. Even their birth order remains a mystery. Most of them were probably born at Snowden, though there are strong indications that Nancy went home to Buck Island to have at least her first baby. Several records state that Thomas Jefferson, Jr., mentioned in the above letter, was born in Albemarle County, and Thomas, Jr. may be Randolph Jefferson's firstborn. On April 10, 1782, Randolph paid personal property tax on himself and three slaves in Albemarle's District #1 in the northern part of the county. That spring, he, Anne, and three servants likely had traveled to Buck Island to await the birth of the Jeffersons' first child.[2]

The order of the children remains arguable. To date, no family Bible has been found and recordings of their births in public records vary considerably. Five sons and one daughter are known to have survived to adulthood:

	range	census information
Anna Scott "Nancy" Jefferson	1782-1785	1820: bet 26-45
Thomas Jefferson, Jr.	1782	1850: age 60; 1860: age 77
Isham "Randolph" Jefferson	1783/84-1791	1850: age 59; 1860: age 69
Robert "Lewis" Jefferson	1786-1788	1850: age 63; 1858: age 72
Peter "Field" Jefferson	1785-1789	1850: age 65; 1860: age 71
James "Lilburne" Jefferson	1789-1796	1830: bet 30-40[3]

Naming practices of the day help suggest a birth order for the Jefferson children. Interestingly, the Jeffersons do not *strictly* observe those practices so common among the Virginia gentry. Frequently, the paternal grandparents were honored first, then the maternal grandparents, followed by the parents themselves. This would have suggested the names Peter and Jane, Charles and Mary, then Randolph and Anne. However, the Jeffersons took another tack.

It is not known if Anna Scott "Nancy" was the firstborn; however, as it turned out, she was the Jeffersons' only surviving daughter. The fact that she carries Anna Scott Jefferson's name, rather than Jane Jefferson's (for Randolph's mother) or Mary Lewis' (for Anne's mother) is almost startling. Her parents were definitely not playing by the naming rules of the day. Jane Jefferson was already dead; however, Mary (Randolph) Lewis was very much alive and remained at Buck Island until her death in October of 1803. Anne may even have gone to her mother's to give birth to her daughter.

Because of the intermarriage of two sets of Jefferson and Lewis siblings, Mary Lewis had just had, or very soon would have, one granddaughter named Jane Jefferson Lewis and another named Mary Randolph Lewis. Perhaps Grandmother Mary graciously suggested the Jeffersons feel free to use another name.

No matter what the couple's reason for not selecting Jane or Mary or even Mary Jane, the inevitable conclusion is that there was a positive, strong connection between the Jefferson twins, Randolph and Anna Scott. It also stands as a reminder that Anna Scott Jefferson must have had many qualities which Randolph and Anne valued and wished to see embodied in their daughter. Additionally, young Anna Scott would be called Nannie or Nancy – the nickname shared with both her aunt and her mother.

As for the boys, the fact that both of Randolph's parents were dead by the time his first son was born was undoubtedly a factor in his choice of names. His brother, on the other hand, had had many opportunities to be a father to him. Already a rising star in the American Colonies, Tom was also quickly becoming a brother of note. So the first born male was named Thomas Jefferson, known for many years as Thomas Jefferson, Jr., the 18th century equivalent of Thomas Jefferson, the younger. The name Thomas, of course, also harked back to Randolph's grandfather, Thomas Jefferson II.

Next, true to practice, Randolph honored his mother's line, claiming the very important Randolph connection. The Randolph line's achievements and reputations may have been a bit uneven, but they were unequivocally a blue-blooded, first family of Virginia, so Isham Randolph Jefferson came next, remembering Randolph's grandfather, Isham Randolph (1685-1742). Of course, Isham Randolph was Anne's grandfather as well. The boy ultimately went by Randolph, Jr.

It is not absolutely certain which boy was born next and any of these children might have been a set of twins; however, if naming practices are applied, then Robert Lewis Jefferson would be the next born, for it was Anne's turn to name a child after her line.

It is notable that there was no Charles Lewis Jefferson, especially since Anne's father died two years after the Jeffersons were married. The name would have honored both him and Anne's brother, who was married to Randolph's sister, Lucy. It is possible the name was given to a boy who did not survive infancy, as happened with Randolph's brother, Peter Field Jefferson. Arguably, there were already too many Charles Lewises in the vicinity, and this may have liberated the Jeffersons to honor another Lewis. They chose Robert Lewis, calling the boy Lewis.

Who was Robert Lewis and what was he to Randolph and Anne? Robert, along with Charles and John, was a favorite Lewis name. Anne had an uncle, Robert Lewis, and a great uncle, Col. Robert Lewis (1702-1765).[4] Col. Lewis lived at Belvoir, a plantation in Louisa County, adjacent to Albemarle to the east. He married Jane, daughter of Nicholas Meriwether, and their children mostly intermarried with Lewis cousins reinforcing this vast kinship web. He deeded a significant percentage of his 1,800 acres on the Hardware River to Anne's father, Charles Lewis, Jr., suggesting the men were closely associated. This land included what became Red Hill Depot and was eventually given to Anne's brother, Isham Lewis.[5]

Additionally, Col. Robert Lewis was the father of Anne's brother-in-law, one of the many Charles Lewises in the neighborhood. Anne's sister, Mary Randolph Lewis, married Charles Lewis, son of Col. Robert Lewis, nephew to Charles Lewis of the Byrd, and her father's first cousin. The lineage looks like this:

JOHN LEWIS M. ELIZABETH WARNER
  1. Charles Lewis of the Byrd m. Mary Howell
    2. Charles Lewis, Jr. of Buck Island m. Mary Randolph
      3. Mary Randolph Lewis m. Charles Lewis of North Garden
    2. Frances Lewis m. Robert Lewis, Jr.
    2. Robert Lewis m. Jane Woodson

1. Robert Lewis of Belvoir m. Jane Meriwether
2. Charles Lewis of North Garden m. Mary Randolph Lewis
2. Robert Lewis, Jr. m. Frances Lewis

The repetition of names in this intensely intermarried family highlights the need for both new names and new blood. Anne and Randolph Jefferson may have particularly admired Col. Robert Lewis and his family. It would still be some time before his grandson, Meriwether Lewis, made a name for himself.

Following the nod to the Lewis family, the Jeffersons alternated between their lines when assigning names. Peter Field remembered Randolph's father and his uncle, Field Jefferson, whose name represented the deeper lineage of the Field family. "Field" or "Fields" was what the boy was called in his youth.

Last born was James Lilburne, sharing the Lilburne surname with Anne's brother, Charles Lilburne Lewis, and her nephew, Lilburne Lewis.[6] Though the children of Randolph and Anne Jefferson would continue the pervasive family tradition – "Lewises marry their cousins" – they at least were bold enough to pick one apparently "new" name: James. It is a curveball, and its inspiration currently remains unknown.[7] He was, however, called Lilburne and not James.

It is not known when Anne Jefferson died; however, James Lilburne was her last surviving child. In 1795, Anne was still alive and may have paid a visit to Monticello. On January 22, 1795, in a letter from Thomas Jefferson to his daughter, Martha Randolph, he mentioned that he was alone at the house, but was "in hopes soon of a visit from my sister Anne." As was the custom of the day, he called his sisters-in-law "my sister," while he identified his own sisters by their married names: "Sister Marks" was Anna Scott (Jefferson) Marks, "Sister Carr" was Martha (Jefferson) Carr, etc. Anne may have been in the neighborhood, visiting her own family. In 1795, her mother, Mary (Randolph) Lewis, was still alive and living at the Lewis homeplace, and her sister-in-law, Lucy (Jefferson) Lewis, was living nearby at Monteagle, both in the neighborhood of Monticello. Lengthy visits to the Buck Island collective of Lewises would have been common events.

In September of 1798, Anne Jefferson was alive when she and Randolph, along with her siblings, Charles L. Lewis and Molly (Lewis) Wingfield, brought a complaint in the Fluvanna Chancery Court against the executors of the estate of their brother, Isham Lewis.[8]

A last possible reference to Mrs. Randolph Jefferson comes from Martha (Jefferson) Randolph, in a letter written to her father on February 8, 1799. In it, Martha reports her busy social life with "Carr, Trist, &c. &c. . . ." during which time she went to a ball in Charlottesville, "*danced* at it and returned home fatigued and unwell." This event was topped by a later one: "The visit from Mrs. Jefferson with preparations for a second ball where I accompanied her and the girls, added to the cares of the house hold some what increased by so long an absence from them. . . ."[9] While this could have been another "Mrs. Jefferson," it certainly might have been Anne, whose daughter, Nancy, was just the right age for "coming out" and going to balls in early 1799. In the spring of 1801, Nancy would become engaged to Charles Lewis. If "Mrs. Jefferson" of the letter is indeed Anne (Lewis) Jefferson, she died after February 8, 1799.

Additionally, notations in Thomas Jefferson's memo books point to a possible death date for Anne Jefferson in the fall of 1800. In October and November, the following expenses were recorded:

October 8: Expenses at Albemarle old C. H. 1.D.
October 10: Ferrgs &c. at do. 1.1.
October 11: Lent my brother 6.5
October 12: Pd. sm. exp 2.20
November 13: Pd. the Revd. Mr. White for performg. funeral service 4.D.[10]

It appears that in conjunction with a trip to Albemarle Old Courthouse (at Scott's Ferry), Thomas took the ferry to Snowden where he saw Randolph and lent him six shillings, five pence. The expenses on October 12th might or might not relate to expenses at Snowden. Anne Jefferson may have been seriously ill at this time, prompting Thomas' visit. Or he may have come to attend her funeral.

A month later, Thomas paid Rev. White $4.00 for a funeral service. As in so many cases, the payment could be long delayed from the service rendered, but it might relate to Thomas' October visit to Albemarle Old Courthouse and Snowden. Reverend Hugh White (d. 1827) was a Baptist minister. He owned a lot in Milton, near the Lewis lands. His marriage to Nancy Dawson made

All of Randolph Jefferson's children potentially spent some time with their uncle at Monticello. Thomas Jefferson, Jr. may have stayed there while he attended Benjamin Sneed's school and James Lilburne Jefferson may have studied there. (LIVES OF THE PRESIDENTS, BY HARRIET PUTNAM, 1903)

him the brother-in-law of Rev. Martin Dawson, who performed marriages for at least two of Randolph Jefferson's children. Anne Jefferson and others in the family may have already converted to the Baptist church by the time she died, making a Baptist minister an appropriate choice for her funeral service.[11]

With the exception of Thomas, Jr. and James Lilburne, there are no surviving details of the education of Randolph Jefferson's children. In both cases, their uncle, Thomas, stepped in. Thomas, Jr. received some formal education which echoed his father's experience, and James Lilburne was offered an opportunity to study at Monticello.

Despite the lack of direct evidence of their education, Randolph's children clearly learned at least the basics of reading, writing, and arithmetic, and they did well for themselves. Nancy married a gentleman, Zachariah Nevil, and one of their sons went on to the University of Virginia to study law. All were literate, and samples of their writing skills survive in a few scant letters, their wills, and advertisements placed in the Virginia newspapers. They were savvy in their use of the court systems. Those who chose to become planters ran successful farms, and Peter Field Jefferson developed into an astute businessman.[12]

As Randolph Jefferson's children reached maturity, their Uncle Thomas took an interest in some of them, contributing to their education and refinement, as evidenced in his memorandum books and correspondence. First, he contributed to the education of Thomas Jefferson, Jr., who may have lived away from Snowden for an extended period of time prior to his marriage. He certainly resided in Albemarle during 1799 and 1800, either at Monticello, old Buck Island, the Charles Lilburne Lewis' home, Monteagle, or some combination thereof. Uncle Thomas provided him with dancing lessons from Thomas W. Vaughan and paid Mrs. Sneed for six months of young Thomas' schooling.[13]

Indeed in 1799, the same Benjamin Sneed who had instructed Randolph Jefferson and his sisters was still living near Shadwell and Buck Island, teaching the "English School."[14] While Thomas, Jr. studied with Sneed, it is possible he was living at Monteagle with his aunt and uncle, Lucy (Jefferson) and Col. Charles Lilburne Lewis, and their children. If so, like his father before him, this experience of boarding with the Lewises would cement his relationship with his Lewis-Randolph cousins and, again, as with his father, it would end in matrimony. In 1808, Thomas, Jr. married his double first cousin, Mary Randolph "Polly" Lewis.

During this period, Uncle Thomas Jefferson purchased boots for his namesake from Mr. Wertenbaker (a.k.a. Wirtenbaker) and paid Mr. Burk for Thomas, Jr.'s tailoring. The boots, suit, and dancing lessons indicate that elegance, as well as education, may have been part of Thomas, Sr.'s plan for his nephew.[15]

In 1801, Thomas Jefferson included his niece, Nancy, in a group of family members and slaves who were inoculated for smallpox at Monticello.[16] That year, he also paid several accounts for Nancy, as well as his nephew, Thomas Jefferson. These included twenty-one shillings for "things furnished Nancy

Jefferson."[17] This may have been in preparation for her engagement to her first cousin, Charles Lewis, which did not ultimately end in marriage and possibly ended in public scandal.

Despite the fact that Uncle Thomas had been inaugurated President of the United States on March 5th, he showed ongoing interest in Nancy's upcoming nuptials. In a letter dated April 8, 1801, written to his son-in-law, John W. Eppes, Thomas commented, "Nancy Jefferson is said to be about marrying Charles Lewis."[18] Thomas was clearly in communication with someone about the affairs of his niece, Nancy, and it is interesting to note that he took time to share this bit of family news. He was, after all, a rather busy man.

Sometime after April of 1801, Nancy's engagement to Charles Lewis was broken. The reasons are unknown; however, the Lewis fortune had been steadily decreasing and, perhaps, Randolph questioned whether or not Charles could be a good provider for his only daughter. Some have speculated that the existence of Charles Lewis' illegitimate, "yellow" child was the reason for the broken engagement.[19] The public extent of Nancy's embarrassment is currently unknown. By January 1, 1803, she had married Zachariah Nevil, making an excellent match. Baptist minister, Rev. Martin Dawson, performed the ceremony.[20]

Despite Nancy Jefferson's experience with Charles Lewis, Thomas, Jr. was not deterred from marrying Charles' sister, Mary Randolph "Polly" Lewis. They, too, were wed in Albemarle County. Again, the minister was Martin Dawson. The bondsman was John Peyton, who affirmed that Polly was "upwards of 21." She was the daughter of Charles Lilburne Lewis and Lucy Jefferson of Buck Island. Shadrack S. Lively, a friend, testified that Thomas Jefferson, Jr. was "upwards of 21."[21]

It would be many years before Thomas Jefferson again took an active interest in Randolph's children. In 1813, Thomas offered his nephew, Lilburne, an opportunity to study at Monticello. Following an extended conversation with the young man during a visit to Snowden, Thomas wrote to Randolph on May 23rd:

> ... In conversing with your son Lilburne, I found that he would prefer employing himself in reading and improving his mind rather than in being idle. It is late for him to begin, but he has still time

enough, to acquire such a degree of information as may make him a very useful and respectable member of society. I formed a favorable opinion of his understanding. If both you and he approve of it, I think he had better come and pass some time here. I can put him on a course of useful reading adapted to his age. This would be of geography, history, agriculture, and natural philosophy: as soon as you and he can make up your minds, he had better come without delay, as he has not a day to lose. He can pursue his reading as well while I am absent in Bedford, as when here.[22]

Randolph responded quickly to this offer, writing to Monticello just three days later. "I will endeavor to fix Lilburne as soon as possible and send him agreable to your request and hope he will endeavour to improve him self by applying closely to his book."[23] Apparently, Lilburne was delayed by going into the Volunteers, the Virginia Militia. Still hopeful that his son would accept Thomas' offer, on June 21, 1813, Randolph wrote to Thomas that Lilburne would be over by the middle of July.[24] Did Lilburne eventually go to Monticello as Randolph's letter suggests? The record is quiet concerning Lilburne from the summer of 1813 until August 1814. While there is no evidence among Thomas Jefferson's memos that covered expenses for Lilburne at this time, he may have lived there at no extra expense to his uncle.

Some records of James Lilburne Jefferson's involvement in the War of 1812 survive, which indicate he was in active service from 19 August 1814 - 22 February 1815.[25] He served as a private in the 7th Regiment of the Light Infantry in the Virginia Militia under Capt. Boaz Ford of Buckingham. Lilburne was stationed, at least some of this time, at Camp Carter near Richmond.[26] According to Stuart Lee Butler, "Buckingham county soldiers participated in the defense of Norfolk and Richmond." The Pay Roll for 29 August 1814 - 22 February 1815 for the 7th Regiment under Boaz shows the group of Buckingham neighbors and friends who served with Lilburne. This list credits him with five months and twenty-eight days of service.[27] These were significant months in the war, especially in eastern Virginia. Between August 24 and 25, 1814, the British burned Washington, D.C., and President James Madison fled the White House.[28]

James L. Jefferson

Pvt.

| 7 (Gray's.) | Va. Militia. |

{ Captain Boaz Ford's Co. of Light Infantry, 7 Reg't Virginia Militia.

(War of 1812.)

Appears on

Company Pay Roll

for Aug 29 to Oct 28, 1814.

Roll dated _____

_____ Oct. 28, 1814.

Commencement of service, 29 Aug, 1814.

Expiration of this settlement, } Oct 28, 1814.

Term of service charged, 2 months, ____ days.

Pay per month, ____ 8 dollars, ____ cents.

Amount of pay, ____ 16 dollars, ____ cents.

Remarks: _____

(572)   E. B. Thompson
          Copyist.

Randolph Jefferson's son, James Lilburne Jefferson, served as a private in the Virginia Militia during the War of 1812. Later, he served in the Fluvanna County Militia. (NATIONAL ARCHIVES AND RECORDS ADMINISTRATION)

THOMAS, JR., PETER FIELD AND
ROBERT LEWIS JEFFERSON, SLAVEHOLDERS

After Thomas, Jr., Peter Field, and Robert Lewis Jefferson left Snowden, they all placed advertisements for runaway slaves. Their language reveals their personalities. Thomas Jefferson, Jr. offered a highly detailed description and was potentially a tough task master. Peter Field Jefferson was matter-of-fact and to the point. Robert Lewis, by contrast, expressed himself in more colloquial language; his tone is softer, lacking Thomas' forceful prose and Field's briskness. Runaways were a universal "problem" in Virginia and care should be taken not to conclude too much about the Jefferson brothers simply because a slave ran away.

Twenty Dollars Reward – Runaway from the subscriber, on the fifteenth day of September last, a negro man slave, named Gary, aged about twenty-six years, light black complexion, height about six feet, stands quite erect, and when walking handles himself quite nimbly, his weight is one hundred and sixty or seventy, countenance cast down when spoken to – the white of this (sic) eyes is unusually large, on being apprehended is very surly, he is a rough Cooper. On examination his shins will be found injured – he has probably obtained a free pass. About twelve years ago, Gary was brought from Brunswick county to Surry county in this state, by Mr. Travis Harris, where said boy lived for eight years, and was in the year 1808 purchased by myself of Thomas Edwards and Reuben Butlers ex'ors of said Harris, dec'd out of Surry, and brought to Buckingham, where said negro got a wife, and lived with me about two years on James river, not far from Scotts ferry, and was then brought by myself to Albemarle county, near Milton, from whom said negro eloped – said slave is probably harbored in Surry or some of the above places. I do hereby forewarn all person(s) from harboring said negro. Alas, I do command all captains of vessels at sea, not to grant said slave his passage, or take him on board.

The above reward, with all reasonable charges, will be paid to any person taking said slave within this state, and bringing him to me, or else for confirming him in any jail in this state, so that I get him again. An additional reward of ten dollars will be given to any person taking said slave out of this state, and for acting with him as above states.

      Thomas Jefferson, Jr., Albemarle county, Virginia.
       [*Richmond Enquirer*, 18 February 1812]

  Thirty Dollars Reward. Runaway from the Subscriber in this City, a Negro Man named Ned Henderson. He is a spare made raw-bone negro, about 5 feet 8 or 9 inches high, had on when he went away a suit of yellow home spun clothes, and has been engaged in running the river for several years. I suspect he is gone to Fluvanna, where he has a wife, belonging to Mr. Henderson near Hardware river, in said County. I will give the above for the delivery of said Negro at Scott's Ferry, in Albemarle, or for securing him in any jail so that I (may) get him again.

           Peter F. Jefferson.
        [*Richmond Enquirer*, 8 June 1816.]

  Runaway, from the Subscriber, living at Scott's Ferry, Albemarle County a negro man, named David, but has been in the habit of altering his name, sometimes passing by the name of Billy Logan. He is a small trim-made fellow, of a dark complexion, about five feet three or four inches high, small sunken eyes, and his forehead tolerably void of hair. This fellow has passed for a free man in August, but at this time it is likely he may be in the neighborhood of Buckingham Court house, or Bent Creek, as he has formerly lived with Mr. Cheek in that neighborhood. Any person who may secure the said fellow so that I may get him, shall receive a reward of Fifteen Dollars.

           Robert L. Jefferson
        [*Richmond Enquirer*, 18 June 1811.]

Part Three
THE JEFFERSON BROTHERS

CHAPTER TEN

The Jefferson Servants

There is nothing I would not sacrifice to a practical plan of abolishing every vestige of this moral and political depravity.
Thomas Jefferson to Thomas Cooper, 1814

The absence of towns, even hamlets, near Snowden certainly isolated Randolph Jefferson and his family, making the farm, as so many plantations were, a world unto itself. In this microcosm, the enslaved population played an intimate role in day-to-day life. From the raising of the Jefferson children, to the cooking and the cleaning of the dwelling house, to the tending of the crops and animals, blacks and whites lived together in close quarters. When there were various experiments in small, supplemental industry at Snowden, the Jefferson slaves were key to success or failure. When the Revolutionary War came to the Horseshoe Bend, Randolph's servants joined the cause, with at least one man risking his life canoeing the military stores on the James River. When the white Jefferson family was faced with demands from creditors, the black families of Snowden faced separation and, possibly, the auction block, inextricably entangling the fortunes of the two populations.

In 1764, when Peter Jefferson's estate was divided between Thomas (who had turned twenty-one) and Randolph (who was only eight years old), Randolph was assigned twenty-four of his father's slaves. There is no list naming Randolph's inheritance, though included was his personal valet, Peter, who had been with him virtually since birth. During the years between this distribution and his majority, it is known that Hannah died at Snowden as a result of a beating by overseer, Isaac Bates, and that Randolph received little Rachel, a gift from his mother, Jane Jefferson.

The first list of Snowden slaves appears on the Buckingham County Personal Property Tax record in 1782, when Randolph paid tax on six horses, forty-two head of cattle, and thirty individuals living in Buckingham. In April of 1782, he paid tax on himself and three slaves in the Albemarle County's District #1, which was comprised of the northern part of the county. There, he, Anne, and three servants were probably at Buck Island awaiting the birth of the Jeffersons' first child.[1] It is not known how long the Jeffersons stayed in Albemarle during 1782, but Randolph did not pay a head tax on a white overseer at Snowden that year, indicating that he may have gone back and forth between the counties, leaving a senior black man in charge in his absence.

The servants in Albemarle were not named on the tax list. It is likely that the thirty men, women, and children living at Snowden all came from Shadwell, for there is no record of Randolph purchasing a slave prior to the closing of Peter Jefferson's estate in January of 1790, though Thomas purchased Phill from Randolph. The following individuals were recorded by the Buckingham assessor:

Nimrod	James	Squire	Peter	Adam	Hanibal
Lucy	Jane	Flora	Effy	Edy	Phillis
Dinah	Orange	Milly	Pat	Daphney	Juno
Dilce	Mary	Sally	Betty	Will	Jupiter
Cyrus	Jack	Frank	Syller (Sully?)	Thornton	Jacob[2]

The next year, 1783, Randolph reported thirty-two slaves:

Nimrod	James	Adam	Orange	Squire	Slatey
Flora	Dinah	Effie	Phillis	Juno	Tinny (?)
Esther	Edy	Luce	Milley	Pat	Jenny
Dilcy	Mary	Will	Cirus	Jupiter	Sully (Sally?)
Thornton	Jacob	Frank	Isaac	Jenny	Lucy
Cary	Peter	Elijah	Perkins[3]		

Allowing for misreading of the sometimes cryptic 18th-century script, Hanibal, Jane, Daphney, Betty, Jack, and "Syller" who appear in 1782 are missing from the 1783 list, while Slatey, Tinny, Esther, Luce, two Jennys, Isaac, Cary, Elijah, and Perkins appear. Cary is a recent birth at Snowden. Perkins might also be an infant, possibly named in honor of Randolph Jefferson's neighbor, Hardin Perkins. Three of the "new" individuals were likely in Albemarle during part of 1782. Even considering that infusion, the list implies surprising, significant, and inexplicable changes in the African-American population in just one year.

When compared to Monticello during Thomas Jefferson's so-called "first retirement," 1794-1796, during which 100 slaves at Monticello worked 5,000 acres, Randolph's holdings are somewhat modest.[4] During this period, written exchange between the brothers is silent until 1796. Thomas is engaged intensely at Monticello, experimenting, improving, building, while Randolph is contently occupied with a comfortable life, a moderately-sized family, and a prosperous farm.

Cary and Snowden Ben

In 1796 and 1797, Thomas Jefferson purchased two of Randolph's young male slaves, Ben and Cary. Their value was to be determined by a third party, after which Thomas would be responsible for paying several of his brother's creditors. Thomas' initial memorandum concerning the agreement reads:

> Sept 3 1796   Recd. from Rand. Jefferson a negro boy Ben, Peter's son who is to be valued by John Coles & James Cocke, and I am to pay the valuation to Donald [Scott] & co. in discharge of their acct. agt. him.[5]

This unusual arrangement between the brothers kept two potentially valuable young slaves in the Jefferson family. The plan also kept African-American families more or less together, something that may have concerned particularly Randolph given his subsequent efforts not to part with his slaves, which inevitably meant breaking up families. Peter's son, Ben, was at least a third generation Jefferson slave. He likely had relations at Monticello as well as at Snowden. Peter, Randolph's manservant, was Myrtilla's son and had been willed to Randolph by his father in 1757.⁶ Peter was approximately Randolph's age, born c. 1755, making him about thirty when Ben was born. Ben's mother was also Randolph's property, since a woman's "issue" belonged to her owner. Currently, there is no clue to her identity.

Ben does not appear on either the 1782 or 1783 slave lists at Snowden, indicating he was born shortly thereafter. Indeed, records at Monticello give his birth as 1785. There, Peter's Ben became known as "Snowden Ben" (or Ben Snowden) to distinguish him from another young Ben living at Monticello and working in Thomas Jefferson's nailery.

When Snowden Ben joined the teenage workforce at Monticello's nailery, he became part of an industrial experiment designed and very closely managed by Thomas Jefferson. Approximately twelve to fifteen boys lived and worked together on Mulberry Row near the mansion house, where they made tens of thousands of nails for consumption on the plantation and for sale. Their work was intense and highly competitive, testing Thomas' theories for creating a system of labor incentives based on rewards and distinctions.⁷

In the autumn of 1796, after Thomas agreed to take Ben to cover Randolph's £80.13.10½ debt to Donald, Scott & Co., it became clear that Ben's value did not quite cover the obligation. Five months later, in February of 1797, Ben was joined by Cary, whom Thomas purchased from Randolph under the same conditions he had acquired Ben.⁸ When Cary was added to the agreement, Randolph managed pay off other debts with some funds left over.

While the loss of two boys at Snowden was significant both for their own families and for the Jeffersons, the boys may have been better off having each other as familiar companions. There is no proof they were brothers, though they certainly could have been. Cary, who was about two years older than Ben, first appears on Snowden's 1783 slave list. Absent on the 1782 list, he was apparently born during the intervening year.⁹ It also seems likely that Cary's parents, or at least his mother, were among Randolph's slaves; however, no information has surfaced concerning them.

Given this unusual, if not unique, opportunity for both Ben and Cary, Randolph could assuage guilt he might have had as the result of separating families, knowing he was giving his boys an opportunity potentially to learn a trade and better themselves in the more complex world of Monticello.¹⁰

Ultimately, Colonels Coles and Cocke valued Ben and Cary together at £155, and, in March of 1797, Thomas Jefferson executed a bond with Donald, Scott & Co. for £80.13.10½, covering Randolph's debt to that establishment. Concerning this payment, Thomas described it as "a debt due them from my brother Randolph and is to go in part paiment for the two boys Ben & Cary recieved (sic) ante Sep. 3. and Feb. 9." The terms of the bond were 5% from October 1, 1796, payable on July 1, 1798. Donald, Scott & Co. also carried accounts for Jane (Randolph) Jefferson, Anna Scott (Jefferson) Marks, and Thomas. This sizable debt to the Glasgow firm indicates that Randolph Jefferson purchased numerous imported items, as did his mother and his siblings. Thomas' purchases from this Scottish merchant included one of his Dollond telescopes.[11]

In September of 1797, Thomas paid a man named Anderson £9.6.0 towards an order for Randolph.[12] Then, in October of 1798, Thomas cleared Randolph's debt for £11.10 to David Anderson, assignee of Christopher Hudson.[13] Two small debts were also paid to Will Beck, the man who long ago accompanied Randolph to Williamsburg.[14] Additionally, Thomas may have sent some of the proceeds from the sale of Ben and Cary directly to Randolph, including £13 that went to Snowden via Squire on September 8, 1798.[15]

~

In early 1797, Cary was assigned work along with Ben in the nailery, which Thomas had established in 1794. Cary was about fourteen years old. Ben was about twelve. They were both at ideal ages for Thomas's already successful training ground for adolescent male slaves. In his "Farm Book," he described the work assigned to his youngest laborers. "Children till 10. years old to serve as nurses. From 10. to 16. the boys make nails, the girls spin. At 16. go into the ground or learn trades."[16] Making nails tested these young men, setting up a competition among them for efficiency and productivity. The fastest most accomplished were selected for trades and more complicated training. The slower, less accomplished were sent to work in the fields.

The nailery was a core part of the plantation reforms and experiments Thomas Jefferson devised during 1794-1796. Having accepted that slavery would not be abolished in his lifetime, he absorbed himself with the amelioration of both the living and working conditions of slaves at Monticello. In particular, he devised the nailery and weaving operations as opportunities for his slaves to prove and better themselves. There were financial rewards, offering slaves opportunities to earn more spending money. Fueled by the ideals of the Enlightenment, Thomas was determined to eliminate waste, increasing efficiency and productivity in all areas of life on the plantation. Once these reforms were in place and honed, Thomas returned to public life, leaving Monticello once again in the hands of overseers and managers.[17]

Ben and Cary continued in the nailery. During 1802, Cary was promoted, gaining additional work burning coal. His job was to fuel cords of wood into kilns for the production of charcoal. His co-worker was Jame Hubbard. Cary was apparently performing well, and Thomas Jefferson noted that he paid both young men for their work. Then, a horrible incident occurred. In response to a prank played by a fellow nailer, the now eighteen-year-old Cary brutally attacked another Monticello slave named Brown Colbert (b. 1785). Cary struck Brown in the head with his hammer, fracturing Brown's skull. After living for five presumably uneventful years at Monticello, why Cary had suddenly erupted with such a violent action at so little provocation is unclear.[18]

In July of 1803, Thomas Jefferson paid Thomas Wells $13.00 for Cary's imprisonment. Then, he swiftly decided to sell Cary to a Georgia slave trader. Cary's purchaser is unknown; however, Gabriel Lilly, the overseer at Monticello from 1800 to 1805, handled the sale, which brought $300.00. Jefferson chose to sell Cary out of a desire to make an example of him "in terrorem to others." By selling him to a distant buyer, this banishment from friends and family would be "as if he were put out of the way by death."[19] This strategy no doubt worked, and the slaves of Monticello learned that their owner was fully capable of effectively eliminating any one of them from the life they knew. In the future, the cautious among them would conceive of less violent ways to express their anger or protest the system which held them in bondage.

On May 30, 1803, Thomas Mann Randolph wrote to his father-in-law in Washington, D.C., detailing the situation:

> [It] presented a shocking prospect at first but promises now an issue very different from the dismal end at first expected. The boy Cary, irritated at some little trick from Brown, who hid part of his nail rod to teaze him, but restored it as soon as he found him angry, took a most

barbarous revenge; approaching him by stealth he struck him with his whole strength upon the skull, very near the longitudinal suture, on the left side, midway between the horizontal & perpendicular faces of the skull bone, when the body is erect. The skull yielded to the face of the hammer in its whole circumference but was driven in only about 2/3 of it, bending the other part & fracturing no where else. The instantaneous suspension of life did not continue longer than a minute: for an hour no damage was suspected but at the end of that time violent convulsions took place which were quickly succeeded by Coma & its usual symptoms, the leaden eye and apoplectic stertor: the patient was sensible of what was done to him and answered reasonably yet to my astonishment when the pressure was removed had no recollection at all of any circumstance from the blow to that moment. Wardlaw & myself arriving nearly at the same time I acted as his assistant in the operation which he performed by means of the trephine (the saw which works both ways or with the motion of the wrist only) with the greatest boldness, steadiness and skill. The boy is as well as we could have expected today, and will, no doubt I think, be as well as ever. The other, I committed to jail till Browns fate is determined.[20]

Thomas Jefferson responded a week later:

Should Brown recover so that the law shall inflict no punishment on Cary, it will be necessary for me to make an example of him in terrorem to others, in order to maintain the police so rigorously necessary among the nailboys. There are generally negro purchasers from Georgia passing about the state, to one of whom I would rather he should be sold than to any other person. If none such offers, if he could be sold in any other quarter so distant as never more to be heard of among us, it would to the others be as if he were put out of the way by death. I should regard price but little in comparison with so distant an exile of him as to cut him off compleatly from ever again being heard of.... In the mean time let him remain in jail at my expence, & under orders not to permit him to see or speak to any person whatever.[21]

Snowden Ben, on the other hand, proved a satisfactory worker, moving from the nailery to the fields as a farm laborer. This may not have been a particularly desirable occupation, but compared to Cary's fate, it was relatively peaceful. Over the years, Ben earned extra money by cleaning the sewers at Monticello.[22]

Later in life, Ben became the second husband of Lilly Hern. In 1818, he was leased for five years to Thomas Jefferson Randolph, Thomas Jefferson's grandson, who then resided at Tufton. Following Thomas Jefferson's death, Ben was sold as part of the dissolution of the estate. He and his wife, Lilly Hern, were purchased on January 1, 1829, by Professor George Blaettermann of the University of Virginia. Together, their price was $385. This sale was no punishment, simply the all-too-often result of the closing of a Virginia planter's estate.[23]

The sale of Ben and Cary away from Snowden, and their divergent stories, is an unusual link between the Jefferson brothers. These two enslaved boys, presumably born at Snowden, spent their early years with their families, only to be uprooted and taken to Monticello. It must have been almost like moving from a village to a small town. Monticello offered them more possibilities for their lives, yet also contained more distractions, threats, and temptations. Cary's fate, in particular, also may have had a strong impact on Randolph Jefferson. In the future, he would show great reluctance to sell a single slave away from Snowden.

The Black Doctor

A mysterious and complicated story surrounds several slaves at Monticello, "Uncle Randolph," and, possibly, a slave at Snowden. A letter, dated January 30, 1800, from Martha (Jefferson) Randolph to her father, who was in Philadelphia at the time, reveals the crux of the mystery:

> ...To your enquiries relative to poor Jupiter he too has paid the debt to nature, finding him self no better at his return home, he unfortunately conceived him self poisoned and went to consult the negro doctor who attended the Georges. He went in the house to see uncle Randolph who gave him a dram which he drank and seemed to be as well as he

had been for some time past, after which he took a dose from this black doctor who pronounced that it would *kill or cure*. 2 ½ hours after taking the medecine he fell down in a strong convulsion fit which lasted from ten to eleven hours, during which time it took 3 stout men to hold him, he languished nine days but was never heard to speak from the first of his being seized to the moment of his death. Ursula is I fear going in the same manner with her husband and son, a constant puking, shortness of breath and swelling first in the legs but now extending it self. The doctor I understand had also given her means as they term it and upon Jupiter's death has absconded. I should think his murders sufficiency manifest to come under the congnizance of the law.[24]

Jupiter was Thomas Jefferson's lifelong servant, devoted and deeply trusted in the Jefferson household. "The Georges" were, likewise, a very significant family at Monticello. Great George, Ursula and their son, Little George, ultimately all died, perhaps poisoned, at the hands of this black doctor who might have been anything from herbalist to a conjurer.[25]

Martha begins her story of Jupiter's death with his trip to "consult the negro doctor who attended the Georges." The sentence that follows, stating that Jupiter "went in the house to see uncle Randolph who gave him a dram," indicates that doctor may have lived at or near Snowden. Apparently, Uncle Randolph's "medicine" had a positive effect on Jupiter; however, the slave later took a potion from the African American which led to his death.

Jupiter occasionally traveled from Monticello to neighboring plantations, making deliveries for Thomas Jefferson. He certainly knew his way to Snowden. Cash was often exchanged between the two men, and Jupiter easily could have had funds to hire a doctor. He also would have known of him through the George family.

Whether or not the black doctor was one of Randolph's slaves remains unclear. There is no surviving correspondence between the Jefferson brothers mentioning this episode. If one of Randolph's slaves had been responsible for the death of four of Thomas' most valuable servants, a discussion of this certainly would have followed, though it might not have survived in correspondence.

In 1800, there were also free Blacks living in Buckingham County. The doctor could have been a free man living at or near Snowden, for whom Randolph Jefferson had no claim or responsibility. It has also been suggested that the doctor may have been owned by a Snowden neighbor.

An entry in Thomas Jefferson's memo books dated November 25, 1798, offers a possible clue. Jefferson writes, "Gave Perkins's Sam order on Isaac Millar for 10.D in full of attendance on smith George."[26] Perkins's Sam would apparently be the slave of a Mr. Perkins.

Hardin Perkins, Sr., Randolph's friend and neighboring planter, had lived adjacent to Snowden all of his adult life and had died three years earlier, in 1795. He left a will, a copy of which survived in a Henrico County Chancery suit. He named several slaves, though there is no mention of one named Sam. The bequests of the slaves read as follows:

> I give & bequeath to my daughter Polley Price Moore the four following Negroes, to wit, Billey, Eve, Fanny, and John . . .
> I give and bequeath to my son Price Perkins, the following Negroes, Abram, Isaac, & Peter. . . .
> I give and bequeath to my Grand Daughter Salley Price Lewis a Negroe Girl Frank. . . .
> I give and bequeath to my Grand Daughter Salley Price Key a Negroe Boy Major. . . .
> I give and bequeath to my wife during her natural life and after her death all the said property with the increase of Negroes (except my Negroe man Abraham) to be equally divided among all my Children. My Negroe man Abraham after the death of my wife, I give to my son Daniel Perkins to him & his heirs forever.
> I give and bequeath to my son Hardin Perkins three hundred and fifty pounds . . .[27]

It is possible that a man named Sam was already the property of Hardin Perkins, Jr., who received no slaves by the will. Price Perkins, who also remained in the neighborhood, might have already owned a slave named Sam. If this

Warren Mill. (Fred Briggs Collection, Albemarle Charlottesville Historical Society)

potentially dangerous slave "absconded," Randolph Jefferson or one of the Perkins brothers might have advertised for a runaway slave. No such advertisement has been discovered.[28]

James and Mr. Jefferson's Carp

In April of 1812, Thomas Jefferson wanted some fish for his stocked ponds at Monticello – one of which was specifically for carp and the other designed for chub. That year, Thomas sent one of his servants, named James, to John Achlin's Fluvanna plantation in search of fish.[29]

The following year, in April of 1813, Thomas was ready for more carp. On April 26th, he wrote memos distinguishing between two men named James. Foreman James appears to be in charge of that year's carp acquisition. The entries read like this:

Apr 26: To James, $2.00 for expenses to Bedford
Apr 26: "Left 5.D. with Mr. Bacon for foreman James to pay for carp."[30]

Then on May 15th, Thomas Jefferson visited Snowden, giving $.75 in gratuities to Randolph's servants and spending $1.00 on fish. While at Snowden, Thomas Jefferson had discovered the possibility of purchasing carp from his brother, Randolph. Once back at Monticello, on May 25th, Thomas penned a letter to Randolph which was delivered by Ned's James, who took $2.00 with him to purchase fish from Randolph.[31] Thomas instructed:

> Supposing the shad season not to be quite over, and that in hauling for them they catch some carp, I send the bearer with a cart and cask to procure for me as many living carp as he can to stock my fishpond. I should not regard his staying a day or two extra, if it would give a reasonable hope of furnishing a supply. He is furnished with money to pay for the carp, for which I have always given the same price for shad, should he not be able to lay out the whole in carp he may bring us 3. or 4. shad if he can get them.[32]

The following day, May 26, 1813, Randolph replied to Thomas:

> I received your friendly letter by the boy they catch no shad a tall at this time so that I have sent James up to Warren to try and procure some carp for you and have wrote to Mr. Brown a bout them if it is in his power to git any to fernish your boy with what you derected him to bring in the barril a live I have understood they cetch a number there every night in the mill race.[33]

Thomas' memo book indicates that Ned's James returned $1.50 of the $2.00 that he was advanced on the 25th.[34] Ned's James had carried Randolph's letter as well, indicating that he returned immediately to Monticello with no fish, while Randolph's James went up to Warren with the note to Mr. Brown.

About a month later, on June 20th, one of the Jameses carried a letter from Monticello to Snowden along with a gift from Thomas to Randolph's wife, Mitchie. Randolph responded the following day, "I received your letter by James and also the book which you sent . . . I wrote very pressingly to Capt. Brown by your boy in respect to the carp for you but found it was all in vane from what James tells me he got none."[35] At this point, keeping track of which James was which is virtually impossible. In the end, as Randolph observed, it was all in vain. What started out as a simple errand for carp ended up reading like a comedy of errors: Too Many Jameses and No Carp At All.

Fanny, The Spinner

In May of 1813, Thomas Jefferson offered Randolph a spinning jenny that had been operating at Poplar Forest, the Bedford County plantation. It is not clear if Randolph asked his brother for the second-hand Jenny or if Thomas had "prescribed" it for the advancement of industry at Snowden. The letter, with the offer, came after Thomas had paid a visit to Snowden earlier in May. In the letter, Thomas also made unsolicited suggestions about how Randolph should manage his crops. During the visit, he also had a conversation with Lilburne Jefferson. Finding the young man eager for education, Thomas seized the opportunity to begin designing his nephew's studies. It is equally possible that bringing the spinning jenny to Snowden was Thomas' idea, not his brother's notion. On May 25th, Thomas wrote to Randolph:

In this sketch, folk artist, Lewis Miller, captures the work of Virginia slaves. He identifies the woman as: "Our next door neighbor. A little black girl spinning wool." Miller also noted the value of corn in Virginia, where "much is raised." (LEWIS MILLER, "SKETCHBOOK OF LANDSCAPES IN THE STATE OF VIRGINIA, 1853-1867," ABBY ALDRICH ROCKEFELLER FOLK ART MUSEUM, THE COLONIAL WILLIAMSBURG FOUNDATION, WILLIAMSBURG, VA., GIFT OF DR. AND MRS. RICHARD M. KAIN IN MEMORY OF GEORGE HAY KAIN)

> . . . I shall be able to give you the spinning Jenny which I carried to Bedford. It is a very fine one of 12, spindles. I am obliged to make a larger one for that place, and the cart which carries it up shall bring the one there to Snowden on it's return. You will have to send a person here to learn to use it, which may take them a fortnight. But that need not be till I return from Bedford, for which place I shall set out the day after our court, this day fortnight, or very soon after.[36]

Randolph responded the following day, expressing sincere gratitude for his brother's generosity. He wrote, "We are extreemly oblige to you in respect to the spining ginney as letting your boy come by and leaveing it with us as it was more then we could of asked of you at any rate or expected."[37] It would be some time before the Jenny actually arrived at Snowden, and, in the meantime, Thomas offered to train one of Randolph's slaves to use the equipment. Almost a month later, in late June, Thomas wrote again:

> The unexpected difficulties of bringing water to my saw mill and threshing machine, and the necessity of doing it before harvest, have obliged me to put off my visit to Bedford till after harvest. The spinning Jenny for Bedford is now ready but will not be sent until I go. While it is here it offers a good opportunity for your spinner to learn upon it. After it is gone there will be no idle machine for a learner to practice on. I send the bearer therefore to inform you of this, that you may not lose the opportunity of getting the person taught whom you intend to employ in that way. I should think she had better come immediately, as it will require a month or more to become perfect in roving and spinning. By the time she is taught, the machine will be off to Bedford, and the cart which carries it will return by Snowden and leave the 12. spindle machine there, on which she may go to work immediately. This will be early in August. I do not know whether I can call on you as I go. I will if I can, but certainly will as I return. Is your road cleared out?[38]

One model of the spinning jenny. (HAND-BOOK OF THE USEFUL ARTS, BY THOMAS ANTISELL, 1852)

James carried the letter to Snowden. He returned to Monticello on June 21st with the news that Randolph would send young Fanny to Monticello to learn to use the spinning enny. Randolph explained that she was so young because, "We have not a woman except a girl of twelve or fourteen years old but what has children." He noted he would send Fanny over in "the course of a week."[39]

It took a bit longer, however. On July 11th, Randolph sent Fanny with Squire to stay at Monticello and learn how to operate the spinning jenny. He apologized to Thomas for the delay but explained they had been very busy "about" the wheat. Clearly, Randolph needed Squire in the field, if not Fanny as well. He wrote, "I could not spare a hand out of the field to bring her and would be very much oblige to you to put her under one of the grone hands to keep her in good order. I suppose we may send for her in three or four weeks."[40] The next day, Thomas responded:

                                          Monticello July 12. 13.

Dear brother

Your's is received by Squire, and the girl begins this morning the first necessary branch, which is roving, or spinning into candlewick to prepare it for the spinning Jenny. This will take her some days, more or less, according to her aptness, and then she will commence on the Jenny. As she appears rather young, it will probably take her a month or 6. weeks to learn well enough to be relied on for carrying it on herself where she can have no further instruction. However I will by any opportunity which occurs let you know her progress and when you may send for her. It will be near a month before I shall be able to set out for Bedford, and uncertain whether I can go by Snowden or not; but if I do not, I will certainly return by there, and the machine will go to you at the same time; about which time I imagine it will be best that your girle should meet it there, continuing to spin here until then, that she may be more perfect. With affectionate salutations to my sister and yourself accept my adieux.

TH: Jefferson[41]

Eager for news of Fanny's progress, Randolph wrote on August 8th inquiring, "... would be glad to heare how Fanny comes on."[42] And, once again, Thomas responded immediately:

Your girl comes on tolerably well. She was some time learning to rove, for without good roving there cannot be good spinning. She has been sometime spinning, and by the time of my return from Bedford, when the machine will be carried to you, she will be able to spin by herself.[43]

The carts set out on September 8th from Poplar Forest, bearing the now long-awaited spinning jenny towards Snowden. Thomas promised that the Jenny was "in perfect order," and he hoped it would be delivered "safe from accident." He followed, headed for his second visit to Snowden in less than four months.[44]

Here, their correspondence ends on the subject, leaving many loose ends to the story. Did the jenny arrive safe from accident? Who brought Fanny home? Did she prove competent? How did the presence of a spinning jenny affect the community at Snowden?

In the end, no matter whose idea it was to bring the jenny to Snowden, Thomas' gift and the education of Fanny, the spinner, stands out as what may have been Thomas Jefferson's largest contribution to life at Snowden. At the very least, young Fanny had quite an adventure, living at Monticello for nearly two months, learning an important skill, and seeing the larger world of Albemarle County.

A Man Named Squire

Between 1801 and 1815, Randolph Jefferson's servant, Squire, is frequently mentioned in Thomas Jefferson's memos and in correspondence between the Jefferson brothers. Despite the commonness of his name, there is every reason to believe he is the Squire listed among Snowden's slaves in both 1782 and 1783. He may, in fact, be one of the Squires listed on Peter Jefferson's 1757 inventory. A boy named Squire, valued at £27.10, impersonally listed between Ceasar (£30) and Jammey (£17.10), was unknowingly poised to become the life-long servant and companion to Randolph Jefferson.

Based on his value, Squire was a bit older than his future owner.[45] By 1776, when Randolph took over operations at Snowden, it is possible that the somewhat senior Squire could be relied on for accumulated knowledge of plantation workings. If Squire had skills beyond agricultural labor, they are unknown, and the surviving details of Squire's errands are deceptively mundane. They elucidate the comings and goings of a mature, trusted man, riding between Snowden and Monticello year after year, carrying communications, goods, and monies between the two Jefferson households.

In 1801, Squire delivered white clover seed to Monticello. He also carried mail. One letter written in 1809 from Randolph to Thomas Jefferson, is addressed to "Mr. Thomas Jefferson; Pr. Squire." In August of 1813, he was selected to go to Monticello with a request for a loan. Thomas Jefferson complied, sending Squire back to Snowden with $40.00.[46] Randolph had written on August 8, 1813:

> I have sent Squire over to see whether I could borrow forty dollers of you as I am compelled to have as much at Court if it is possible to borrow as much of you which shall certainly be replaced a gane in three weeks which will be about the time I shall dispose of my crop of wheat and will take extreemly kind of you if it is in your power to help

me at this time which I shall feel my self under many obligations to you for the loan of. Be pleased to discharge Squire as soon as possible . . . ."[47]

Due to the loss of the Buckingham County records, the reason Randolph needed the $40.00 has been obscured; however, Thomas wrote back, "Your letter of yesterday found me unprovided with the sum you desired; but I have been able to borrow it among our merchants who are not much better off than others, all business being at a stand. We are experiencing the most calamitous year known since 1755."[48] A severe drought was responsible for very poor crops. Thomas also noted in his memos on August 8, 1813, "Inclosed to my brother 40. D. of which he asks a loan."[49]

In February and April of 1815, Squire remained Randolph's choice to make deliveries and carry letters to Monticello, particularly to retrieve a female puppy Thomas had promised him. In April, Randolph reported to his brother that the James River was very high due to spring rains. As a result, they were not able to catch any fish; however, Randolph added, "as soon as the river gits down so as they can ketch any we will immediately send over a parcel by Squire."[50] Despite the elevated waters and undoubtedly very muddy roads, Squire made more than one trip back and forth to Monticello that spring. Assuming this is the same Squire who was a boy in 1757, he was now at least in his sixties.

One of Squire's last duties for Randolph Jefferson was to carry a letter to Monticello on August 1, 1815. Written by Randolph's wife, Mitchie, it informed Thomas Jefferson that Randolph was seriously ill. When Randolph died on August 7th, Squire's fate fell into the hands of the Buckingham Court as the Jefferson family squabbled over Randolph's property. What next happened to Squire and his family, who had spent their entire life with "Mass Randall" and the Jeffersons, remains unknown.

Orange and Dinah: A Marriage Across Plantations

When Peter Jefferson died in 1757, some of his slaves had long served the Jefferson family. Myrtilla's Peter, for example, who was bequeathed to Randolph Jefferson in his father's will, was born at Shadwell about the same time as Jane

Jefferson's twins. In other cases, slaves were relatively new to the Jeffersons. Such was the case with Orange's family, who were purchased from Edward Spencer of Orange County.[51] In 1755, Toby, Juno, and their children, Nanney and Toby, Jr., traveled south to their new home at Shadwell. There, not long after their arrival, Orange was born, making him roughly the same age as the Jefferson twins.[52] Orange may have been conceived in Orange County before Toby and Juno were sold, and his "place" name remembers the family's former home. It has also been suggested that his name described his skin color. Whatever the origin of Orange's name, it would become a familiar one at Snowden.[53]

The family's previous owner, Edward Spencer, Gent., a pioneer peer of Peter Jefferson, settled in Orange County before 1736.[54] Spencer was a commissioned Captain in the Orange County Militia, later promoted to Major. By the 1740s, Spencer was associated with the area around Wilderness Run where he supervised the building of a bridge; he also owned a mill, an ordinary, and a ferry. Precisely how Jefferson and Spencer became acquainted is not known, but Spencer's affiliation with the ordinary and ferry at Orange County courthouse probably holds the key.[55]

Orange County was formed in 1734 from Spotsylvania County, making it Albemarle's neighbor to the north. Its courthouse was located on the Rapidan River, at Germanna, central to the sprawling county. Near the newly laid out county seat, Edward Spencer owned 104 acres which included an ordinary; he also held the licence for a ferry near Orange County's courthouse.[56] In 1744, he went to court and agreed to "keep a Sufficient vessell to transport foot people cross the Rappidan at Court time coming & going for which its ordered that he be paid 500 pounds of tobacco."[57]

In 1747, Spencer's long-term lessee, John Branham, was "allowed for finding Beer taking Care of The Justices Horses & keeping the ferry the same that was allowed Edward Spencer Gent & to begin from the Last Novbr Court."[58] As Magistrate and surveyor for Albemarle, Peter Jefferson likely had business at the Orange County courthouse where he, along with other Justices, utilized Edward Spencer's ferry, stables, and ordinary.

Spencer's involvement in the growth of Orange County neatly parallels Peter Jefferson's role as a "developer" in Albemarle. The men had remarkably similar profiles; each owned a mill, as well as a ferry and an ordinary near their county seat. Friendship based on affinity is easy to imagine. Orange County's courthouse, a day's ride from Shadwell, is a long way for Peter to go to purchase a family of unfamiliar slaves. However, as an acquaintance of Edward Spencer's, Peter may have come to know Toby, Juno, and their family for several years prior to Spencer's death in 1753.

When Edward Spencer died intestate, his wife, Elizabeth, was appointed administrix of his estate and an inventory of his property was recorded in Orange County, on June 28, 1753. Among his twenty-nine slaves were Toby (£30), Juno and Toby (£50), and Nan (£15). Additionally, there was a mature man named Orange valued at £35. Any relationship between Juno, Toby, and the adult Orange is currently unknown, but they certainly may have named their next child in honor of him.[59] While Peter Jefferson's Orange never knew life in Orange County nor the man named Orange who was left behind, his name carried his family's experience and history forward to Shadwell and then to Snowden.

∼

For nearly ten years, Orange and his family lived and worked at Shadwell. Their immediate community consisted of ten individuals: a woman named Phyllis; her son, Goliath; a single woman named Lucey; a couple, named Gill and Fanny; and Toby and Juno, with their children, Nanney, Toby, Jr., and Orange.[60]

In 1764, the family was divided between the Jefferson brothers. Juno and Toby, Jr. went to Monticello, while siblings Nanney and Orange were assigned to Snowden – all part of the group of slaves who were "not herein otherwise disposed of" in Peter Jefferson's will. Because Randolph Jefferson was still years away from taking charge of his property, these family members may have remained together in the interim, continuing to work the Rivanna lands, waiting for Randolph to grow up and claim his patrimony in 1776.

By 1774, Orange was a grown man, running long-distance errands for Thomas Jefferson. On October 26, 1774, Thomas gave Orange 1 shilling, 3 pence to pay "ferriages to Bedford," the location of one of John Wayles' previous properties. The Thomas Jeffersons had just inherited the land in Bedford County, and at the time, there was a drama in process on the farm between the overseer, Mr. Stanly, and the plantation's slaves, who had charged Stanly with stealing.

There were discrepancies among amounts of butter, corn, and hogs. The slaves insisted it was Stanly who had done the pilfering.[61]

Randolph took possession of Snowden in 1776 and brought his bride there in 1781. Two slave lists exist for the farm for 1782 and 1783; Orange appears on both.[62] In his late twenties, Orange had grown to be a trusted and mobile servant, running errands for Randolph Jefferson as he had for Thomas. In February of 1790, for example, he carried apricot and plum stones from Monticello to Snowden. He was strongly motivated to visit Monticello, as his wife, Dinah, lived there as the property of Thomas Jefferson.[63]

In 1773, Thomas Jefferson became one of several men responsible for the estate of his father-in-law, John Wayles. Among the slaves who became Jefferson's property was a thirteen-year-old girl on the verge of womanhood named Dinah. She had lived at the Wayles' property, The Forest, in Charles City. In 1774, she was listed in Thomas Jefferson's "Farm Book" as "Dinah 1761." She stayed at Monticello and in the Farm Book was designated with a "+," indicating that she was a titheable person who followed an occupation other than field work. Trained as a house servant, once at Monticello, Dinah continued her work as a domestic.[64]

When Dinah arrived at Monticello, Orange was probably still living at Shadwell where he may have stayed until 1776, thus giving the young couple time to get acquainted. Orange also spent time at Monticello, as revealed by Thomas' memo. Then, when Randolph turned twenty-one, Orange permanently settled at Snowden, but he did not forget Dinah. His mobility, especially between Snowden and Monticello, made a marriage possible, if not entirely practical.

Dinah was a young wife and mother when little Orange was born in December of 1777. She probably was separated from her husband when she gave birth, but not isolated, surrounded by his extended Shadwell family.[65] In May of 1780, Dinah gave birth to their daughter, Sally. A few years later, Dinah's life changed dramatically. While Thomas Jefferson lived abroad and did not occupy his house on the mountain, from 1784-1792, Dinah continued to live at Monticello, but as fewer house servants were necessary in Thomas' absence, she was sent to the fields to work.

Then, having returned to the United States, Thomas wrote to his brother on September 25, 1792, that he was contemplating the sale of Dinah and two

Thomas Jefferson to Randolph Jefferson, dated 25 September 1792, concerns the sale of Dinah and her children. (Courtesy of Library of Congress) of her children. The funds from their sale would be used to help satisfy some of John Wayles' remaining debts. Thomas' efforts to keep Orange's family near to him are laudable.[66] The letter reads in part:

> Finding it necessary to sell a few more slaves to accomplish the debt of Mr. Wayles to Farrell & Jones, I have a thought of disposing of Dinah and her family. As her husband lives with you I should chuse to sell her in your neighborhood so as to unite her with him. If you can find any body therefore within a convenient distance of you who would be a good master, and who wishes to make such a purchase, I will let her and her children go on a valuation by honest men either there or here. One half the money to be paid within a year, the other within two years, and if not paid at the day interest to run from the date of the bond. Good security would be required. Dinah is 31 years old and two children are to go with her, to wit Sally 12. years old and Lucy whose age I do not know. Dinah is a fine house wench of the best disposition in the world and tho' she has worked out ever since I went to Europe, she would still suit any person for house business....[67]

Over the years, despite the distance between Snowden and Monticello, Orange certainly had many opportunities to visit his wife, as they had at least three children together. Dinah may have visited Snowden as well, though enslaved women were generally less mobile than men. Ultimately, Dinah and the girls would not live in Snowden's immediate neighborhood, though they remained in Albemarle County.

On January 15, 1793, Thomas Jefferson sold Dinah and her daughters to a planter named James Kinsolving, who lived about fifteen miles west of Monticello.[68] Jefferson notes in his "Farm Book" that Dinah, Sally, and Lizzy were sold for £139-15-0. "Lizzy" is probably the "Lucy" of Thomas Jefferson's letter; however, it is unclear which name is actually correct. Young Orange was not sold from Monticello with his mother and sisters. He may have died prior to 1793, hence his disappearance from the plantation's records.[69] Obviously, the sale deeply affected the lives of Orange and his family.

Kinsolving was an established planter in the area west of Charlottesville, near Ivy. In about 1783, he married Elizabeth "Betsy" Leigh. In 1788, he began purchasing land on Mechums River, at Mechums Depot, just east of the foothills of the Blue Ridge Mountains. Circa 1790, he built his mansion house at what he called Temple Hill. It is probable that Dinah became the housekeeper. Given Dinah's talents and disposition, it seems likely that she was Betsy's domestic help and may have assisted in raising the twelve Kinsolving children, including a Kinsolving son named Jefferson. Most of the Kinsolving children were born after Dinah and her daughters came to live at Temple Hill.[70]

A successful planter, Kinsolving owned approximately 1,400 acres at his death in 1829. In addition to planting, Kinsolving operated a blacksmith shop manned by his slave, Bartlott.[71] In 1807, Kinsolving housed a "famous high blooded English horse," making the stallion available for breeding in Albemarle.[72]

Despite these various enterprises, Kinsolving was slow to repay Thomas Jefferson for the purchase of Dinah, Sally, and Lizzy. Thomas Jefferson eventually brought suit against Kinsolving for non-payment. Between 1795 and 1800, Kinsolving made ten additional small payments; the debt was finally met on September 1, 1800, when Jefferson noted: "Recd. of J. Kinsolving 11/ in full of our old acct."[73]

During this period, Dinah naturally disappeared from the Jefferson records. Orange appears for a last time in Thomas Jefferson's memos on August 14, 1802, when Thomas noted, "Sent by Orange to Jesse Moore 9/ for ferrge. of plank from Snowden."[74] Orange had long served Randolph Jefferson and his family. At that time, he was approaching fifty. His fate after 1802 remains unknown. The continuation of Dinah's story unfolds in the wake of James Kinsolving's death, who died intestate on March 21, 1829.[75]

An inventory of Kinsolving's personal property was recorded on June 1, 1829. Among his slaves was "Dianna, an old woman." The two women immediately following the entry for "Dianna" may be her daughters, Eliza (Lizzy) and Sarah (Sally). Their children are listed with them. Eliza, Melinda, and Preston were valued at $500, while Sarah, Rachel, Nelson, and Pheby were valued at $525. "Dianna," who was about sixty-eight years old, was indeed an old woman by slave standards, and was given a value of $000.[76]

Six months later, on December 2nd, four men convened at Kinsolving's mansion house "for the purpose of dividing and allotting the negro property." The slaves, the commissionaires concluded, were worth a total of $10,485. That day, some were again listed and valued, including "Dinah a very old woman either the mother or the grandmother of the whole flock $ —."[77]

Did the commissioners, Messrs. John Jones, [--] Gillum, Francis McGehee, and Thomas L. Skelton, exaggerate? Or did Dinah take a new husband and have children beyond Sally and Lizzie? Purchased by Kinsolving when she was thirty-two years old, Dinah still had approximately ten childbearing years ahead of her in which she conceivably could have had several more children. Since her "increase" belonged to James Kinsolving, it seems reasonable he would encourage her to find a mate close to or at her new home.

While it is unlikely that Dinah was the mother, grandmother or even great-grandmother of *all* of Kinsolving's slaves, which numbered thirty-eight on the June inventory, she may have been related to the majority and specifically to those mentioned in the December 2nd report, which read:

> We have ascertained the whole amount of negro property belonging to said estate which amounts to $10485. From which we allotted a larger portion of the older negroes by consent for the widow viz Bob an old man at the price of $100. Ned, do. somewhat advanced $150. Ben a younger man $345, Hannah a woman advanced $150, Mary a woman $200. Maria a young woman $275. Phill a lad $350, Harry, a boy $275. Sarah a woman & child $550, Lucinda a girl $175. Dinah a very old woman either the mother or Gr mother of the whole flock $ —.[78]

A record of the sale of James Kinsolving's personal property was submitted to the Albemarle Court on September 5, 1831. For the most part, the slaves stayed in the family. Two women named Elizabeth were sold, making it unclear where Dinah's daughter "Lizzy/Eliza" and one child ultimately lived. Dinah's probable granddaughter, Melinda, was sold separately to Jefferson Kinsolving for $151.00. Her brother, Preston, is not mentioned by name and may be attached to one of the Elizabeths. One Elizabeth and her child were sold to

Patsy Wood for $363.00.[79] The other Elizabeth and her child went to G.W. Kinsolving for $100.[80]

Sarah's daughter, Rachel, was sold to William Ballard for $100.00, while her brother Nelson, described as a child, went to James Kinsolving for $103.00. Pheby apparently stayed with her mother and became the property of Mrs. Kinsolving.[81]

One other name stands out among James Kinsolving's human property. On the June inventory, a man named "Orrange" is valued at $400. A coincidence, or was he the grandson of Orange and Dinah? James Kinsolving, Jr. purchased the very valuable Orrange, paying his full value of $400.00.[82]

In 1829, Dinah and at least some of her family members were allotted to Betsy, the widow of James Kinsolving, whom Dinah had served for decades. By 1830, the Kinsolving slave population was distributed among the family. Betsy Kinsolving owned thirteen slaves, while her sons owned others: George W. owned ten slaves, James owned ten slaves, and Jefferson B. Kinsolving owned four slaves, representing many of the remaining individuals. It appears that Dinah did not survive until 1830, as there was no elderly black woman living with Elizabeth Kinsolving that year. Her sons James, Jefferson B., and George W. Kinsolving were enumerated as separate heads of households, though George was likely living with his mother.[83] Betsy lived until late 1837, and George W. Kinsolving became squire of Temple Hill.[84]

At her death, Besty Kinsolving's property included the following slaves: Bob, Ben, Ned, Phil, Lad, Harry, Hannah, with her daughter Mary, Lucinda, boys named Edmund and Spencer, a boy named Richmond, Phebe, her daughter Frances, and Maria. The mothers of Edmund and Spencer are not identified. They may be, however, descendants of Orange and Dinah, their names referring back to Edward Spencer who long ago owned Orange and his family.

From John Wayles' The Forest, to Thomas Jefferson's Monticello, to James Kinsolving's Temple Hill, Dinah knew at least three homes over her long lifetime. In all those years, she never lived with her husband, Orange, and may never have visited his home at Snowden, despite the fact that she had "the best disposition in the world" and refined skills, making her a "fine house wench." How different the lives of Dinah and Orange would have been if Randolph Jefferson had purchased her and their daughters in 1793, uniting them all at Snowden.

In the late 1950s, Albemarle-based architect, Floyd E. Johnson, expanded Temple Hill.
(COURTESY OF K. EDWARD LAY, ALSO PUBLISHED IN THE ARCHITECTURE OF JEFFERSON COUNTRY: CHARLOTTESVILLE AND ALBEMARLE COUNTY, VIRGINIA, 2000)

~

In April of 2010, Temple Hill, the surviving and elaborated Kinsolving mansion house, was open to the public for Virginia's annual garden week. The current owners, Kirby and Laura Farrell, described their home and gardens:

> TEMPLE HILL. From Nichola Farm, take a left onto Gillums Ridge Rd. (Rte 787) (gravel road), go 0.2 mi. Take a left on Broad Axe Rd. (Rte 682), go 0.6 mi. James Kinsolving began purchasing land along the Mechums River in 1788. At his death in 1829, he had more than 1,400 acres. Built around 1790, his house is a fine example of a well-preserved early Virginia home. The one-story frame house sits on a knoll with a terraced lawn and enjoys lovely views of the Blue Ridge Mountains. Charming parterre gardens filled with perennials flank the front entrance.
>
> Originally, Temple Hill had an English basement with an exterior bulkhead for entrance to the cellar. Outbuildings included a smokehouse and an icehouse. The basement now has a polished brick floor and houses various rooms at times used as kitchen, dining room, den and wine-cellar. A porch was added in the 1870s and an attractive wing with a master bedroom suite and large country kitchen

added in the 1950s. Improvements also include masonry walls, terrace, walks, driveway, garage, pool and a small barn. Original brickwork was Flemish bond. The gabled roof, once shingle, is now metal. Exterior finishing is wide-beaded weatherboarding.

A cemetery behind the house contains five existing stones, dating to 1854. Four are finished marble and one is a flat fieldstone. Large boxwoods surround the markers.[85]

The Kinsolving cemetery is located about a hundred yards behind the dwelling house on the northwest slope of a high hill, overlooking the Mechums River Valley. In the distance to the west are the foothills of the Blue Ridge, a view no doubt often enjoyed by the Kinsolvings and their slaves alike.

The graves include that of George W. Kinsolving, who commanded the cavalry and was officer of the day when the Marquis de Lafayette visited at Monticello in the autumn of 1824. That year, Dinah was still alive to hear tales of the exciting reunion of Thomas Jefferson and his old comrade. The stones in the family graveyard include:

Annie B., Wife of Geo. W. Kinsolving
Died Aug. 2 1854,
Aged 66 yrs. 7 mo. 14 days

Geo. W. Kinsolving
died April 16, 1857,
Aged 72 yrs. 6 mo. 22 ds.

Vusulia B., daughter of F.W. & A.B. Kinsolving
Aged 43 yrs. 1 day

Mrs. Verturia A. Clark
departed this life Sept. 13th 1872,
Aged 47 yrs. 4 mo. 28 days

J. Kinsolving
21 May 1899
Aged 69[86]

The grave sites of Dinah, her daughters, and their children are unknown.

Chapter Eleven

The Patient Randolph Jefferson

> *Patience will bring all to rights.*
> Thomas Jefferson to Horatio Gates, 1797

Randolph Jefferson's life was a fairly simple and straightforward one, especially when compared to the whirlwind of mental and physical activity that swirled around Thomas Jefferson and Monticello. As a planter, Randolph's life was tied to the seasons, the weather, his family, and his slaves. His concerns were mundane, regular, and likely lacked excitement. His frugality meant little change at Snowden. A healthy new colt or a particularly good tobacco crop constituted his small pleasures. The frustrations of drought, the threat of pestilence, and accumulation of debt were his constant worries. Such is the life of all farmers.

The Jefferson-to-Jefferson correspondence offers a rare window into two episodes in Randolph's life during which his emotions are revealed. Like all men, he suffered the pangs of separation and loss, enjoyed the hopeful anticipation of change, expressed sincere gratitude for gifts, and, most of all, demonstrated the patience of Job when the excruciatingly slow workings of an agrarian, pre-industrial age kept him from the thing he desired.

The Jefferson Pups

Initially, the Jefferson-to-Jefferson discussion concerning "the bitch" and "the pups" seems an odd one; especially, Randolph Jefferson's *dogged* eagerness to acquire a pup from his brother. However, in the context of Thomas Jefferson's decades-long interest in the importation and breeding of the French sheepdog, Randolph's enthusiasm takes on a new relevance. These were not pets, they were highly prized working dogs and, when Thomas returned from France, Randolph demonstrated an immediate interest in acquiring one of the imported animals.

In November of 1789, Thomas brought home to Monticello sixty European trees which included figs, cork oaks, and larch trees to introduce to North America as well as three French sheepdogs: a bitch, Bergère, and her two nascent pups, born trans-Atlantic.[1]

Initially, Thomas was interested in the sheepdog, not to tend sheep at Monticello, though they ultimately would, but to introduce the breed to North America just as he was introducing new, and hopefully valuable, botanical species to central Virginia. Trees were one thing; dogs, however, were generally despised in Virginia, with the exception of hunting dogs, which were trained and controlled by the ruling class. Lord Fairfax had introduced the foxhound to Virginia in 1738, a breed particularly admired by George Washington, who owned these dogs, giving them overly amorous names such as Sweet Lips, Venus, and True Love.[2]

Keeping sheep in rural Virginia was a risky investment. They fell prey to both wolves and wild dogs. Legislation attempting to protect sheep dated back to 1755, giving Justices of the Peace the power to eliminate sheep-killing dogs stating, "Be it further enacted, by the authority aforesaid, That it shall, and may be lawful, for any justice of the peace, upon due proof made to him, of any dogs killing sheep, to order such dog to be destroyed forthwith."[3]

Slaves frequently were blamed for the increase in feral dogs, and the May 1755 statutes put legal limits on the number of dogs that Negroes could keep.

> [N]o more than two dogs, shall be kept at any negroe quarter, in the counties aforesaid, at any one time; and in case more dogs shall be found to belong to the same quarter, it shall be lawful for any person or persons to kill and destroy every dog kept at such quarter above that number, by order of any justice of the peace; and the several constables at the time of their viewing tobacco fields . . . are hereby required and impowered to examine into the number of dogs kept at the several

negroe quarters in their precincts, and to kill and destroy every dog kept thereat, exceeding two. And whereas dogs frequently ramble from home and destroy great numbers of sheep, and some persons are so unneighbourly as to refuse their being killed. . . .[4]

Slaves were expected to keep *their* dogs under control and at home plantations. Dogs owned by the Virginia gentry were another matter. Trained hunting dogs, which included spaniels, pointing or setting dogs, could be transported legally in any number either by the owner or the slaves of that plantation. It was the rambling dog that threatened sheep and created strain among neighbors.[5]

The problem of sheep-killing dogs did not abate. In 1769, King and Queen County petitioned the Virginia House of Burgesses announcing "the great injury and loss that we sustain in our flocks of sheep, by the dogs which are suffered to run at large." At least in some parts of Virginia, dogs had superseded the native wolf as a threat to livestock. Twenty years later, a roaming dog remained a controversial addition to a Virginia plantation and even a reliable sheepdog could be suspect.[6]

Monticello slave, Isaac Granger Jefferson, remembered the constant threat of wild animals to both man and sheep:

> No wildcats at Monticello; some lower down at Buck Island. Bears sometimes came to the plantation at Monticello. Wolves so plenty that they had to build pens round black peoples' quarters and pen sheep in 'em to keep the wolves from catching them. But they killed five or six of a night in the winter season; come and steal in the pens at night. When the snow was on the groun', you could see the wolves in gangs runnin' and howlin', same as a drove of hogs; made the deer run up to the feedin' place at night.[7]

In late 1789, Thomas Jefferson was bucking a long-standing trend when he attempted to introduce the idea of mixing dogs with sheep. During November and December, Thomas enthusiastically informed his friends about the welfare of Bergère and her puppies. On November 21st, he wrote to William Short, "My

The precise appearance of Thomas Jefferson's French sheepdog is unknown. It may or may not have resembled the Barbet which is still bred today. This etching of the "Shepard's Dog" represents the English breed at the turn of the 19th century. (Published by W. Darton and J. Harvey, 1808)

plants and shepherd dogs are well."[8] Infatuated, he called the *cheinne berger* and her pups *à merveille* – marvelous![9] By mid-December, Thomas received news that another bitch and four puppies had been procured in Normandy.[10] Thomas' breeding program was off and running.

～

On December 23, 1789, the Jeffersons at last arrived at Monticello after a four-year absence. Thomas and his daughters, Patsy and Polly, were accompanied by the new "family members," Bergère and her pups. A "joyous" reception was given "Mass Thomas" by his slaves. Gradually, during these first weeks home, Thomas' immediate family paid their "welcome home" calls. On December 30th, Thomas repaid his sister Lucy Lewis six shillings. On January 11th, Thomas gave his friend and brother-in-law, Dabney Carr, twelve shillings. Most importantly, on February 23rd, Martha "Patsy" Jefferson wed her cousin, Thomas Mann Randolph, Jr.[11]

Among the many guests, Randolph Jefferson likely visited Monticello at least once since the first of the year, seeing Bergère and her pups. Born in November, the puppies were now weaned and could safely be removed from their mother. Perhaps influenced by his older brother's enthusiasm or out of genuine intrigue about the possibilities of owning a sheepdog, Randolph asked for a puppy. His brother apparently rewarded his request. On February 28th, Thomas wrote to Randolph at Snowden saying, "I will give the orders as you desire to George, relative to peach stones and the puppies."[12] On March 1st, Thomas left Monticello and there is no memo or surviving letter concerning either the peach stones or the puppies. No ferriage money was left for a Monticello slave to go to Snowden, so there is every reason to believe that Randolph or one of his servants journeyed to Monticello to pick up one of Bergère's pups.

Thomas's slave, Isaac, remembered that Bergère's puppies were called "Armandy" and "Claremont," which may refer to "Clermont," one of the oldest cities in France. Isaac recalled that they were both black, with stump tails. A dog named Ceres was also suggested as one of Bergère's pups and may have been named after the ship which first took Thomas Jefferson to Europe in 1784.[13] Which one lived at Snowden is unknown.

Randolph's enthusiastic embrace of the sheepdog is a rare peek into his goals for Snowden. Was he merely influenced by his brother's hardy endorsement of the dog as a positive thing for Virginia planters, or did Randolph himself see the usefulness of the breed if Virginians were to prove successful with sheep?

Sheep probably already grazed at Snowden, and, like many central Virginians, Randolph struggled to save his flock from decimation by wolves and stray dogs. Now he faced the exciting prospect of owning a dog which was bred to guard and manage his flock. He doubtless hoped that a highly instinctive sheepdog would, at least in part, free his shepherd for other work and expand the area in which the sheep might range. As Thomas wrote to William Thornton, physician, inventor, painter, and architect of the United States Capitol, "persons now to follow my sheep, and with the aid of the bitch I received from France, perfectly trained, they have the benefit of fine pastures in which they could not run but for the facility she gives of keeping them from the grain in the same fields."[14]

By giving puppies to various family members, Thomas Jefferson extended the breed into the larger community. Recipients included his daughter and son-in-law, Martha (Jefferson) and Thomas Mann Randolph. In 1816, he wrote to John Wayles Eppes, "I send you by Francis a female puppy of the Shepherd's dog breed. The next year I can give you a male. The most careful intelligent dogs in the world. Excellent for the house or plantation."[15]

Details of Randolph's success or failure with his dog are lost to history; however, his interest in the breed continued for years, as did his brother's enthusiasm for spreading the sheepdog throughout Virginia and beyond. During 1809-1810, Thomas was still acquiring new animals, including two sent to him from the Marquis de Lafayette. The female was ready to go to work in Monticello fields, where rows of peach trees provided the only fencing.[16]

Unsurprisingly, numerous requests came to Jefferson for puppies, especially those descended from Lafayette's pair of dogs. Eventually, inquiries came from afar asking about specimens of the breed. Judge Harry Innes of Kentucky requested a pair. Responding to Innes, Thomas extolled the breed: "Their extraordinary sagacity renders them extremely valuable, capable of being taught almost any duty that may be required of them, and the most anxious in their performance of that duty, the most watchful and faithful of all servants."[17] Jefferson also delineated his own experience:

> I have 4. pair myself at different places, where I suffer no other dog to be; and there are others in the neighborhood. I have no doubt therefore that from some of those we can furnish a pair, or perhaps two...they learn readily to go for the cows of an evening, or for the sheep, to drive up the chickens, ducks, turkies every one into their own house, to keep forbidden animals from the yard, all of themselves and at the proper hour, and are the most watchful house-dogs in the world.[18]

A decade after Randolph acquired his first puppy, he requested another, indicating sheep had become a regular part of the landscape at Snowden, and he was eager for another working dog.

In the fall of 1811, Martha (Jefferson) Carr died, and Randolph, who was recovering from an excruciating attack of kidney stones, was detained in Buckingham, unable to go to Albemarle and be with his family at that time. Apparently, not long before, one of Thomas' sheepdogs had a litter of puppies and he was holding a young male for Randolph. On October 6, 1811, Randolph wrote to Thomas, addressing both their sister's death and the anticipated dog. Randolph intended to visit Monticello the following month and asked Thomas to "take care of my puppy if you have not given him a way to any one I expect by this time he must be large."[19] Randolph's long delay in getting to Albemarle indeed may have resulted in Thomas giving the fast-growing dog to someone else.

Based on Randolph's subsequent letter, written four months later on February 8, 1812, it appears that Thomas did not hold the puppy. In it, Randolph asks that a pup be saved from a forthcoming litter. "If your shepards bitch has any more puppys I must git the favour of you to save me one dog puppy."[20] Weeks passed, and apparently no dog arrived from Monticello.

Having eagerly waited months for a new dog, Randolph repeated his request on April 13, 1812:

> ... we shall be over early in May [at] which time the roads will be in good order to travil and as soon as they are I shall set of[f] over I have one request to ask of you and that is if your bitch has any more puppys by her at this time I would thank you to save me a dog if you have not ingaged them to any other person ... [21]

It remains unknown if Randolph received his much-anticipated pup in 1812; however, the following year, Randolph wrote to Thomas asking to borrow a Monticello ram before the weather got too cold. Clearly, Randolph's flock was flourishing and ready for an infusion of Thomas' stock.[22]

Not long before Randolph's death, the Jefferson brothers had a last exchange concerning the valuable sheepdogs. On December 29, 1814, Randolph wrote to his brother referring to the bitch "you were so good as to give me when I was over." Clearly, Randolph had recently visited Monticello, and Thomas had made him a present of one of the sheepdogs; however, Randolph did not immediately bring her back to Snowden and he soon wrote, "I have a great desire to see her." Unable to leave Snowden at that time, he added that he had "waited with all the patience I am master of" for the dog to be delivered.[23]

By February 13, 1815, the dog had still not appeared, so Randolph sent Squire to collect her. In Randolph's letter to Thomas, it is revealed that she was expecting puppies.

I have concluded to send over Squire, after the bitch that you was so good to give me, when I was over as I should be extreemly hapy to git her, if she has not pupt, or if she has and he can make out to bring her and some of the pupies. I can send over for the rest at Esther, without Mr. Randolph will let old Stephen come over and bring the rest for me. If the bitch has no more [than two] Squire can bring them him self, I have waited expecting to see Stephen every day, but the reason I suppose [he] is not coming is that Mr. Randolph has not reternd yet from below.[24]

In an afterthought, Randolph added, "Do pray Sir give Squire such derections in respect to the bitch as you think most necessary and you will very much oblige."[25]

On February 16th, Thomas wrote a prompt but disheartening response, which likely saddened Randolph, who had waited so long and so patiently for an adult dog, complete with a litter of puppies.

. . . with respect to Stephen mr Randolph got rid of him long ago. I am told he stays now at North Milton or somewhere there. he talks of going down the country to live. . . . the bitch I had given you was caught in the very act of eating a sheep which she had killed. she was immediately hung. . . .[26]

As Thomas had warned others, even sheepdogs will kill if they are hungry and neglected. "Their sagacity," he noted, "renders them the most destructive marauders imaginable. You will see your flock of sheep and of hogs disappearing from day to day, without ever being able to detect them in it."[27]

All was not lost, however. Thomas reassured his brother, "as we had a fine litter at the same time from another bitch, I preserved one of them for you, which Squire is now gone for and will carry over to you."[28] At long last, there would be another *chien de berger* puppy at Snowden and Randolph's patience was rewarded with a new pup to enjoy and occupy him in his last days.

This engraving of the "South east corner of Third, and Market Streets" was published by W. Birch & Son in 1799 and depicts Philadelphia as it appeared when Thomas Jefferson purchased a silver watch for his brother, Randolph. (Courtesy of American Philosophical Society)

Randolph Jefferson's One Indulgence

In 1815, Randolph Jefferson was taxed on a silver watch. It was undoubtedly the watch Thomas acquired for him from Philadelphia. In October of 1806, Thomas paid Henry Voigt $66.00 for a Swiss-made watch. Ultimately, the timepiece cost nearly $80.00, including chain, seal, and key. In addition to being a master watchmaker and engraver, Voigt was acting as the Chief Coiner in Philadelphia's U.S. Mint.[30] An extravagant expenditure for Randolph, this price tag was equivalent to nearly two years' worth of taxes on Snowden.[29]

The correspondence between the Jefferson brothers reveals how intimate and delicate a timepiece was to men of this era; they referred to the watch as "she."[31] It required winding, polishing, and other mechanical attentions. Randolph had been a widower for at least five years when he made the decision to order the watch; it promised to be a new bosom companion.

The saga of the acquisition of Randolph's treasure begins in 1806, when he sent $70.00 to his brother, asking him to be his agent to purchase a watch. The funds were sent via Martha's husband, Thomas Mann Randolph, and received in Washington, D.C. by Jefferson on July 7, 1806. Presumably, the order for the watch was placed before he left the capitol on July 21st and arrived home at Monticello on the 24th.[32]

In November, Thomas settled a bill with watchmaker Voigt, which included Randolph's order. Ultimately, the delivery was long delayed and, on November 12, 1806, Thomas wrote from Washington to his daughter Martha (Jefferson) Randolph at her home, Edgehill.[33]

> . . . I expected ere this to have received a watch for my brother from Philadelphia and to have sent it by Davy. But it is not yet come. I mention this, as it is possible my brother may send or come to Edgehill in expectation of finding it there. I am afraid it may be long before another opportunity occurs of sending it from here by a safe hand.[34]

In February of 1807, Thomas made a memo that the "seal, chain & key" cost an additional $1.50.[35] A letter to Randolph details the situation:

> Washington Feb. 9. 07.
> Dear Brother
> It was not until Mr. Randolph went home in December that I received your watch from Philadelphia. The keeping her here has given me an opportunity of proving her and of being able to assure you that a better watch I believe was never made. I set her by a most accurate clock on New year's day, and she has varied but a minute and a half in that time, which is at the rate nearly of a minute a month. She cost something more than you limited. I paid 79 ½ Dollars for the watch, the chain, seal and key. You will have to make up the difference by some white clover seed. As I shall be at Monticello by the middle of March, I conclude to carry her myself so that you may receive her there. This is the more necessary as it will be requisite to give you some information as to her management. In hopes I shall have the pleasure of seeing you then at Monticello. I conclude with my affectionate salutations and assurances of constant attachment.
> Th:Jefferson[36]

In April, Thomas traveled home to Monticello where he took care of typical household business, including paying Ben $.50 for cleaning the sewers. Randolph made his way to Albemarle, where he picked up a very fine and accurate timepiece, including instructions from his brother on how to care for such a delicate item. On April 22, 1807, Thomas noted that Randolph had paid him the "balance above cost of his watch &c. 1.50."[37]

It had taken over ten months to settle the business, but once Randolph had his watch, he quickly became attached and dependent. Understandably, the watch was highly prized by him and presumably ran well for a couple of years. Then "she" began to reveal herself as the rare and delicate instrument she was. Repair and/or maintenance became an annual event.

In late 1811 or early 1812, Randolph sent her, via his brother, to Richmond for repairs. Separation from the watch caused Randolph concern, even anxiety. On February 8, 1812, he was distressed when news reached him that the watchmaker's shop in Richmond had gone up in flames, and he feared the precious watch had been lost in the fire. He wrote this post script to Thomas: "If you sent my watch to Fast Bender it is more than probable that she went to the flames with the rest of the watches in his shop as his shop were burnt about the eighteenth of Jany. – – –"[38]

Then, by April 13th, news reached Randolph that the watch was safe. He wrote again to Thomas:

> I have bin informed by Mr. R: Patteson who has just got up from Richmond a day or two past that my watch is safe and in the possession of Mr. Fass Bender will you be so good as to send down for her by some person who will be going down shortly that can be depended on to bring her up safe as I expect we shall be over early in May which time the roads will be in good order to travil. . . .[39]

Randolph was able to take advantage of the somewhat regular traffic between Richmond and Monticello. However, it did put him in the position of being dependent on Thomas' schedule and generosity. Progress can be slow waiting for someone to do you a favor. Randolph had given the watch to Thomas prior to January 18, 1812, and did not expect to be reunited with it until early May.

On February 3, 1814, Thomas made a memo that he "Exchanged a watch with my brother & gave him the 40.D ante Aug. 8 in boot." It is not clear which way the exchange was going – to or from Thomas.[40] Did Thomas loan Randolph a watch while his was once again went off to Richmond for repairs, or had the timepiece just returned from one of its "annual" sojourns? Whichever was the case, the watch certainly traveled more than Randolph did!

In late 1814, the delicate piece was off to Richmond again, and Randolph wrote the following to his brother: "I would be greatly oblige to you if Mr. Randolph had reternd home from Richmond if you will be so good as to ask him to send old Stephen over with my watch as I am at the greatest loss in the world for want of her...."[41]

On February 13, 1815, Randolph was still anticipating the return of his beloved watch. He wrote to Thomas, frustrated by the long wait, "As for my watch I have bin without her so long that I am intirely weand from her, however if Mr. Randolph should of reternd and brought her, I should be extreemly happy once more to receive her agane."[42] Then Randolph had second thoughts; realizing he was not truly weaned from his watch, he added later in the letter:

> I am at the greatest loss immanegnable for the want of my watch. If Mr. Randolph has not reternd yet I shall be oblige to send down to Verina but I am still in hopes there will not be any occasion to do that, as he must certainly have reternd long before this time.[43]

Thomas wrote back immediately, on February 16, 1815, putting Randolph's mind at ease concerning the watch. He began:

> After several disappointments in getting your watch from Richmond, I received her a week ago. I sent for Stephen, who came to me and pretended to be sick. Finding he didn't mean to go to Snowden I had concluded to send her to you in a day or two, when Squire arrived. She appears to have gone well since I have had her, except a little too fast.[44]

Thomas went on to explain that "Old Stephen" no longer worked for Thomas Mann Randolph, though he was still in the vicinity. Stephen's age, poor health, sloth or some combination of these, had added to the delay in the delivery of Randolph's watch.[45] However, now that reliable Squire had delivered Randolph's letter, Thomas could send his response and, finally, safely transport the watch to his brother.

As with the "dog puppy," Randolph demonstrated much patience with long delays, and his affection for the watch is earnestly expressed. This was not a frivolous concern; the watch was an outstanding luxury in Randolph's life. It also represented a significant investment. Beyond his dependence on his brother to do him a favor, in this instance Randolph was also dependent on his niece's husband and that man's ex-employee to help him get his watch back and forth to Richmond. There were plenty of anxiety-producing factors to make him feel out of touch with the process of the repairs. Additionally, Randolph may have felt the pressures of ill-health and wanted his valued things around him.

The story of the timepiece does not conclude in the Jefferson-to-Jefferson letters, but Squire undoubtedly returned to Snowden with the precious item. The presence of the silver watch on the 1815 personal property tax indicates it was likely in Randolph's hands when he died that August. One of his sons may have ultimately received the cherished and fragile watch, though no trace of that legacy has been discovered.

Chapter Twelve

The Second Mrs. Randolph Jefferson

*I am now in a tranquil situation which is my delight,
with all my family living with me, and forming a delicious society.*
Thomas Jefferson to Angelica Schuyler Church, 1795

As the century turned and the first decade of the 19th century progressed, Snowden experienced the death of Anne Jefferson, the maturation and marriage of some of her children, and ultimately, the second marriage of Randolph Jefferson. In the absence of other documents, Virginia's personal property tax records help trace the Jefferson boys as they gradually turn into independent men. Unlike others in the Lewis clan, they did not marry exceptionally young. Thomas, Jr. was the first of the boys to marry and that was not until the autumn of 1808. Thomas, Isham Randolph, and Robert Lewis were at least twenty-five years old when they married. Peter Field may have been closer to thirty years old. There are many indications that their initial steps toward independence were made close to home, working with their father and/or other extended family members.

In 1800, Randolph turned forty-five, making him solidly middle-aged. Looking to the future, he was still young enough to think of new partnerships and business ventures beyond farming. At the same time, he saw his sons quickly growing old enough to desire independent lives. The records suggest that, as Randolph aged, he made at least two attempts to establish new partnerships while grooming a successor for Snowden. The death of his wife, Anne, may have escalated his efforts. The first known attempt to bring new perspectives to Snowden was the arrival of his kinsman, John Jefferson, Jr.

The son of John Jefferson of Cumberland County, John, Jr.'s exact role and relationship to the Snowden household is speculative, but evidence supports the following narrative. Born in Cumberland on April 14, 1779, John Jefferson, Jr. was the son of John and Elizabeth (Broome) Jefferson. John, Sr. was Randolph Jefferson's first cousin, the son of his uncle, Field Jefferson.[1] Like Stephen Perkins and Thomas Ware before him, John was an ideal age to spend a year or two at Snowden as overseer. Not a stranger to Randolph Jefferson, they might have enjoyed a friendly, working relationship. In 1800, at age twenty-one, John was listed with Randolph on that year's personal property tax record, appearing as overseers had in the past.[2]

John Jefferson came with his bride, Sarah "Salley" Smith (Criddle) Jefferson, whose sister, Anne "Nancy" Criddle, married Thomas Ballowe, giving Salley family ties to Buckingham County's Horseshoe Bend, as well. During their stay in Buckingham, the John Jeffersons started what would become a large family, beginning with Elizabeth Broome Jefferson (b. 1801) and Sarah Suggett Jefferson (b. 29 May 1802).[3]

In 1802, John Jefferson, Jr. became involved in land transfers in Cumberland County. That year, on February 12th, John's father deeded him 4 ½ acres in Cumberland for the price of £150; however, it does not appear that John intended to give up his residence in Buckingham. A year later, John and Sarah Jefferson transferred the 4 ½ acres to John J. Reynolds of Cumberland for £10. In the deed, John and Sarah are described as "of Buckingham," indicating they are still in place in the county, if not still residing at Snowden.[4]

If John worked as Randolph Jefferson's overseer during 1800 and 1801, he had other ideas by 1802. That year, Randolph once again paid tax on two white males at Snowden; however, this time, a note reading "Th jr" appears next to Randolph's name. Presumably, his son, Thomas Jefferson, Jr., had turned twenty-one years old and may have stepped up as Snowden's newest overseer. Randolph and Thomas, Jr. are listed on the tax record on "May 13th," and John Jefferson, still in Buckingham, is listed separately on "August 11th." It is unclear what might be concluded about the location of John Jefferson's domicile based on the separation of dates.[5]

That year, John paid personal property taxes on himself, one slave over twelve years old, four horses, and one ordinance license for an impressive tax liability of $13.42. A plausible scenario is that John had convinced Randolph to allow him to operate the tavern at Snowden's ferry house; though, currently, there is no direct evidence linking the license to Snowden.

The year of 1802 may have been a successful one for all the Jeffersons, encouraging John (and Randolph) to expand his interests. In 1803, John not only paid taxes again for an ordinary license but also for a stud horse (taxed at 30 shillings), his personal property tax obligation soaring to $22.98. Interestingly, he no longer paid for a slave or for riding horses. Randolph Jefferson may have had capital invested in both the ordinary and the stud service, though young John was responsible for the tax burden.[6]

Whatever the agreement had been between Randolph Jefferson and John Jefferson, Jr., things did not develop smoothly. By 1804, John Jefferson had removed from Buckingham and had returned to Cumberland County, where he paid personal property taxes that year. Significantly, in 1806, Randolph's dissatisfaction with his cousin was so deep that he initiated a Buckingham Chancery Case against him for damages amounting to £32.[7] The details were lost in the Buckingham Courthouse fire, so the particulars of their disagreement are unknown. Upon his return to Cumberland, John Jefferson, Jr. found himself involved in litigation there, as well.[8]

Following this significant and likely disheartening four-year relationship with John Jefferson, Randolph attempted another family collective business, this time including his son-in-law, Zachariah Nevil. Born before 1764, Nevil was a mature man with significant property of his own to bring to operations at Snowden. The tax records for 1805-1809 show a swell in slaves while Nevil resided there.[9] A company styled as "Nevil and Jefferson" was formed; however, just what the company did is a bit of a mystery.

As early as 1807, perhaps much earlier, Zachariah Nevil, Randolph, and some of his sons joined together in a business venture. Store ledgers at Stony Point, located just above Scott's Ferry where William Moon, Jr. operated a store, reveal the partnership of "Nevil and Jefferson." A surviving ledger and journal from 1808 includes charges by Randolph Jefferson, his sons, Thomas, Lewis, and Randolph Jefferson, Jr., Zachariah Nevil, and "Nevil & Jefferson." Just which Jeffersons formed the partnership and the nature of their business is not entirely clear, though it does not appear to concern agriculture.

At Moon's Stony Point store between June and December of 1808, purchases for "Nevil and Jefferson" included: shoe brushes, 27 [units?] iron, a blacking ball, quires of paper and pencils, "copperass," pit and hand saws, two orders of German steel, and black pepper.[10] Orders for "Nevil and Jefferson" were placed by Randolph, Zachariah, Thomas, Jr., and Randolph, Jr.

What was the new endeavor? Copperass is an antiquated term for ferrous sulfate which was used in the manufacture of inks, most notably iron gall ink. This process dated from the middle ages and was used until the end of the 18th century. It was also used in the dyeing of wool and in a material called harewood, used in parquetry since the 17th century.

Significantly, in 1808, Thomas Jefferson, Jr. purchased a skilled slave named Gary, who was a cooper. Gary may have been making barrels or specialized containers for "Nevil and Jefferson."[11] Gary's skill may offer a clue to changing operations at Snowden, indicating an increased need for barrels, tubs, casks, or other containers made of wood on the plantation. Hogsheads, large wooden barrels used to transport tobacco from farms to ferry landings and to tobacco warehouses, were always in demand. Also, distilleries sprang up from time to time in Buckingham, some more successful than others. Did the young Jeffersons plan to make whiskey? Later in life, Randolph's son, Peter Field Jefferson, would make a great deal of money selling spirits in Scottsville – with and without a proper license.

One thing is clear, at the end of the decade, as Randolph Jefferson's days as an unmarried widower drew to a close, he was surrounded by his children, their spouses, and his first grandchildren. It is currently unknown if they all lived in a single dwelling at Snowden or if there were multiple households spread across the farm. By 1809, the following family members were collected at Snowden:

Thomas Jefferson, Jr. & wife, Mary Randolph "Polly" Lewis
Anna Scott "Nancy" & Zachariah Nevil,
    grandson: James "Lilburne" Nevil (b. 24 May 1806)
    granddaughter: Louisa A. Nevil (b. abt. 1808)

Isham Randolph, Jr.
Robert Lewis Jefferson
Peter Field Jefferson
James Lilburne Jefferson

Then, during 1809, the prospects for the "Nevil and Jefferson" enterprise were altered when Randolph chose to marry a young woman still in her teens. Visions of an enlivened middle-age and a potential second family may have sparkled in Randolph Jefferson's eyes; however, the realities of dashed inheritance may have reflected simultaneously in those of his children and his only son-in-law.

No matter how Randolph's children felt initially about his bride and their new "step-mother," it is known that at least two of his sons grew to dislike her. It may have been quickly apparent that the future which had been projected for Snowden had just taken a left turn. As a result, the group that had coalesced there gradually went their separate ways, charting new courses.

Zachariah and Nancy Nevil returned to his home on the Rockfish River in what had been Amherst County and was now Nelson County. There, they built a good life for their growing family.

Thomas, Jr. and his new wife, Polly, removed to Albemarle, where he was taxed in 1810. They likely joined her extended family at Buck Island and the town of Milton before they settled on a farm just across the Albemarle County line in Fluvanna County.

By March of 1810, when it was tax time again at Snowden, Randolph Jefferson paid for just one other white male who is not identified by name.

Enter Mitchie B. Pryor

If Buckingham County had published a newspaper in 1809, a banner headline might have prominently announced: "Mitchie B. Pryor to Marry President's Only Brother." However, this sparsely populated, rural community boasted no such newspaper and, to date, no record of the marriage of Mitchie B. Pryor and Randolph Jefferson has been located, not even a Bible record which might offer details of the nuptials.[12]

Dolley Madison (1768-1849) was raised a modest Quaker; however, by the time she was First Lady (1809-1817), low-cut gowns were in fashion and Dolley flaunted her shoulders and breasts. Former First Lady, Abigail Adams, remarked that Mrs. Madison looked like "a nursing mother." When Randolph Jefferson met Mitchie Pryor, this revealing Empire style of dress was in fashion, potentially heightening any young woman's allure. (COURTESY OF LIBRARY OF CONGRESS)

Randolph and Mitchie (pronounced "Mickey") were likely married in Buckingham County, probably at her home, Woodlawn, not far from Buckingham Court House.[13] An exact date is unknown; however, we do know that Randolph was unmarried when he drafted a "last will" in May of 1808 and that he wed Mitchie sometime before December of 1809, when he wrote to his brother, Thomas, asking if he could borrow a gig harness to visit her ailing brother in Charlotte County.

Mitchie came from an upstanding, "middling," Buckingham family, with a long presence in central Virginia. She undoubtedly had lived all her life at Woodlawn. Her parents, David and Susannah (Ballow) Pryor, were residents in the county during the Revolutionary War and roughly of Randolph Jefferson's generation. Pryor may have been a blacksmith or had access to a smith's shop. After the war, he put in a public claim for "making an axeltree for public waggon," valued at five shillings.[14]

While the Pryor family had a long presence in central Virginia, it is not a particularly common English surname in Buckingham, and it is currently unknown when David Pryor settled in the county or if he was born there.[15] In 1773 and 1774, there is a David "Pryer" on the Buckingham tythes lists.[16] In 1782, a David Pryor paid Buckingham County Personal Property tax on himself and one taxable slave named Jacob. It is also unknown when David Pryor married Susannah Ballow; however, their son, Zane, was born on October 27, 1781. Indications are that he was their first child.

Following the Revolutionary War, David Pryor steadily expanded both his family and his land holdings, indicating some prosperity. The Pryors likely had at least two sons by the time David surveyed a tract of 300 acres on the west side of Rock Island Creek in 1784. Two years later, he was granted 550 acres in northern Buckingham described as "beginning &c. in Joseph Cabells Courthouse Road," which in subsequent tax records was located on "the head (waters) of Walton's fork."[17] Ten years later, in 1796, he acquired an additional 195 acres by patent. It was described as adjoining Thomas Anderson, James Stanton, William Holdwright, and John Couch.[18]

Pryor's 230-acre farm on Ripley's Creek, called Woodlawn, was "new" land and not inherited by either David or Susannah. Though David owned and likely developed other land in his lifetime, this is the farm his widow and children retained decades after his death. There are no extant descriptions of Woodlawn, which was located four to five miles north of Buckingham's courthouse. However, in 1815 the structures there were valued at $700, indicating a very solid, but not showy, dwelling house. Susannah would remain at Woodlawn until she removed to Tennessee circa 1818-1819.[19]

In 1802, David Pryor was taxed on these three tracts of land in northern Buckingham, and there are indications that he sought to buy an additional 300 acres from a man named Boyd in 1803. The sale may not have been completed due to Pryor's death the following year.[20] When he died on September 24, 1804, David Pryor's anticipated land expansion drew to an abrupt halt.[21]

In addition to his farmland, he left many sons to care for his widow and his young daughter, Mitchie. Few records have been found to identify, or even roughly estimate, the births of the Pryor children. Mitchie is, however, the only known daughter among eight Pryor sons. Zane was born on October 27, 1781, a peer of Randolph Jefferson's oldest son, Thomas Jefferson, Jr. Zachius/Zachariah B. and Banister Pryor were born before 1783.[22] William S. Pryor was born before 1785. John C., Langston S., and Nicholas Ballow Pryor were all born before 1789. Leonard was likely the youngest Pryor son, and Mitchie may have been younger than all of Randolph's children, with the possible exception of James Lilburne Jefferson.[23]

Several of Mitchie's brothers demonstrated interests and abilities beyond farming in Buckingham County. Their accomplishments, as well as those of their children, also indicate advanced education, at least for some members of the Pryor family. Two of Mitchie's nephews practiced the law, while her nieces married lawyers and doctors as well as farmers. One nephew, Benjamin P. Pryor, served on the city police force in Richmond, Virginia.[24]

Perhaps the most accomplished of Mitchie's brothers was Dr. William S. Pryor, who eventually settled in Hanover County, Virginia. William married into the prominent Pollard family, several of whom served as clerks of the Hanover County Court and the Circuit Court over several generations. He clearly valued education, and his daughters were schooled in Richmond.

Brother Banister S. Pryor became a physician, appearing in the Prince Edward County records as Dr. Pryor. Sometime before 1813, Nicholas Ballow

Pryor removed to Tennessee, eventually settling in the city of Nashville. There, he served as a trustee of the Nashville Female Academy. Another brother, Judge John C. Pryor, removed to Mississippi and later went to Tampa, Florida.[25]

Langston S. Pryor, who remained in Buckingham County, was a Lieutenant during the War of 1812, serving with the 5th Virginia Militia. He was also a Freemason in Buckingham's Union Lodge No. 1, achieving the degree of Master Mason by 1803.[26]

The Pryors and the Jeffersons were not close neighbors and no documents have surfaced directly linking the families before Randolph and Mitchie married. Woodlawn and Snowden were at least fifteen to eighteen miles apart, and in the early 1800s, that was a considerable distance on unpredictable country roads. How, then, did Randolph and Mitchie meet and come to know each other?

The strongest possibility is that Mitchie's mother, born Susannah Ballow, was connected to a plantation very close to Snowden, and that Randolph Jefferson had long known her and her extended family. Thomas Ballow, who married Chloe Battersby in 1757, was the long-term neighbor of Snowden. His family intermarried with Anthony Murray, son of Richard Murray, one-time ordinary keeper at Snowden. Additionally, Thomas and Chloe Ballow had a daughter "Micah" Ballow (b. 3 August 1769), who certainly might have inspired Mitchie B. Pryor's name.[27]

Randolph's marriage to Mitchie coincides with another significant change in his life, the removal of the Charles Lilburne Lewis family to Kentucky. In the autumn of 1807, Randolph's sister, Lucy, his brother-in-law and cousin, Charles Lilburne, and most of their children left Albemarle County and their plantation, Monteagle, behind, never to return. Financial embarrassment and severe diminishment of wealth had driven them to the more affordable West. There are many indications that the Randolph Jeffersons and the Lewises were closely connected. When Anne (Lewis) Jefferson died about 1800, the two families may have drawn even closer.

The Pryor family – Susannah, Mitchie, and the Pryor brothers – offered Randolph a new extended family, and one with local benefits. Prolonged visits at Woodlawn took the now middle-aged Randolph Jefferson into the heart of Buckingham County, putting him in close proximity to the county seat, which was filled with the activity of county business and the comings and goings of commerce, particularly on court days. Likewise, the youthful Mitchie was drawn to the activity at Snowden on the James River. There, she would have access to shops both in Scott's Ferry and in Warren, as well as the more sophisticated culture of the plantations along the river. Additionally, Randolph's comparatively elevated status in Buckingham, both through his wealth and his family's broader reputation, offered Mitchie a remarkably promising future.

When Randolph married Mitchie B. Pryor, her mother, Susannah, had been a widow for six years. While some of Susannah's sons would eventually establish homes elsewhere, the majority resided in Buckingham during the next few years. By 1810, Langston and Zane may have been living independently in the county, though at least two Pryor sons still resided at Woodlawn.[28]

Randolph apparently liked his mother-in-law, enjoyed the atmosphere of her household, and, in return, provided the presence of a mature man on the farm. Mitchie's youth may have encouraged a slow transition from her childhood home to Snowden, and the newlyweds took a leisurely approach to their first months of marriage, spending time paying lengthy calls on Mitchie's family, including her brother in Charlotte County. In Randolph's absence from Snowden, overseers, possibly in conjunction with some of Randolph's sons, operated the farm.[29]

The bulk of the surviving correspondence between Thomas and Randolph Jefferson was written after Randolph married Mitchie Pryor. The letters suggest Randolph's comfortable acceptance of and easy integration into his new family. At some point, Randolph and "his family" lived with Susannah Pryor, indicating that at least James Lilburne Jefferson joined his father at Woodlawn.[30] In October of 1811, one of Mitchie's brothers carried a letter for Randolph to Monticello. The following February, Randolph wrote to Thomas from Woodlawn, mentioning the fact that he felt good enough to ride down to Snowden with no adverse consequences. Struck with an attack of kidney stones the previous October, Randolph spent some time recuperating at Woodlawn. A year later, he was living at Woodlawn for another extended period due to Susannah Pryor's illness.[31]

As Mrs. Randolph Jefferson, Mitchie eventually established herself as the new mistress of Snowden and letters between the Jefferson brothers reflect

Mitchie's presence there. By 1813, eager to improve and expand her kitchen garden, Mitchie asked her brother-in-law's advice concerning seeds and plantings unfamiliar to her. He kindly responded by sending her both seeds and instructions.[32]

Perhaps overconfident in her role, Mitchie also approached Thomas Jefferson with concerns about Randolph's management of the farm. In 1813, Mitchie indicated that she doubted her husband's ability to manage his own affairs, felt he was not getting top dollar for his crops, and hoped that Thomas would counsel and advise her concerning the plantation business. She went so far as to ask Thomas' consent that she might take a part in "directing" Randolph's affairs. Thomas declined to help her make this step.[33]

While letters from Thomas to Randolph contain only the most cordial references to Mitchie, she was not popular among Thomas Jefferson's extended family and friends. Mitchie's personality clearly rubbed the Albemarle establishment the wrong way. There is every indication that she was a very young woman, possibly not yet twenty years old when she married Randolph, and no doubt heady with her role as sister-in-law to the newly retired President of the United States.[34] Joseph C. Cabell found her coarse referring to her as a "Jade of genuine bottom."[35] Years later, following Mitchie's death, Martha (Jefferson) Randolph commented that the clergyman who pronounced Mitchie's funeral oration must have been hard put to deliver a complementary eulogy.[36]

Despite these disparaging opinions of Mitchie, Randolph stood to gain from the alliance and there are many reasons why a man might marry a woman young enough to be his daughter. Randolph had been a widower for some years and, undoubtedly, her youth was invigorating. Very much at home with the Pryors, Randolph added a second household and farm to his sphere. As his vitality became uncertain and his health unpredictable, Randolph had a potential nurse in Mitchie. In 1812, when Randolph wrote to his brother about his renewed health, he pointedly mentioned that he had not had a "drop of any kind of spirits" since they were last together. It is possible the doctor advised him that alcohol aggravated his kidney stones; it is also possible that Mitchie had a positive influence on Randolph's drinking.

Additionally, Randolph and at least three Pryor sons shared involvement with the Buckingham County Militia. Langston and Zach both served in the

5th Virginia Regiment during the War of 1812. Langston continued his service in Buckingham's 100th Regiment. In 1815, he was promoted to Captain and his older brother, Zane, was recommended as Ensign. Their experiences may have influenced Randolph's son, James Lilburne, in his decision to join the militia.[37]

This was not the first time the Jeffersons had witnessed a family member's second marriage to a much younger woman. In 1790, the marriage of Thomas Mann Randolph and Gabriella Harvie had drastic consequences for Martha (Jefferson) Randolph and her children. When Martha's father-in-law, Thomas Mann Randolph, Sr. remarried, he was fifty years old. His bride was a very youthful seventeen. The marriage, which only produced one living heir, completely upset Thomas Mann Randolph's household, the children of his first marriage, and their inheritance. Adding insult to diminished inheritance, Thomas Mann and Gabriella Harvie, named their only surviving son, Thomas Mann Randolph, Jr., effectively erasing the first "Jr." from the record.[38]

Significantly, Randolph did not seek Thomas' opinion or counsel before he married Mitchie, presenting his second marriage as a *fait accompli*. Did Randolph fear his brother would disapprove, or did he feel this union was his decision to make, with no need to seek approval?

Thomas Jefferson never criticized his brother's remarriage, at least not in his correspondence with Randolph. Generally, it was not in Thomas' personality to criticize his family; rather, he tended to see the best in them and look the other way when it came to their foibles. His attitude toward his second sister-in-law appears to be no exception. Contrary to the comments about Mitchie made by his friends, Thomas' letters indicate that he was kind and welcoming toward her, particularly in the beginning.

On December 7, 1809, Randolph asked his brother if he could borrow a harness, so that he and Mitchie could take their gig to Charlotte County to visit her brother, who was gravely ill, perhaps dying.[39] The following day, Thomas sent the harness to Snowden via Squire; his correspondence included an invitation to Monticello:

Monticello Dec. 8. 09.
    Dear brother
        I send by Squire the gigg harness, and shall be very happy if after your return, instead of sending it you would avail yourself of it to pay

us a visit here with my sister. She promised me a visit in the spring but the distance is too short to require it to be put off to so remote a period. Perhaps you might find an absence from home during winter less inconvenient than after the operations of the farm and garden shall have been begun in the Spring. However we shall be happy to see you both at your own best convenience. In the mean time accept for both the assurances of my affectionate esteem.

Th: Jefferson[40]

In February of 1813, Randolph extended a similarly warm invitation to his brother to visit Snowden. While thanking Thomas for some garden seeds he had promised to send to the Buckingham plantation, Randolph wrote, "we are now living at home and would be happy to see you and your familey when ever convenient."[41]

In March, Thomas responded with a long letter detailing the seeds he sent. Thomas Jefferson's enrichment of the kitchen garden at Snowden is particularly interesting given his reputation as an "amateur" botanist.[42] One wonders why Randolph required such a complete collection of vegetable seeds and had not saved his own from the previous fall. Had he lost his stock to fire, flood or some other natural disaster, or, as with Thomas' gardener, had Randolph's household failed to gather enough seed the preceding season?

In the same letter, Thomas mentions his desire to visit Snowden and his disappointment that Randolph had not come for the seed himself, sending Squire instead.

...I do not expect to go to Bedford [Poplar Forest] till late April. If by that time I learn that the road from Mr. Hudson's to Scott's ferry is passable I will certainly call on you. I meant to have done it as I returned in November last, but learned that the boat at the ferry was gone, and that I would not be able to cross there. All here are well and salute my sister [in-law] and yourself, disappointed that you did not come for the seeds. Adieu

Yours affectionately
Th: Jefferson[43]

MARCH. 37

Capsicums.

Sow capsicums the middle of this month in the same manner as love apples; they make excellent pickles; you may transplant in May or April.

Garlic and Shallots.

Plant them separately in drills nine inches asunder, the roots six inches apart in the drills; cover them about two or three inches deep.

Salsify, Scorzonera and Skirrets.

Sow the seeds in an open situation, thinly, on separate beds, and rake them in. When the plants are up, thin them to stand six inches apart; their roots will be fit to use next fall.

Large rooted Parsle

Sow the seed in an open situa face, and rake it in; when the pl strong, thin them to stand six inc large root like a parsnep is esteem sons.

Skirrets, by John Gerard. (DOVER PUBLICATIONS)

THE
AMERICAN GARDENER:
CONTAINING AMPLE DIRECTIONS
FOR WORKING A KITCHEN GARDEN
EVERY MONTH IN THE YEAR,
AND COPIOUS INSTRUCTIONS
FOR THE CULTIVATION OF FLOWER GARDENS, VINEYARDS, NURSERIES, HOP YARDS, GREEN HOUSES AND HOT HOUSES.

BY JOHN GARDINER, AND BY DAVID HEPBURN,
Late Gardener to Gov. Mercer, & Gen. Mason.

A NEW EDITION, MUCH ENLARGED.

TO WHICH IS ADDED
A TREATISE ON GARDENING,
BY A CITIZEN OF VIRGINIA.

ALSO,

A FEW HINTS
ON THE
CULTIVATION OF NATIVE VINES,
AND DIRECTIONS FOR
MAKING DOMESTIC WINES.

THIRD EDITION.

WASHINGTON CITY,
PUBLISHED BY WILLIAM COOPER, JR.
1826.

The American Gardener, by John Gardiner and David Hepburn, remained a very popular reference for many years.

The following June, Thomas described a gift he sent for Mitchie in his letter to Randolph:

> ... My sister desired that when I should send her seeds of any kind I would give her directions how to plant and cultivate them. Knowing that there was an excellent gardening book published at Washington, I wrote for one for her, which I now inclose. She will there see what is to be done with every kind of plant every month in the year. I have written an index at the end that she may find any particular article more readily: and not to embarrass her with such an immense number of articles which are not wanting in common gardens, I have added a paper with a list of those I tend in my garden, and the times when I plant them. The season being over for planting everything but the Gerkin, I send her a few seeds of them. She will not find the term Gerkin in the book. It is that by which we distinguish the very small pickling cucumber. Affectionate salutations to you both.
> Th:Jefferson[44]

This gesture reveals several things about both the giver and the receiver of the "excellent gardening book." Thomas Jefferson's urge to instruct corresponds immediately with young Mitchie's urge to learn and improve herself and her garden. She is clearly literate, which safely might be assumed, but it is nice to have proof. Jefferson's care and generosity to create an index for her is charming, as is his helpful gesture to make the book manageable by listing the common local plants. Randolph wrote back on June 21, 1813, "My wife is extreemly oblige to you for your present and is very much pleased with it." What a treasure of a book, with its index and notes, that simple guide would be today.

Thomas was not the only member of the Jefferson family who exhibited kindness toward the second Mrs. Randolph Jefferson. In the summer of 1814, Mitchie was seriously ill, and Randolph's twin, Nancy Marks, spent an extended stay at Snowden. A note from John Nicholas to Thomas Jefferson explained the situation.

Augst 12th 1814
John Nicholas also presents his Comp^{lts} to Mr Jefferson, & informs him that he saw his Sister, Mrs Marks, a few days ago, who requested him to ~~inform~~ desire M^r J — to postpone sending for her untill he heard from her; Mrs R. Jefferson being then very sick & Mrs Marks not wishing to leave her untill her entire recovery — [45]

John Nicholas, of Seven Islands in Buckingham County, was either visiting Snowden or encountered Nancy Marks in the neighborhood. No more is known about Mitchie's illness.

Over two weeks later Thomas' memos reveal that Wormley was given $1.00 for ferriage to Snowden on August 27th. He brought Nancy back to Monticello on August 29th and returned ".125" to Thomas Jefferson. Without the note from John Nicholas, the purpose for Wormley's trip would be unknown.[46]

The Gig and the Harness

The story of the gig and the harness is a fairly simple one, but somewhere in it lurks the frustration of a young bride who is eager to be out and about in her husband's gig – visiting family and shopping in Scottsville or Warren – but who is thwarted, first by a borrowed, worn-out harness, and then by a demolished buggy.

Gigs were common among Randolph's "set" in Buckingham, while fancier carriages, coaches, or landaus were rare. The gig may have been acquired in conjunction with his young bride, as he first paid tax on one in 1809.

Early in Randolph and Mitchie's marriage, one of her brothers was ill – lying like to die – in Charlotte County, south of Buckingham. Randolph had a gig, but apparently lacked a functioning harness so he wrote to his brother on December 7, 1809, requesting to borrow one. Randolph sent the letter by his servant, Squire. Thomas wrote back the next day, sending Squire home to Snowden with a loaned harness. Thomas encouraged Randolph and Mitchie to return the harness themselves and pay him a visit at Monticello.

Years passed and Randolph never returned the harness. Then, in the summer of 1815, Thomas wrote the following:

Monticello June 23.15.

I lent you some years ago the harness of our family gigg, until you could get one made for your own. Mrs. Marks tells me your gigg is now demolished and out of use. Mine has been used with one of our chariot harness. A neighbor asks the loan of it to go on a journey, and if we let one of our set of harness to go, we shall not be able to use the carriage until his return which will be very distant. Under these circumstances I send the bearer to ask the return of the harness I lent you, in order to accommodate my neighbor. Present my respects to my sister and be assured of my best affections.

Th: Jefferson[47]

Later that same day, Randolph wrote back to his brother, telling the sad story of the harness which dated back to 1809. He began, "I received yours by the boy the harness in which you were kind enough to lend us was intirely worn out so that they did not scarce las(t) us over to Prince Edward and back."[48]

The weak harness, however, was not returned to Thomas and was apparently strong enough for Randolph to loan to Mr. Patteson, who borrowed both it and Randolph's gig to go "over to the springs." Heading west, Mr. Patteson got as far as Staunton, where the harness gave out. He bought a "new set," which he gave to Randolph and Mitchie, when he returned from the springs.[49]

During the intervening years, Randolph and Mitchie must have made use of the gig and Mr. Patteson's replacement harness. It seems logical that the gig traveled to Monticello for visits. Apparently, the harness was neither noticed nor discussed. Perhaps, after a while, Randolph considered it forgotten. After all, Thomas' harness was long gone, having "died" in Staunton.

Over the years, Randolph even wrote to his brother, mentioning the gig. In June of 1810, the gig required repairs and delayed Randolph visiting Monticello. It needed a new axletree, and Randolph was waiting for iron.[50] In early 1812, once Randolph recuperated from his attack of kidney stones, he wrote to his brother that he could once again enjoy a ride in his gig with no discomfort. In 1813, Randolph was still paying a "luxury" tax on the buggy; however, in 1814 and 1815, the gig was no longer accounted for on Randolph's personal property tax record, indicating it may have been inoperable for some time before Thomas wrote in June of 1815, asking for his harness to be returned.

To this request, Randolph freely admitted that now that their gig was indeed demolished, "the harness is of no service to us." He immediately sent Mr. Patteson's no longer new set back to Monticello. The harness was carried by Thomas' servant, Israel, whose expenses to Snowden were $.25.[51]

By the time the buggy was demolished in 1813, Randolph's costs (especially his annual tax obligation) were rising. He may not have had ready funds to replace it. There may also have been another motive: the absence of a gig made it easier to keep the freely spending and gallivanting Mitchie at home. Of late, in Randolph's opinion, his wife had been frequenting the stores at Warren and Scott's Ferry a little too often.

In the end, the gig and harness incident reads like a very typical family story. On a visit to Snowden, sister Anna Scott (Jefferson) Marks noticed that Randolph's gig was no longer operable. How it was demolished and by whom remains a Snowden mystery; however, Anna Scott may have meddled a bit, gossiping to Thomas about the gig, having known full well that Randolph had never returned Thomas' harness.

Whether or not Thomas and his family were ever truly inconvenienced without their second harness is not known. If it was as worn as Randolph describes, there was not much life left in the set when it was originally loaned and, in the beginning, Thomas really did not care if it was returned; he makes it clear, though, that he would have been happy for his brother and new sister-in-law to use the excuse of returning the harness to pay a call at Monticello. Indeed, if Thomas was aware of its shabby condition, he might have been privately surprised that Randolph was getting so much use out of the old thing. Of course, Randolph was enjoying the new set purchased by Mr. Patteson, benefitting from the replacement, as did Thomas, ultimately.

It is commonly accepted that Thomas Jefferson did not like confrontation and would have been hesitant to ask his brother to return the harness. In fact, in August, Thomas returned the harness to Randolph, writing to Mitchie, "I return by squire the harness he was so kind as to lend me. It has answered the only occasion I [...] had for it, and may on some other be useful to yourselves."[52]

This engraving of Thomas Jefferson, by J. B. Longacre, copied the Gilbert Stuart portrait (1805). In the 19th century, mass-produced engravings, like this one of Jefferson, guaranteed that art reached a wider audience. (MEMOIR, CORRESPONDENCE, AND MISCELLANIES, FROM THE PAPERS OF THOMAS JEFFERSON, VOL. IV, BY THOMAS JEFFERSON RANDOLPH, 1830)

Mr. Jefferson Stops at Snowden

It would be wonderful if a lengthy letter from Thomas Jefferson to Randolph survived describing his delightful stay at Snowden. There is, however, no such document. Instead, their letters contain invitations and mentions of desired visits, any of which might have taken place when Thomas was going to and from Poplar Forest in Bedford County. We may never know which, or how many, of these materialized.

There are indirect references that the brothers met, both at their respective plantations and while abroad in the countryside. One curious loan from Thomas to Randolph, dated September 14, 1812 and noted in his memorandum books, was for "Mrs. Pryor's vales .25." At the time, Thomas was heading home to Monticello from Poplar Forest. That day he spent $.50 on the ferry at Warren and $2.75 at H. Flood's on dinner and lodging. Had the brothers crossed paths in Warren and dined together at Flood's Tavern? Was Randolph heading toward Buckingham Courthouse to his mother-in-law's home, and, being short of cash, asked his brother for $.25 for Mrs. Pryor's servants?[53]

It is certain that Thomas Jefferson visited Snowden twice in 1813, once in May and again in September. The memorandum for May reveal that he stopped for a visit on his way back from Poplar Forest. The entries read as follows:

May
- 6. Pop. For. gave him [James] for his expences back to Monto. 1.D.
- 12. Settled with Jer. A. Goodman. See settlements.
    Also with Mrs. Goodman weavg. & teachg. Sally £9-10.
- 13. Gave Billy for his expences back to Monto. .50.
    Pop. For. Pd. debts & vales 2.2.
    H. Flood's dinner 1.625.
- 14. Noah Flood's lodging 1.5.
    Gibson's brkft. .625.
- 15. Snowden vales . 75 fish 1.D.
    ferriage. 75 cash on hand 32.25 D.[54]

On this visit, Thomas and Mitchie discussed what she now perceived as Randolph's mismanagement of the farm, particularly his "disadvantageous sales" of crops. Mitchie expressed concern about Randolph's ability to manage his own affairs. What reason she gave is unknown. Did she feel that at age fifty-seven, he was losing his judgment due to mental confusion? Perhaps she was intimating that he had never had a head for business and now she was observing it firsthand.[55]

Mitchie sought more involvement in Randolph's business affairs and hoped that Thomas would "consul and advise" her. She found him unwilling to do so. Whether it was Thomas' belief that the business of the farm was not a wife's business or that he was careful not to triangulate his relationship with Randolph and Mitchie, his unwillingness to get involved personally was certainly in keeping with his personality.

Otherwise, the Snowden household seemed to be as it had been, without any appearance of new extravagances on the part of the second Mrs. Jefferson.

Her recent expenditures may have been on personal items or they may have been exaggerated in Randolph's mind. It might also be concluded that there was no sign of decline from gross mismanagement on Randolph's part.[56]

Additionally on this spring trip, Thomas found that the condition of the roads made for a difficult and unpleasant side trip to Snowden. Indeed, the county roads and the James River sometimes conspired to keep the brothers apart. There was, however, at least one road that Randolph had some control over and his brother expressed concern about its viability. On May 25, 1813, Thomas wrote from Monticello: "I shall set out [for Bedford] the day after our court, this day fortnight, or very soon after. I will go by Snowden if I can; but certainly will return that way, on condition you will previously re-open the old road to the old smith's shop. You will never find more leisure time for your people to do that than the present."[57]

Randolph wrote back the next day, stating, "I will do my best to have the rode put in better order aganist you come a long as fare as the shop on the rode." He further reassured his brother in a "PS." Its casual, colloquial content is

This sketch by Lewis Miller, dated May 19, 1853, evokes Randolph Jefferson's gig headed up the long and winding road to the dwelling house at Snowden. (LEWIS MILLER, "SKETCHBOOK OF LANDSCAPES IN THE STATE OF VIRGINIA, 1853-1867," ABBY ALDRICH ROCKEFELLER FOLK ART MUSEUM, THE COLONIAL WILLIAMSBURG FOUNDATION, WILLIAMSBURG, VA., GIFT OF DR. AND MRS. RICHARD M. KAIN IN MEMORY OF GEORGE HAY KAIN)

typical of Randolph's written expression: "Dont be in dred of the old rode I will have that put in good order against you come a long for you – "[58]

Despite Randolph's phonetic spelling, his letter of May 26, 1813, demonstrates that he still possessed a reasonably clear and organized mind. Its contents would also seem to argue against the fact that he was slipping into a state of dementia at that time. Responding to Thomas' long letter of May 25, 1813, Randolph addressed the following points: 1) catching some carp for Monticello; 2) sending Lilburne to study at Monticello; 3) fixing the road for Thomas' visit; 4) the arrival of the spinning jenny; 5) thanking Thomas for his agricultural advice; and 6) commenting on their sister Marks' visit.[59]

In anticipation of the 1813 visit, various delays at Monticello kept Thomas from leaving for Bedford. When he wrote to Snowden again in June that he might not arrive until August, Thomas inquired, "I do not know whether I can call on you as I go. I will if I can, but certainly will as I return. Is your road cleared out?"[60]

As the summer progressed, things eventually fell into place. In July, Randolph's slave, Fanny, went to Monticello to learn the spinning jenny as planned. In August, Thomas traveled to Poplar Forest and the usual entries appear in the memorandum books: ferriage and vales at Warren, more vales for the servants at Gibson's ordinary, Noah Flood's, Henry Flood's tavern, and Hunter's ordinary. Then, on September 8th, Thomas wrote from Beford:

> Poplar Forest Sep. 8.13.
> Dear brother
> The cart sets out this morning with your spinning Jenny in perfect order, and will deliver it I hope safe from accident. According to present appearances I may leave on this Saturday morning, and if in time to get to Noah Flood's I may be with you to dinner on Sunday, but if I only get to Henry Flood's I shall dine at Gibson's and be with you on Sunday evening; and it is yet possible I may be detained here till Sunday. My best affections to my sister and yourself.
> Th: Jefferson[61]

At last, on the way back from Poplar Forest to Monticello, both Mr. Jefferson and the 12-spindle spinning jenny arrived from Bedford County at Snowden, though a couple of days later than anticipated. The memorandum books recorded the event long-awaited by both brothers, undoubtedly. The familiar pattern is reversed beginning on Sunday, September 12th: vales paid Hunter's ordinary, paid bills at Henry Flood's tavern, lodging at Noah Flood's tavern, on the morning of Monday, September 13th breakfast at Gibson's ordinary, then the mid-day arrival at Snowden. Thomas spent the night at the farm, for he gave his gratuity to the Snowden servants on Tuesday before heading home to Monticello.

> Sept. 14.   Snowden vales .25 ferrge. cart and self 1.25.
> Cash on hand 6.75.[62]

No other details were noted.

However, when the cart with the spinning jenny indeed arrived, we can easily imagine Mr. Thomas Jefferson freely giving advice about its proper use and value to the industry of the farm. That alone would have provided quite a lot of excitement. When the dust settled and Thomas Jefferson headed for Scott's Ferry, it would be his last visit to Snowden during his brother's lifetime.[63]

Other Visitors to Snowden

Randolph's twin sister, Anna Scott "Nancy" (Jefferson) Marks, made extended trips to Snowden, at the very least in her later years, if not earlier. She did not marry until her thirties, lived with her sister Martha and her brother Thomas after their mother's death, and was well-loved in the family as an "old maid aunt." In the years before her marriage, she also spent extended stays with her sister Lucy at Monteagle, and there is every reason to believe she did the same at Snowden. In August 1807, before the Lewises left for Kentucky, Nancy paid a visit to Monticello and Randolph received an invitation to visit her there. Thomas wrote on August 12, 1807: "Our sister Marks arrived here last night and we shall be happy to see you also."[64]

To the surprise of at least some members of the family, she married Hastings Marks in October of 1787 and lived for many years in Louisa County.

In Mitchie Jefferson's day, women in Virginia looked to London and Paris for the latest fashion. This watercolor depicts a typical summer evening dress of 1815. (COSTUME OF THE LADIES OF ENGLAND 1810-1823, ART & ARCHITECTURE COLLECTION, MIRIAM AND IRA D. WALLACH DIVISION OF ART, PRINTS AND PHOTOGRAPHS, THE NEW YORK PUBLIC LIBRARY, ASTOR, LENOX AND TILDEN FOUNDATIONS)

Following her husband's death, Anna S. Marks came to live at Monticello in January of 1813.[65] Apparently quite ill when she arrived, she fully recovered and out-lived both Randolph and Thomas, dying in 1828.

While residing at Monticello, Sister Marks, as she was called by her brothers, was more convenient to Snowden and could freely visit Randolph and his family. By 1812, the household may have been reduced to Randolph, Mitchie, and James Lilburne, leaving plenty of room to accommodate Anna Scott. Some of these visits are mentioned in the Jefferson-to-Jefferson letters.

In May of 1813, Randolph invited his sister to Snowden. In a letter to Thomas, he wrote, "You will be so good as to tell my sister Marks that we shall be extreemly happy to see her hear and that I will retern with her if she will come over." In June, Randolph again noted, "We shall certainly expect my sister Marks over this summer." In August of 1814, Thomas mentioned in a letter to Lancelot Minor that Mrs. Marks had "been for some time absent on a visit to our brother."[66]

On December 18, 1813, Thomas Jefferson's memorandum book reveals that he paid "Wormly ferrges to Snowden for Mrs. Marks .50."[67] This time, apparently, Randolph could not accompany her back to Monticello, so servant Wormley Hughes was sent to escort the widow Marks.

In the summer of 1814, Nancy spent an extended stay at Snowden when she helped care for Mitchie Jefferson, who was very ill that August. Again in April of 1815, Randolph wrote that Sister Marks had arrived very fatigued from her journey.[68] Randolph, himself, was recovering from an illness and was scarcely able to walk. The pair of middle-aged, weak twins gave Mitchie Jefferson an opportunity to act as nurse, if she was so inclined.

Other members of the Jefferson family undoubtedly paid calls to Snowden over the years, but little direct evidence has been found. One note, however, indicates that Thomas Jefferson Randolph, Randolph Jefferson's great-nephew, and Jeff's brother-in-law, Wilson Cary Nicholas, Jr., were in the habit of calling. On April 2, 1815, in a P.S. to Thomas, Randolph noted that the young men "took a ride to see" him; but for the first time, they "made no stay of account" at Snowden.

Chapter Thirteen

"My brother Randolph Jefferson died this morning."

When we have lived our generation out, we should not wish to encroach on another.
Thomas Jefferson to John Adams, 1816

"My brother Randolph Jefferson died this morning." That's all Thomas Jefferson wrote in his memorandum book on August 7, 1815.[1] There is more detail in his ledger concerning his purchase of a horse on June 14, 1815. "Purchased a horse of David Isaacs, bright bay, without any white except a few white hairs in the forehead where there is usually a star. Said to be 6. but believed 7.y. old."[2]

This likely does not reflect the relative importance of the animal, which was a carriage horse named Tecumseh, but rather the deeply personal nature of the death of Thomas Jefferson's only brother. Fortunately, Jefferson's personal papers and a few other memos shed a bit more light on Randolph's last months and the urgent last days of his life.

~

In February of 1815, Randolph seemed to be in good health. A letter dated February 13th, written from Snowden, indicates that he expected to be summoned to the March Court in Albemarle to testify in a suit between his son, Isham Randolph Jefferson, and Craven Peyton.[3] From this reference, it would seem that the nearly sixty year old Randolph was going about his normal business, riding to Charlottesville for Court Days, etc.

Likewise, he was busy planning for his spring planting, asking Thomas for fruit tree cuttings, especially apple and cherry. He also requested seeds for the kitchen garden: cabbage and ice lettuce. At the end of March, however, Mitchie wrote to Thomas that Randolph's condition had deteriorated. Dated Sunday, March 26th, this letter is lost, but in it Mitchie apparently painted a dire picture and Nancy Marks desired immediately to visit Snowden.

On Tuesday, March 28th, Thomas gave Wormley Hughes $1.00 for ferriage to Snowden.[4] A letter written on March 31, 1815 from Thomas to his grandson, Thomas Jefferson Randolph, explains what happens next:

> Ellen's visit to Warren has been delayed by an unlucky accident. On Monday we heard that my brother was very sick. Mrs. Marks wishing to go and see him I sent her the next morning in the gig with a pair of my horses, counting on their return the next day so that Ellen and Cornelia might have gone on Thursday according to arrangement. After Mrs. Marks had got about 7. miles on her road, one of the horses (Bedford) was taken so ill that she thought it best to return. He died that night, and no pair of the remaining three could be trusted to draw a carriage. Mrs. Marks going off again to-day to Snowden, I make Wormley take Seelah and direct him to return by Warren and exchange him with you for the mare Hyatilda . . . .[5]

On Saturday, April 1st, Thomas Jefferson Randolph and his brother-in-law, Wilson Cary Nicholas, Jr., paid a call on Uncle Randolph. The young men no doubt visited in response to Thomas Jefferson's March 31st letter.

The following day, in a letter dated April 2nd, Randolph wrote to Thomas revealing that their Sister Marks had indeed arrived safely. She was fatigued from her journey, which, of course, had been complicated and protracted by the death of Bedford. While Randolph was well enough to greet her, visit with the young men, and write to his brother, he was, by his own account, still extremely weak and could hardly walk. It is unclear if Randolph was suffering an acute illness or something chronic or debilitating. He added in his letter, "I will try to take a ride over some time this summer if my health permits."[6] This might suggest a chronic condition or that Randolph expected a long recuperation.

In a postscript to the letter, Randolph mentions that "Jefferson and young Wilson Nicholas took a ride to see me on saterday but made no stay of account with us for the first time."⁷ He was clearly disappointed that they did not stay longer; however, the men wisely may have kept the visit short, acknowledging the general fatigue of the household.

The April 2, 1815 letter from Randolph to Thomas also indicates that Randolph had some debts which were bothering him. Unlike so many Virginia planters who often created debt with abandon, Randolph hesitated to carry obligations. That spring, he was determined to see to his business affairs and settle his accounts. His recent sale of 70 acres of low grounds to Charles A. Scott at $100.00/acre would accomplish this. In fact, Randolph told his brother, Scott's first payment on April 12th would "take me out of debt with every man that I am involved with and which will enable me to keep all my slaves as long as I live."⁸

An active concern about indebtedness had plagued Randolph for some time. Beginning in the autumn of 1813, in conversation with his brother, Randolph expressed that he was increasingly distressed about Mitchie's spending habits. While Thomas did not observe "any appearance of extravagance" in the household when he visited Snowden in May and September of 1813, Randolph firmly believed something had changed about Mitchie's behavior.⁹

What would have prompted this sudden extravagance on Mitchie's part after almost four years of marriage? Where was the money going? Given Mitchie's youth, she may have been bored, which outings and purchases temporarily assuaged.

No accounts have surfaced to reveal the extent to which she was actually draining Randolph's wealth, and it is impossible to discern to what degree the actual spending might have been compounded by fears mounting in his mind. He particularly complained to his brother of his accumulating debts at stores at Scott's Ferry and Warren. By the fall of 1814, Randolph told Thomas that he must sell some land to raise money to pay these obligations.

Determined to stop Mitchie's spending, Randolph and the merchants involved agreed that no sales would be made to Randolph's account without a written order from him. Soon, though, he was convinced that forged orders were being given to the merchants, so he devised a "secret mark," which was eventually imitated. These imitations, he told Thomas, were made by his wife, Mitchie.¹⁰

This atmosphere of mistrust and growing financial stress was no doubt debilitating for the household, and it was compounded by quarrels between Randolph and at least two of his sons, Thomas, Jr. and Field. Even though Randolph called Mitchie's honesty into question, he was intolerant of his sons' criticisms of her. Additionally, at least one of the boys had reacted to rumors that Randolph had re-written his 1808 "last will," which he denied. All of this mounting tension certainly could have contributed to the declining health of the inherently kind-tempered man.

Adding to Randolph's financial concerns, the War of 1812 had brought various pressures to Virginia farmers, including steadily rising taxes. Just before his death, Randolph's personal property tax obligation alone nearly doubled in one year to $20.83½. Still, Randolph managed to hold on to most of his land and maintain his structures.

Between 1810 and 1815, Randolph paid the following personal property taxes:

1810:	2 white males; 13 slaves above 16; 2 slaves 12-16; 7 horses; 1 gigg	tax: $ 8.30
1812:	1 white male; 14 slaves above 16; 0 slaves 12-16; 5 horses; 1 gigg	tax: $ 7.62
1813:	1 white male; 12 slaves above 16; 0 slaves 12-16; 5 horses; 1 gigg	tax: $ 8.38
1814:	1 white male; 12 slaves above 16; 0 slaves 12-16; 5 horses; --	tax: $10.53
1815:	2 white males; 21 slaves above 12; 1 slave above 9; 0 slaves 12-16; 6 horses; 25 cattle	tax: $20.83½[11]

Between 1810 and 1815, Randolph also paid the following in land taxes:
1810:	1827 acres	$34.11
1812:	1827 acres	$45.48
1815:	1827 acres	$60.38[12]

In April of 1815, the sale of land to Charles A. Scott seemed to put Randolph's mind at rest, and though his unpredictable health may have kept him at Snowden that spring, come June, he was busy with the harvest. A letter dated June 23, 1815 is the last surviving correspondence from Randolph written to Thomas Jefferson. In it, Randolph laments the death of Mrs. Carr.[13] He mentions the busy harvest, and closes: "My Wife Joins Me in love and respect to you and family. – I remain your most affectionately. — Rh: Jefferson."[14] All appeared to be well at Snowden.

July passed and Randolph became deathly ill. As Randolph himself philosophically put it, death "is what we may expect to come to – either later or sooner." His came on August 7, 1815. Likely at his side were his twin sister, Nancy, his son, Randolph, Jr., his wife Mitchie, and no doubt others, including his son, James Lilburne, who still resided at Snowden.

To the upset of the majority of his family, Randolph left a recently written will favoring his second wife, Mitchie, over the children of his first marriage. His sons wasted little time in opening a chancery case in Buckingham contesting the new will, which was no doubt quickly recorded by Mitchie and/or whoever else might have been named Executor(s). Her brother, John C. Pryor, may have been among them and the one to engage William Wirt as his sister's attorney. The Jefferson brothers would argue that their father's will dated May of 1808 remained his true intention for his estate at the time of his death and that, while in the process of dying, he was unduly influenced by Mitchie.

Currently, only a few disparate items survive to give any idea of what took place in Buckingham County's courthouse. The argument continued for months, if not years. The future of Snowden and the African Americans enslaved there, who were part of the estate as Randolph's personal property, hung in the balance of this decision. While we know, ultimately, what happened to Snowden, the plantation and its dwelling house, the facts concerning the fates of more than twenty-two black men, women, and children turned to ashes with the lost documents in the 1869 Buckingham Courthouse fire.

Buckingham County: A Contest of Wills

Thomas Jefferson was not at his brother's side when he died on August 7th, though he did make an effort to be there. On Tuesday, August 1st, Thomas received a communication from Mitchie, addressed to him and Mrs. Marks, informing them of Randolph's condition. He had been very ill, but was stable when Mitchie wrote her letter, which Squire delivered. Thomas quickly responded, explaining that Anna Scott Marks would have come immediately; however, there was much sickness at Monticello. On August 2nd he wrote:

> Dear Sister
> your letter of yesterday gave us the first information of my brother's illness. we learn it with great concern mrs Marks would have visited him, but that we have now in our family, both in doors, and out, more sickness than I have ever had since I was a house keeper. Cornelia, Virginia and both of mrs. Bankhead's children. all of these are on the recovery except Virginia. without doors two or three are taken of a day, so that all the houses of the negroes are mere hospitals requiring great and constant attendance and care; all of an epidemic dysentery now prevailing thro' the neighborhood. your letter gives us a hope that my brother is mending. we shall be very glad to hear of him by the mail which furnishes a weekly-opportunity.
> I return by Squire the harness he was so kind as to lend me. it has answered the only occasion I had for it, and may on some other be useful to yourselves. affectionately Yours.
> TH: Jefferson[15]

The cause of Randolph Jefferson's death is unknown. He may have been suffering from the same "epidemic dysentery" which was prevailing in the neighborhood of Monticello. If so, it is entirely possible that he died of complications of dehydration.

A second letter arrived at Monticello on Friday, August 4th, written by Zach Pryor, Mitchie's brother. His letter may have warned that Randolph had taken a turn for the worse and was now failing fast. More may have been conveyed, however, both it and the letter from Mitchie are lost.[16]

On Saturday, August 5th, in response to Zach Pryor's likely more urgent letter, Nancy Marks headed for Snowden. Thomas' servant, Wormley Hughes, accompanied her, taking $1.00 for their ferriage expenses.[17] After their departure, that same day, Randolph Jefferson, Jr. arrived at Monticello informing his uncle of "the extreme danger of his father's situation." Randolph, Jr. told Thomas of a new will, signed by his father, about which he was uneasy and which he believed his father did not understand. The new will favored Mitchie over Randolph's sons. Obviously, Randolph, Jr. wished the 1808 will to remain valid.[18]

In 1815, Randolph, Jr. lived nearby in Fluvanna County. That he had anxiously ridden to his uncle's for help indicates that he understood all too well the implications of the new will – that it left him and his brothers with far less property than if the 1808 will stood after Mitchie received her dower right. Randolph, Jr. and his brothers desperately wanted this new will rescinded, so much so that Randolph, Jr. risked not being present at his father's deathbed in his urgent desire to thwart Mitchie's inheritance.

Thomas understood the gravity of his nephew's plea and assured him that he would visit Snowden "the moment his horses returned from carrying their sister, Mrs. Marks, who had gone that very day to see Randolph." Thomas also assured Randolph, Jr. that he "would immediately prepare a short instrument for revoking the will recently made, and reestablishing the former one." If Randolph, Sr. chose to sign it, he would be indicating that such an instrument was his wish. Thomas prepared this, and Randolph, Jr. returned to Snowden with the paper of revocation.[19]

On Sunday August 6th Wormley returned with the horses and $ .25, which he gave to Mr. Jefferson.[20] When Wormley left Snowden, Randolph was still alive and Nancy had reached her twin's side in time to say her goodbyes.[21] While the horses rested, Thomas prepared to leave for Buckingham the next day.

On Monday morning, Thomas set out for Snowden, but at Scott's Ferry he learned that his brother was already dead. Returning home to Monticello, he quickly sent a copy of the May 1808 will to fellow executor, Hardin Perkins, who soon attested to the will at Buckingham Courthouse.

Randolph Jefferson apparently died early in the day, and while at first it might seem odd that Thomas did not continue on to Snowden to see his grieving sister, nephews, and in-laws, he had a job to do as the formerly appointed executor of Randolph's estate. He only had Randolph, Jr.'s word that there was a new will and did not know whether or not his brother had signed the document revoking this alleged new will. Time was of the essence for the administration of the estate, so he put aside any familial duties, assuming his role as Counselor and Executor.

During the emergency surrounding Randolph's last hours, Thomas had received a letter from Wilson Cary Nicholas dated August 6th, but did not address it until he returned from Scott's Ferry. It contained news of Randolph's dire state and the rumored details of the second will. Nicholas had good reason to take the liberty to send Thomas his concerns. He was not only Jefferson's friend and political colleague, but the previous March he also had become father-in-law to Thomas Jefferson Randolph when he married Jane Hollins Nicholas. The letter reads:

Wilson Cary Nicholas (1761-1820). [© THE CLEVELAND MUSEUM OF ART. WILSON CARY NICHOLAS, C. 1805. GILBERT STUART (AMERICAN, 1755-1828). OIL ON CANVAS, 72.8 X 59.8 CM. THE CLEVELAND MUSEUM OF ART, GIFT OF MR. AND MRS. THOMAS W. MOSELEY 1992.305]

Warren, Aug 6, 1815

My Dear Sir

I have this moment heard that the doctor, who attends your brother left his house last night, under a belief he cou'd not live many hours. It is reported he has lately made a will; by which he has given the whole of his property, except, about six hundred acres of his back land, and eight or ten negroes to his wife in fee simple. I thought I owed it to you to give you this information as it came to me in such a way as to induce me to fear there was some reason to believe both that he is in great danger and that he may have made the disposition of his property I have mentioned. I am sure you will pardon the liberty I have taken.
I am Dear Sir
yours most respectfully   W.C. Nicholas[22]

These are the only known details concerning Randolph's second will and what it implied for the integrity of Snowden. Instead of being sold whole, as Randolph initially envisioned, the second will deemed it would be divided. About 1,200 acres would go to Mitchie, with 600 acres of "back land" assigned to the boys. If none of his sons were excluded, about 120 acres of the less desirable land would go to each of his sons. At the time of Randolph's death, there were at least twenty-two slaves living at the farm, more than half of whom were to go to Mitchie. This added up to considerably more than her dower right, giving her closer to 2/3 of Randolph's estate as contrasted with the 1/3 of his property, which the law decreed.

Thomas took Tuesday, August 8th to consider the situation. Interestingly, he made no memorandum between August 7th and August 11th. On August 9th, Thomas responded to Nicholas, explaining how he started out for Snowden to see his brother, but when he reached Scott's Ferry, learned that Randolph was already dead. In his role as Executor, he returned to Monticello to retrieve his copy of Randolph's will. On the 9th, he wrote back to Wilson C. Nicholas:

I duly received your two favors of the 3. & 6th. I was engaged in the moment in preparing some necessary orders before my departure to see my brother, and could not therefore immediately answer them. . . .

Thomas Jefferson to Wilson Cary Nicholas, 9 August 1815. (Courtesy of Library of Congress)

I learnt on the 6th for the first time the imminent danger in which my brother was, and never till the day before that had any suspicion that he had made any disposition of his estate but by a will in my possession. I put into the hands of one of his sons a paper which if executed might have set all to rights. but they failed to use it. I set out early on the 7th to see him, but at Scott's ferry met the news of his death, and returned. some members of his family had behaved very undutifully to him and under the impressions from that he must have been led to the act which involves, I fear, the whole in ruin.[23]

On August 11th, Thomas wrote to Capt. Hardin Perkins, who was also appointed an executor in the 1808 will. His formality indicates that he is not well-acquainted with Randolph's neighbor and friend in Buckingham County.

                                                Monticello Aug. 11.15.

Sir

My brother lately deceased having some years ago made me the depository of his last will and testament, wherein you are named an executor, I think it a duty to put it into your hands, and now therefore inclose it to you. altho' also named an executor, my age disables me from undertaking it, and the pursuits of my life have not been such as to qualify me for any useful interference in the affairs of the estate. the will having been all written with his own hand will need no other proof: but if that could be doubted the witnesses who attested to make it more sure, can all attend when called on. Accept the assurance of my great esteem and respect.

                                                          Th: Jefferson

Capt Harding (sic) Perkins[24]

While some of Thomas Jefferson's associates may have felt that Mitchie had solely influenced Randolph to rewrite his will, Thomas knew that Randolph was disappointed in the behavior of at least two of his sons, particularly Thomas, Jr.

and Field, and that this may have, at least in part, caused Randolph to change his mind about what they deserved to inherit.

Sister Nancy Marks stayed at least another week at Snowden. She may have helped the immediate family, remaining for Randolph's burial. There are no existing details concerning a funeral service or where Randolph was laid to rest. On August 16th, Thomas gave his servant, Gill, $2.00 to cover the expenses -- "going for Mrs. Marks."[25] Thomas did not join her; he was clearly elsewhere. On the 13th and 14th he was about the neighborhood on business at Col. Reuben Lindsay's home and at James Madison's Montpelier. On the 20th, he traveled to Bedford County and Poplar Forest.

While Thomas was at Montpelier, Martha (Jefferson) Randolph was at Monticello and forwarded to Madison's home a letter concerning the August Court in Buckingham. Father and daughter discussed the upcoming contest of Randolph's wills in the following exchange:

View of Montpelier, by Anna Maria Thornton, watercolor on wove paper, c.1802. (COURTESY OF THE MONTPELIER FOUNDATION, JAMES MADISON'S MONTPELIER, ORANGE, VIRGINIA. PURCHASE MADE POSSIBLE THROUGH THE GENEROSITY OF SEVERAL MONTPELIER PATRONS, 2011)

Monticello Aug. 13, 1815

My Dear Father

The emergency of the occasion must apologise for the liberty I took in opening the enclosed. But as to morrow is Buckingham court and not knowing the danger that might accrue from a disappointment I ascertained by opening the letter Whether it was your self or the Witnesses only that were wanting. In which latter case they could have been summoned with out applying to you, but as some thing more seems requisite perhaps you can by writing do the business. Adieu. We are all as you left us the sick mending and no new cases. Yours,

MR.[26]

Montpelier Aug. 13. 15.

The letter you forwarded, my dear Martha, desiring me to attend the Buckingham court of this month, requires an impossibility because that is tomorrow. I know also that the trial of the question cannot be at the same court at which the two wills are presented. Time must be given to summon witnesses, and I suppose I shall be served with a summons notifying the day I must appear. We have had a safe journey, shall return to Mr. Lindsay's tomorrow night and shall be at home to dinner on Tuesday. Yours with unalterable and tender affection, TH: Jefferson[27]

Thomas also wrote to his nephew, James Lilburne Jefferson.

Monticello Aug. 19. 15.

Dear Lilburne

It was at the President's in Orange that I received your brother's letter requesting me to be at the next Buckingham court to give evidence on your father's will. it came to hand the Sunday evening & the next day was that of the court. time and distance therefore rendered my attendance impossible. I set out for Bedford tomorrow morning and shall be there to the 1st October. I do not know that my evidence can be of any importance; but if it be thought so, I could not attend without being subpoenaed; because the testimony of a person volunteering in a cause is not received with the same confidence as [wh]en he attends in obedience to a summons. if it be desired therefore subpoena may be inclosed by mail directed to me 'at Poplar forest near Ly[n]chburg,' and if sent soon I shall receive it in time. Accept the assurance of my friendship and best wishes.

Th: Jefferson[28]

On August 31[st], Thomas wrote from Poplar Forest to his daughter, Martha, concerning his anticipated appearance in Buckingham County's September Court: "We all arrived here without accident. . . . I find the distance from hence to the Natural bridge, by Petit's gap is only 29. Miles, I shall therefore go there on horseback; but not until I return from Buckingham court, where I presume I shall meet Jefferson (Randolph), and hear something from home. . . ."[29]

The contest of the two Jefferson wills had begun in the August Court, without Mr. Thomas Jefferson in attendance. As he suggested, the two wills were no doubt presented. The Court's initial job would be to hear witnesses and depositions, then decide which of Randolph's wills was valid. Once one will was chosen, the executor(s) could go about administering the estate. Because the results of the case are lost, one can only speculate on the Court's opinion based on what did and did not happen in the ensuing years.

No more details are known about the August hearing. In mid-September, Thomas Jefferson went to Buckingham Courthouse as a witness in the "trial" between Randolph's adult children and his widow, Mitchie B. Jefferson.[30] Jefferson's memo book indicates the following expenses at ordinaries and taverns. Bolling Branch and Guerrant Staples were tavern keepers in the vicinity of Buckingham Courthouse.

Sept 11, 1815   H. Flood's vales .25.
Sept 12, 1815   Raleigh. vales .25 Buckingham C.H. Branch dinner .50. Buck C.H. Staples entertainmt. &c. 8.075.
Sept 13, 1815   H. Flood's lodgg. & c. going & coming 6.75.[31]

The single most important surviving document concerning the case is Thomas Jefferson's three-page deposition. Fortunately, a copy remained in his files at Monticello. It reveals that Mitchie Jefferson, represents one party; Randolph's sons by his first marriage – Thomas, Lewis, Field, Randolph, and Lilburne – represent the second party, and that a case was opened in the Superior Court of Buckingham County.[32] Anna S. (Jefferson) Nevil, Randolph Jefferson's only daughter, was not involved. She was not a legatee in the 1808 will and is not known to make any claims in 1815.

Thomas Jefferson's deposition reads as follows:

September 15, 1815
The deposition of Thomas Jefferson of Albemarle aged seventy two years, taken by consent of parties in a controversy depending in the Superior court of the district of Buckingham between M.B. Jefferson the widow of Randolph Jefferson of the county of Buckingham lately deceased of the one part, and Thomas Jefferson, Robert Lewis Jefferson Field Jefferson, Randolph Jefferson and ~~Lewis Jef~~ lilburne Jefferson, sons of the said Randolph Jefferson decd on the other part.

This deponent being first sworn on the holy evangelists deposeth and saith that in the month of May 1808, being at his own house in Albemarle, his deceased brother Randolph Jefferson came to visit him, and while there, requested him to write his will, and stated to him the distribution he wished to make of his estate: that this deponent accordingly made a rough draught, read it to the testator, and by some small corrections, made it conformable to his wish: that the testator, by the advice of this deponent, copied the same with his own hand, but on account of some small inaccuracies, copied it a second time, fairly and wholly with his own hand, had it moreover attested by three witnesses subscribing in his presence, and deposited the same with this deponent for safekeeping: that the testator was, at that date, a widower and married some time afterwards: that, after his marriage, one of the testator's sons came to see this deponent, and conversing on the subject of the marriage, expressed great uneasiness, on a report to which he gave credit, that his father had made a marriage settlement on his wife of the greater part of his estate: that the said Randolph Jefferson coming afterwards to see this deponent, he mentioned to the sd Randolph the uneasiness of his family on the report of a marriage settlement; whereupon the said Randolph declared to him it was entirely without foundation: that he never had had a thought of making a settlement, nor had his wife or any of her family or friends ever made such a proposition to him; that they knew she would be entitled to dower of his estate, and he supposed they deemed it a sufficient provision; that they proceeded to converse on the subject of the will, and both of them considering that in case of his death, the law would so far controul his will as to give his wife dower of his estate, and that the will would be valid as to all the rest, it was deemed by both unnecessary to make any alteration in it, and was still left with this deponent in it's original form: that the testator was always in the habit of consulting this deponent in all (?) cases of importance respecting his ~~affairs~~ interests, and he knows of no such case in which he did not consult him, except that of his last marriage, of which he never spoke to him until after it's consummation; that being on a visit to him, at his house in Buckingham in May 1813 his wife spoke to this deponent concerning her husband's management of his affairs, and particularly the disadvantageous sales he made of his crops, and expressed a wish that this deponent would recommend to him to consul and advise with her on such transactions, and to consent to her generally taking a part in the direction of his affairs; which however he this deponent did not do; that neither on the occasion of this visit nor a second in September of the same year, which was the last he ever made to his brother, did he observe any appearance of extravagance in the economy of his house, or in any other article of his expenses; that he considered his said brother as not possessing skill for judicious management of his affairs, and that in all the occasions of life a diffidence in his own opinions, an extreme facility and kindness of temper, and an easy pliancy to the wishes and urgency of others made him very susceptible of influence from those who had any views upon him: that soon after the period of his last visit beforementioned, his sd brother, in the occasional conversations with him on the subject of his affairs, began to complain of his store debts at Scott's ferry and Warren, he thinks particularly with mr Moon and mr Johnson, & that these were accumulating chiefly by his wife; that these complaints became more and more serious, and at length, in the autumn of the last year, he stated that he should be under the necessity of selling some of his land: that he then spoke more pointedly of the agency of his wife in contracting these debts, he said that he had desired the merchants to furnish nothing but on an order written with his own hand; that after this [?] orders were sent, not written by himself, but so like his hand as to deceive the merchants and produce the articles; that thereupon he used a secret mark in his orders, which had effect at first, but was soon after discovered & imitated: and these imitations he expressly said were by his wife: That as well before, as after his last marriage he expressed dissatisfaction with the undutiful and disrespectful conduct towards himself of some of his sons, particularly of Thomas and Field, but he does not remember his particularizing any other of them: that nevertheless in all the conversations which this deponent had with his said brother on the subject of his affairs, altho' he cannot recollect the particular expressions, yet it was perfectly understood by both parties that the will in the possession of this deponent continued to be that which he meant to continue as his will and that their conversations were founded on this basis; and he does verily believe that if the testator, in his sound & healthy state, had intended to change his will, he would have applied to this deponent to make the change: that on Tuesday the 1st of August he received a letter from the wife of the testator, addressed to himself and Mrs Marks his sister, and on Friday the 4th one from Mr Zach Pryor, which are the two letters deposited in the court of Buckingham; that on Saturday the 5th Randolph Jefferson, son of the testator came to this deponent, informed him of the extreme danger of his father's situation, that he had expressed to him his uneasiness as to a will he had signed, which he did not understand, that his former will in possession of this deponent was the one he wished to stand, and his anxiety to see this deponent and have this effected: that this deponent assured the sd Randolph the younger that he would go to see his brother the moment his horses returned from carrying mrs Marks, who had gone that day to see him but that he would immediately prepare a short instrument for revoking the will recently made, and reestablishing the former one, which if his brother chose to sign ~~it~~ would effect what was said to be his wish whi[ch] instrument he did prepare and deliver to the said Randolph the young[er] that he set out on Monday the 7th of August to visit his sd brother, but at Scott's ferry meeting information of his death, he returned home, and soon after inclosed the will of May 1808, deposited with him, to mr. Perkins, who was named an executor in it, which is the same will which was presented to him in Buckingham court and was there recognized by him. And further this deponent saith not.

TH: Jefferson[33]

No known copy of the second will exists; however, a copy of Randolph Jefferson's 1808 will survived in Thomas Jefferson's papers. Having helped Randolph create the will and being named one of the executors, Thomas maintained a copy. It provides significant insight into Randolph's outlook in 1808. His first draft, dated May 27, 1808 reads as follows:

I Randolph Jefferson of Buckingham county in Virginia being in sound health do make the following testamentary disposition of my estate.

I give all the negroes which I shall own at the time of my death to be equally divided between my five sons Thomas, Robert Lewis, Field, Randolph & Lilburne, each of them to whom I may have given slaves during my life time, bringing the value of those slaves into hotchpot with those to be divided, and ~~draw~~ taking from those to be divided only so much as with those before given shall make his portion of slaves equal to that of each of his other brothers.

It is my will that all my lands and other property whatsoever be sold after my death, and the proceeds be equally divided among my said five sons, my son Thomas bringing into hotchpot with those proceeds the sum of one thousand pounds (at which I estimate the advantages I have given him during my life) and taking from the proceeds only so much as, being added to the £s Thousand pounds, shall make his portion under this bequest equal to that of his other brothers. ~~I give to my son Lewis my violin, to my son~~ Lastly I make my friends Harding (sic) Perkins & Robert Craig, my son Robert Lewis, and my brother Thomas Jefferson executors of this my will, revoking all others heretofore made and ~~writing~~ in testimony hereof I have written the whole with my own hand this 27th day of May 1808.[34]

These two documents combined illuminate Randolph's last years and the general situation among the family members. At the same time, they raise many unanswerable questions.

To begin, there is no clear reason why Randolph excluded his only daughter, Anna Scott "Nancy" Nevil, in his will. In May of 1808, she was married with children, and very likely living at Snowden. There is no reason to believe the families were estranged following the Nevils' departure for Nelson County, and Zachariah Nevil eventually served as Administrator of Randolph's estate. He was probably court-appointed and may not have been the first administrator in 1815. The fact that Nevil was assigned this duty might indicate that none of Randolph's named Executors ever served in their capacities.

It is noteworthy that, in 1808, Randolph still prized his violin, and that he wished Lewis to have it. Music must have remained important to Randolph and at least one of his sons shared his interest in the instrument. His request of a "hotchpot" approach to the division of his personal property, however, discouraged individual gifts, and this may be why the bequest was stricken out.[35]

Why did Randolph request that Snowden be sold and the proceeds divided? Perhaps, he did not wish to see it splintered into five sections. Did he imagine that at the time of his death, the boys would long be settled elsewhere and no longer interested in living on the place? Maybe, he hoped that one of the boys would have the capital to buy the farm in its entirety. Later in life, both Lewis and Field would acquire farms in Buckingham, indicating that they might have wanted to own Snowden. In 1815, however, neither of the young men had the

> I Randolph Jefferson of Buckingham county in Virginia being in sound health, do make the following testamentary disposition of my estate.
>
> I give all the negroes which I shall own at the time of my death to be equally divided between my five sons Thomas, Robert, Lewis, Field, Randolph & Lilburne, each of them to whom I may have given slaves during my life time, bringing the value of those slaves into hotchpot with those to be divided, and ~~draw~~ taking from those to be divided only so much as with those before given shall make his portion of slaves equal to that of each of his other brothers.
>
> It is my will that all my lands and other property whatsoever be sold after my death, and the proceeds be equally divided among my said five sons, my son Thomas bringing into hotchpot with those proceeds the sum of one thousand pounds (at which I estimate the advantages I have given him during my life) and taking from the proceeds only so much as, being added to the 1d thousand pounds, shall make his portion under this bequest equal to that of his other brothers.
>
> I give to my son Lewis my violin, to my son
>
> lastly I make my friends Harding Perkins & Robert Craig, my son Lewis Randolph, and my brother Thomas Jefferson executors of this my will, revoking all others heretofore made, and ~~writing~~ in testimony hereof I have written the whole with my own hand this 27th day of May 1808.

(COURTESY OF SMALL SPECIAL COLLECTIONS, UNIVERSITY OF VIRGINIA)

While visiting Monticello in 1808, Randolph Jefferson decided to write his last will. The first draft (27 May 1808) indicates his desire to leave his violin to his son, Lewis. The second draft (28 May 1808) abandons individual bequeaths in favor of an equally divided "hotchpot." Also note, that Randolph names his sons as they are commonly called by the family.

I Randolph Jefferson of Buckingham county in virginia being in sound health, do make the following testamentary disposition of my estate.—

I give all the negroes which I shall own at the time of my death to be equally divided between my five sons Thomas, Robert Lewis, Field, Randolph & Lilburne, each of them to whom I may have given slaves during my life time, bringing the value of those slaves into hotchpot with those to be divided, and taking from those to be divided only so much as with those before given, shall make his portion or of slaves equal to that of each of his other brothers.

It is my will that all my lands and other property whatsoever be sold after my death, and the proceeds be equally divided among my said five sons, my son Thomas bringing into hotchpot with those proceeds the sum of one thousand pounds (at which I estimate the advantages I have given him during my life) and taking from the proceeds only so much as, being added to the sd thousand pounds shall make his portion under this bequest equal to that of his other brothers.—

Lastly I make my friends Harding perkins & Robert Craig, my son Robert Lewis, and my brother Thomas Jefferson executors of this my will revoking all others heretofore made and in testimony hereof I have written the whole with my own hand this 28th day of may. 1808.

(Courtesy of Small Special Collections, University of Virginia) means to purchase it from the estate, even if there had not been the complication of the contest of wills.

There is little doubt that Randolph's 1808 will expressed his wishes at that time. Randolph desired his brother to be one of his executors and sought his advice in composing the will. Thomas was, after all, a trained lawyer and very experienced in the formal writing of legal documents. In addition to his friends, Hardin Perkins and Robert Craig, Randolph singled out his son, Robert Lewis, who may have been his most *simpatico* offspring. Interestingly, Lewis is not later named as disrespectful.

At the time Randolph's first will was conceived, he likely had no plans to remarry. It was probably sometime after May of 1808 that Randolph decided to wed Mitchie B. Pryor. Even after his marriage, there was little need to change his will, unless he intended to alter the division among his sons or add new beneficiaries beyond Mitchie. Virginia law guaranteed her a dower third and the remainder would be divided equally among his sons according to the original plan.

There could be any number of reasons why Randolph did not consult Thomas prior to his marriage to Mitchie. He might have feared criticism because of Mitchie's youth. She was a great deal younger, although from a solid Buckingham family. Two of her brothers were eventually doctors, one a judge, and a third a Master Freemason. Given the accomplishments of the family, Randolph's initial defense of Mitchie, her family and friends to his brother is certainly understandable.[36]

Thomas indicates that by May of 1813, Mitchie expressed concern over Randolph's handling of financial affairs. He was fifty-seven years old, and, according to his brother, may never have been a particularly good businessman. Diffident, kind, and pliable – easily influenced by the opinions of others – these personality traits might have increased with age. It is possible he was persuaded into risky ventures by John Jefferson and others. His oldest son, Thomas, had convinced Randolph to advance him a great deal of money, £1,000, before 1808. Perhaps, too, where Randolph lacked confidence, Mitchie was bold.

If Mitchie had begun to drain Randolph's purse, Thomas Jefferson did not see any indication in the household as of September of 1813. Yet subsequently, either at Monticello or elsewhere, Randolph expressed increasing concern about Mitchie's spending in Scott's Ferry and Warren. By the autumn of

1814, Randolph sold land to Charles A. Scott to cover the debts. Subsequent tax records indicate that Scott purchased a total of at least 178 7/8 acres of Snowden's acreage.[37]

Sometime before Randolph's last illness, one of the Jefferson boys (which one is unknown) went to Monticello and complained of a new "last will" their father had drawn up which favored Mitchie. Randolph's son had heard a credible report that "his father had made a marriage settlement on his wife of the greater part of his estate." Later, when Thomas suggested to Randolph that his family was uneasy with the settlement, Randolph denied the rumor as "entirely without foundation."[38]

Randolph insisted that "he never had had a thought of making a settlement, nor had his wife or any of her family or friends ever made such a proposition to him." He continued that the entire family knew that Mitchie would be entitled to her dower right of his estate, which in Virginia was 1/3 of his property. Randolph certainly understood how the law worked in this case and that his existing "last will" would remain in force for 2/3 of his estate. Of course, if Mitchie preceded him in death, without issue, the 1808 will would stand as written. Randolph's understanding of basic Virginia law as applied to widows and inheritance, demonstrates more than a rudimentary familiarity with "how things work." This part of Thomas' deposition states that Randolph did not make a second will *before* his last illness.

In fact, based on Thomas' deposition, it would seem that the second will was written close to Randolph's death. Thomas says that it was understood that Randolph did not intend to change his will. If Randolph had wanted to write a new will, he would have consulted Thomas. As stated in the deposition, "he does verily believe that if the testator, in his sound & healthy state, had intended to change his will, he would have applied to this deponent to make the change." Perhaps, however, the very same hesitation that kept Randolph from discussing his pending second marriage with his older brother reasserted itself when he wished to change his will in Mitchie's favor.

In the deposition, Thomas planted the seeds of doubt about his brother's abilities and stressed an easily influenced mind, without directly accusing Mitchie or the Pryors of manipulating a dying man. Was there reason to believe that Randolph's mental faculties were deteriorating?

The story of Randolph's mounting concerns about Mitchie's unauthorized spending during 1813-1815, and his subsequent agreement with the merchants of Warren and Scott's Ferry, could be viewed as a rather bizarre way for a mature man to manage his youthful and, perhaps, frivolous wife. To accuse Mitchie of forgery, to devise a "secret mark" for his orders, then to further accuse his wife of learning and forging his "secret mark," seems to compound the intrigue. Was Randolph becoming increasingly paranoid about his assets and debts, or was Mitchie actually a very mischievous, if not an outright duplicitous, spouse? Unfortunately, there is nothing to compare to Thomas' version of Randolph's story. Thomas Jefferson is, of course, an impeccable source. It is also possible that Randolph was losing his grip on the reality of the situation, and that Thomas was not fully aware of the day-to-day strain his brother was experiencing.

It is not surprising that some of Randolph's sons expressed open disapproval of his marriage to a much younger woman, young enough to be his daughter. It is interesting to note that the "undutiful and disrespectful conduct," as reported by Randolph, came from Thomas, Jr. and Field. Thomas, Jr. was bold in other family matters, particularly in the case of the Lewis family's manipulation of Craven Peyton's property.[39] Of Randolph's sons, it is Field who would become highly successful in business, indicating a strong, assertive personality.

Notably, Robert Lewis Jefferson, who was apparently a trusted, favorite son and appointed as an executor in Randolph's first will, is not named as one of the disrespectful sons. Nor is Lilburne, who was considerably younger than the others and may have been the only one of the brothers to actually live with Mitchie for a protracted period, though his references to her as "the widow" in his correspondence with his uncle Thomas do not evoke a warm relationship.

There was one piece of very significant news that Thomas Jefferson failed to mention in the deposition when considering Randolph's last-minute change of mind in regards to his estate and how it would be distributed at his death. **Mitchie B. Jefferson was pregnant.** Thomas probably did not know this fact when he wrote his September 15th deposition.

Based on the birth of her son, John Randolph Jefferson, on March 8, 1816, she was about two months pregnant when Randolph died. Often, first babies are late, so she may have been a couple of weeks further into her pregnancy.

The big question remains, exactly how many individuals were aware of this on August 7, 1815, including Mitchie herself?

In August, if Mitchie knew or suspected she was pregnant, she surely presented this to her husband. Perhaps the dower third of the estate no longer seemed adequate to her or to Randolph. If Snowden was sold, where would she and her child live? Surely, he would want the unborn child to be considered in the will, although he knew that Virginia law protected the rights of a "posthumous" child. If Mitchie's pregnancy was not apparent when Randolph died, then whatever appeals she made to Randolph to leave her more than her dower right certainly worked.

Since the exact contents of the will which Mitchie B. Jefferson and/or Randolph's appointed executors presented to the Buckingham Court are unknown, we must rely on Wilson Cary Nicholas' description that Mitchie received everything in fee simple, which would give her absolute ownership of the land and personal property, allowing her to pass the property to any heirs she chose. Nicholas went on to note that Randolph had excepted "about six hundred acres of his back land and eight or ten negroes," which he presumably left equally to his sons.

The 1815 Personal Property tax and typical wills of the day help define just what "everything" might have been. Mitchie's inheritance would have included: Randolph's dwelling house, valued at $900; all its household and kitchen furniture; the livestock: horses, any oxen and other cattle, sheep, and hogs; all of the plantation utensils, including wagons and carts. That year he was taxed on twenty-five head of cattle and six horses. Valuable household furniture included the Mahogany pieces and a chest of drawers. There was also Randolph's silver watch and his violin – unless, perhaps, he still gifted the violin to Lewis and named a son to receive the beloved watch. The gig, of course, was demolished.

As for the human property at Snowden, the 1815 tax included twenty-two taxable slaves and there were no doubt additional children under nine years old. If eight or ten were to be divided among the Jefferson brothers, which would leave Mitchie with over a dozen taxable slaves.

In 1815, Randolph was taxed on 1,827 acres on the James River, though some of this was in the process of being purchased by Charles A. Scott. This was an excellent property; that year the tax obligation was a whopping $60.38.

Joseph Carrington Cabell (1778-1856). (UNIVERSITY OF VIRGINIA VISUAL HISTORY COLLECTION, RG-30/1/10.011)

Typical of the best Buckingham plantations, Snowden was a combination of high and low ground near the river. Some fields were developed; other sections were undeveloped woodland timber. What exactly did Nicholas mean by 600 acres of back land? It does not sound very desirable, analogous to the proverbial "back forty." Perhaps, it was undeveloped and further from the river.

On the other hand, if the Jeffersons could convince the court that the 1808 will was Randolph's dying wish, Snowden would be sold, likely at public auction, to raise cash to divide among his sons. Of course, any of the five Jeffersons could attempt to purchase Snowden if the first will was upheld.

On September 13, 1815, Thomas Jefferson's friend, Joseph Carrington Cabell (1778-1856), wrote from his Nelson County home, Edgewood, to his dear friend, John Hartwell Cocke, at his Fluvanna County home, Bremo. Cabell describes the "Jefferson vs. Jefferson" atmosphere at Buckingham Courthouse and mentions the involvement of Richmond attorney, William Wirt.

Wirt is at Buckingham C⁺ house, engaged to assist in establishing Randolph Jefferson's last will, giving nearly all his estate to that widow of his. I stood there two days when the testimony not being gone thro. I came home. The case is not as strong as I suspected, but I should overthrow the will. The widow is a Jade of genuine bottom.[40]

It is completely clear how Cabell felt about Mitchie. "That widow of his" is bad enough, but to call her a "Jade of genuine bottom" paints a very unpleasant picture, conjuring a morally unprincipled character. Seen through Cabell's eyes, she was more than a mischievous young bride on spending sprees; she had taken advantage of Randolph Jefferson and deceived him, perhaps in many ways.[41]

What prompted Cabell's attendance at the September Buckingham Court? Cabell was long an advocate of Thomas Jefferson's political party and, in January of 1815, Jefferson had approached Cabell to support him in his plans to establish what would become the University of Virginia. Did Cabell come to Buckingham Court out of allegiance and friendship to Thomas Jefferson, who gave his deposition in September, or did he have a relationship with Randolph Jefferson and his sons yet to be discovered? Perhaps Cabell was merely curious to see William Wirt in action and that drew him to the courthouse. Interestingly, he chooses to tell Cocke that Wirt is there, not mentioning Jefferson's deposition.[42]

If Cabell came from Nelson County to hear the evidence, who else attended Buckingham Court in mid-September? A contest concerning a former President's brother, including an appearance by Thomas Jefferson, himself, and the famous William Wirt, certainly drew attention. The gossip mill was no doubt very busy. Likely in attendance . . . neighbor, friend and executor for the first will, Hardin Perkins, the Jefferson sons, Mitchie and members of the Pryor family, fans and detractors of both sides. It must have been a very lively week at Buckingham Courthouse.

William Wirt's reputation across the Commonwealth was enough to bring Joseph C. Cabell and many others to Buckingham Court that week. By 1815, he had become one of the best known lawyers in the country:

> Wirt [1772-1834] was conspicuous for his personal beauty, both in youth and manhood. His manly, striking figure, intellectual face,

This engraving of William Wirt (1772-1834) was based on a painting by Alonzo Chappel and published by Johnson, Fry & Co. Publishers, New York. (COLLECTION OF JOANNE L. YECK)

clear, musical voice, and graceful gesture won the favor of his hearer in advance. In his public addresses he was usually calm, self-possessed, and deliberate. His memory was very retentive, and he excelled in felicity of quotation, sometimes retorting upon an adversary with telling effect a passage inaptly cited by him from an English or Latin poet. A pocket edition of Horace was often thumbed in his journeys, but Seneca was his favorite classic author. Wirt's conversation, enriched by multifarious reading, yet easy, playful, and sparkling with wit and humor, was full of interest and charm. Similar qualities pervade his letters. He was a member of the Presbyterian church, and in his last years took great interest in missionary societies, and was president of the Maryland Bible society.[43]

A short biography prepared by the Maryland Historical Society, where many of Wirt's papers are housed, discusses Wirt's life just before the time of Jefferson vs. Jefferson: "Even [Thomas] Jefferson wanted Wirt to enter the national political arena, but Wirt refused. National politics offered too little money and too much pain; a two-year stint in the Virginia legislature, 1808-1810, only convinced Wirt that politics was not his future...."[44]

In fact, the ambitious William Wirt was hungry for both money and fame. The son of immigrants, a Swiss father and a German mother, he was orphaned by the age of eight. Suffering from a debilitating stutter, the strong-willed Wirt overcame it to become a commanding speaker and successful courtroom attorney. As a further strategy to gain acclaim, Wirt defended the unpopular, if not the despised. He was part of the defense teams on the sedition trial of journalist John Callender (1800) and the murder case of George Wythe Sweeney (1806), who was accused of poisoning his uncle and Thomas Jefferson's mentor, Judge George Wythe.[45] In 1807, Thomas Jefferson asked Wirt to serve as the prosecutor in the treason trial of Aaron Burr.

Beyond professional ties between Wirt and Thomas Jefferson, there were strong friendships among their extended families and in-laws. Wirt's first wife was Mildred Gilmer, the daughter of Jefferson family friend, Dr. George Gilmer. Wirt's companion and associate, Dabney Carr, was the nephew of Thomas and Randolph Jefferson; Wirt and Carr read law together. Young Carr was very close to his Uncle Thomas, having lived extensively at Monticello following his father's death. It would be "natural" for Carr to take the side of his Jefferson cousins. Why, then, would his good friend, William Wirt, take the opposing side?

How had the Pryors come to attract the interest of this established Richmond attorney and well-known essayist? Was it simply that Wirt loved rising to the challenge of a difficult defense, particularly one associated with a prominent Virginia family? Did Wirt have a personal connection to Randolph Jefferson or even to the Pryors?[46] Significantly, Wirt took the case without insisting on his usual "money up front." This might suggest a relationship with Randolph or the Pryors, especially following Randolph's marriage to Mitchie. Was Wirt doing a favor for his friend's widow? Did he see something in Mitchie most others did not? Had he believed the marriage beneficial for Randolph? Wirt's seems a lone voice of sympathy towards the widow Jefferson, though he refers to a friendly informer, someone who is not a member of the Pryor family.[47]

In February of 1816, the Pryors were preparing for their next round in Buckingham's April Court. Mitchie, who was less than a month away from delivering her first child, was living at Woodlawn with her mother, Susannah. Brothers Zachariah and Langston also remained at home, as did Leonard. Zane Pryor was still in Buckingham, though he may have been established in a home of his own. John C. Pryor, who was about twenty-seven years old, was there as well and appears to be the family member who communicated with William Wirt. One Pryor family history states he was later a Judge, so he may have had the mindset and/or training for legal thinking.

In a letter written on February 16, 1816, from Wirt to John C. Pryor, Wirt indicates that "Jefferson vs. Jefferson" continued to be heard in the Buckingham Superior Court, which met twice a year in April and October. Wirt had been retained through the autumn months and now warned the Pryors that the Jefferson brothers planned to go to court with new evidence. Wirt, in his mid-forties, wrote with confidence, even fatherly advice, to young Pryor:

To Mr. John C. Pryor
Buckingham C House
(Ric^d February 16 1816)
Dear Sir

I wrote you a letter of general instructions a mail or two ago as to preparations of evidence for the case of Jefferson vs. Jefferson It has since occurred to me that it would be well if it could be done, to repel the charge of offensive extravagance on the part of Mrs. J. as it relates to her dealing at the stores, and the tale of the instruction given to the merchants by Randolph Jefferson not to trust her further On this subject the merchants themselves ought to be questioned and the average state of Randolph Jeffersons accounts before and after his last marriage to be examined, and if they furnish conclusions in our favor to have copies of the accounts before and after the marriage produced – The character of the overseer (Goss I believe was his name) ought to be proven if as I have heard, it be a bad one. The terms on which R.J. and his family lived at your mothers – after his marriage with your Sister (I mean as to supplies etc) ought to be shown to repel the conclusion that the whole family were living on him and on his means. It is also proper that I should mention to you what under any other circumstance it would be highly indelicate I should do, that the party on the other side have announced (as I have been told in confidence) their determination on the trial before the Judge of the Superior Court to offer proof of the Infidelity of Mrs. J. to her marriage bed I may have been and hope I have been misinformed – at all events since this hint was communicated to me in the utmost secrecy and from friendly motives towards Mrs. J. you will perceive the impolicy and impropriety of making any noise about it. I communicate it as her counsel with the view of putting her on her guard and enabling her to be prepared to repel such an imputation if the thing should be attempted. Let there be no threats or noise on this subject, act with prudence, and do not exasperate the prejudice already too strong against your sister's cause.

Finally it is proper and candid that I should state that your sisters is the first cause, in which I ever left the range of my profits and the current profits of my office without receiving the fee before I entered upon the cause – seeing that I have already been five months and shall have been six, without receiving a cent of my fee, she must not think it amiss, that I give notice that I cannot reconcile to the sense of duty which I owe to my other clients and to my family to leave Rich^d again, in her cause til I shall have received the fee in arrears, as well as that for which I have engaged to attend in April. Both the fees must be paid to me before I leave house - and with this length of notice, I hope she will not complain of the condition

       I am Sir respectfully your obt. Sevt.
       Wm Wirt[48]

  Especially given the fact that the Buckingham County Courthouse burned in 1869, Wirt's letter offers extremely valuable insight into the next chapter in "Jefferson vs. Jefferson." This is the only communication between Wirt and the Pryors known to survive, and it is not known if Wirt received his fee or appeared at April Court. In any case, in 1817, Wirt would step up to the national scene when James Monroe named him Attorney General of the United States. At that point, the Jefferson case was far from concluded.

  Wirt's letter is very clear concerning both the evidence he suggests the Pryors collect and the matter of his fee, as yet unpaid by Mitchie. The strategy Wirt suggests for new evidence illuminates some of the Jefferson family's accusations. They have likely charged that, prior to Randolph's death, Mitchie and her family had diminished their inheritance in several ways. First, they repeat the charges that she was spending extravagantly in the local shops. This "tale" was told by Thomas Jefferson, and Wirt reduced it to hearsay. Notably, he is not entirely sure that the merchants can produce evidence in favor of Mitchie's case.

  Second, the Jeffersons are charging not only that Mitchie freely spent Randolph's money, but that her family took advantage of Randolph's financial generosity. To accuse "the whole family" of living on Randolph "and on his means" suggests that Randolph and Mitchie must have spent significant periods of time at the Pryors or *vice versa*. The Jeffersons may have gone as far as to suggest he was spending money on the upkeep or improvement of Woodlawn.

It is clear that Randolph and Mitchie spent extended time at Woodlawn, with Mitchie's unmarried brothers and her mother, but exactly when and for how long? On October 6, 1811, one of the Mr. Pryors delivered a letter from Randolph to Thomas at Monticello. The Jeffersons may have been living with the Pryors at that time. An extended stay is documented in a letter dated February 8, 1812, when Randolph wrote to his brother from Mitchie's home. Again in February of 1813, the R. Jeffersons were at Woodlawn due to Mrs. Pryor's illness.

However, it is also clear from the Jefferson-to-Jefferson letters that Randolph and Mitchie spent significant time at Snowden. Randolph was at home at Snowden to receive his brother in May and September of 1813. Two years later, Randolph writes to Monticello from Snowden in February of 1815, their sister Anna Scott Marks visits in April, and Randolph writes of a busy harvest in June.

Randolph was in his mid-fifties when he married Mitchie, and observing his brother's longevity, had every reason to believe he had many years ahead of him of vigorous living. Perhaps, he had plans eventually to purchase Woodlawn for his bride. It was still in her father's estate when they married. This may have been due to the fact that some of the children, likely including Mitchie, were still minors when David Pryor died. If Randolph had that in mind, he may have "invested" money in the farm, and his grown sons likely did not approve of that choice.

There is every reason to believe Randolph was a kind person and when he married Mitchie, he took on her concerns, whether those were about her widowed mother or her not-as-yet independent brothers. Randolph stayed with his mother-in-law when she was ill and may have paid some doctor's bills. His sons from his previous marriage could easily resent this growing attachment to a new family, but it didn't necessarily add up to full support or make the Pryors leeches. They had property of their own, and while they weren't wealthy, no evidence has surfaced to indicate they were in any way destitute.

The introduction of Mr. Goss, the overseer, is a new wrinkle in the case. If he was a bad character what could he be accused of? Was he left "in charge" at Snowden while Randolph and Mitchie lived with the Pryors? Was Mr. Goss guilty of stealing, skimming profits, or perhaps mis-management of the farm, thus creating losses? Wirt's desire to pursue Goss' rumored bad character indicates that dishonesty, not incompetency, may be behind Goss' actions.[49]

The shocking, but not entirely surprising, section of Wirt's letter is the "highly indelicate subject" that the Jefferson brothers were considering accusing Mitchie of infidelity of "her marriage bed." They were undoubtedly not alone in their suspicions or accusations. Cabell's comment that she was "a Jade of genuine bottom" indicates he did not approve of her behavior, and the term Jade carries with it a sexual connotation. She may have been flirtatious. She may have been young and pretty. She may or may not have been adulterous.

Mitchie conceived her son in early June, based on John Randolph Jefferson's birth on March 8, 1816. By the time Wirt wrote to John C. Pryor in February, Mitchie was very pregnant, possibly confined to her home.

Though married to Randolph Jefferson for at least four or five years, possibly longer, this was Mitchie's first known pregnancy. She could have had others – miscarriages, stillborn babies, infants who died; however, if this was her first public pregnancy after five years of marriage there would be even more basis for gossip. Other behaviors may have fueled the gossip, as well. Wirt makes it clear that Mitchie had many detractors beyond Mr. Cabell. He warned Pryor not to "exasperate the prejudice already too strong against your sister's cause."

While Randolph had periods of sickness which are revealed in his letters to his brother, he appears quite hardy in his June 23rd letter, written just six weeks before his death. In it, he explains they are very busy with the harvest. He did not expect to finish for a few more days. While his closing is formulaic, it may still be very true: "My Wife joins Me in love and respect to you and family. – I remain your Most affectionately. — Rh:Jefferson.

Whatever Randolph's children or Thomas Jefferson's friends thought of Mitchie B. (Pryor) Jefferson, the fact was she was a young widow expecting a child. It is unclear which, if any, of Randolph's sons remained at Snowden to be of any help to her, or if she went to Woodlawn immediately following Randolph's death. We know from Lilburne Jefferson's subsequent correspondence that he was not inclined to appreciate her or her condition, which he does not mention in the surviving letters.

1816: No Squire for Snowden

On the Jefferson brothers side of the case, little is currently known. Who was their counsel? Did Thomas Jefferson, Jr. "take the case" of the plaintiffs as he did in other family matters at other points in time, or did they hire a professional to take on the well-known and very competent William Wirt, Esq.?

After Uncle Thomas gave his deposition in September of 1815, there is nothing in his memorandum books to indicate ongoing involvement in the case. Given Thomas Jefferson's overwhelming desire for familial harmony and avoidance of confrontation, the public deposition itself was a large gesture on his part. He must have felt strongly about the perceived injustice of the second will. However, just because the will was not fair did not mean that it was not legal or not ultimately what Randolph wanted at the time of his death. As Cabell put it, the case against Mitchie was "not as strong as I suspected." He, like others, would have liked to "overthrow the will," not because it was unsound, but because he disapproved of the widow Jefferson.

On January 30, 1816, Thomas Jefferson advanced his slave, Wormley Hughes, $1.00 for expenses to Snowden.[50] Wormley's purpose remains a mystery, though it may indicate that some of the Jefferson sons were still occupying it. On February 10, 1816, a concerned Uncle Thomas wrote to Lilburne.

Monticello, Feb. 10, '16

Dear Lilburne:

My sister Marks tells me you are in want of clothes and other necessaries, and are living at the town at the ferry until the question is decided about my brother's will. I wish you would come and stay with us. I have proposed this on one or two former occasions, and would now press it. You shall employ your time as you please, and as usefully to yourself as you please, in which, and in anything else I can I will render you my best services. But come particularly and let me have you furnished at Charlottesville with all proper and comfortable clothing. In the hope of seeing you I remain affectionately.

Th. Jefferson[51]

It was more than the delays of the settlement of Randolph Jefferson's will that compromised Lilburne's situation. The dwelling house at Snowden had burned, likely taking much of Lilburne's personal property in the blaze. In a letter written from Scott's Ferry, Lilburne describes the abandonment of Snowden and the shocking destruction of the house.

Scott's Ferry Feb 18th 1816
Dear uncle
    I received your letter by Guilley your advice in respect to my situation. I thank you kindly for your advice. I went to Buckingham CH on monday last and spoke to the Curator in respect to my situation and he refused to let me have money out of the estate. I then appealed to the Court for justice. the court would not authorize the curator to let me have money out of the estate unless I would shoose (sic) a guardian. I then choosed a guardian and he will no doubt do justice by me. I have not been in want of clothing but I thought that I was entitled to funds out of the estate I should be very happy to come and live with you but I have rented the ferry and the man that I rented it of wont I am affraid compromise with me but if he will I will come over. My anxiety is to travel and that westardly. The plantation snowden is to be rented out next week and I had a thought of renting a part of it. The widow has moved to her mothers. She had not moved there more than two days before the house caught on fire and bournt everything into ashes. I will let you know in a few days is soon as I can see Mr. Thomas the gentleman that I rented the ferry of he is gone to Richmond.
                 I am Sir your affectionate nephew
                 James L. Jefferson[52]

The following month Lilburne accepted his uncle's offer for clothing and charged to Thomas Jefferson cloth and stockings from Charlottesville merchants James Leitch and Bramham & Jones.[53] It is not currently known whether Lilburne also accepted Thomas' warm and insistent offer to join him at Monticello.

Lilburne's need for a guardian indicates he is not yet twenty-one years old. This would place his birth after February 1795, which is quite possible.[54] His relative youth also explains his uncle's concern for his welfare and the invitation to join him at Monticello. Having served in the militia during the War of 1812, Lilburne was no longer a boy, but he was still not a man in legal terms. The only suggested livelihood Lilburne may have in mind could be connected to the ferry.

Lilburne's "anxiety" to travel westward is intriguing. With his home at Snowden burnt, his father dead, and his family fighting, leaving Buckingham County no doubt had great appeal. How far west did Lilburne imagine going? He watched at least two of the Pryor brothers remove to Tennessee and his Lewis cousins were long gone to far western Kentucky. There is currently no evidence that Lilburne ever left Virginia. By the 1830s, he was living in Lynchburg, which may be as far west as he ever got.

Lilburne tells of going to Buckingham Court on Monday last, which was February 12th. There he approached the Curator of his father's estate to advance him some funds. Neither the identity of the Curator or Lilburne's eventual guardian is currently known. The fact that the court had appointed a Curator, rather than an Executor or Administrator, is revealing. This information, coupled with Wirt's letter, makes it clear that, by February of 1816, the court had not yet committed to an opinion about the two wills. The estate had to be safeguarded while the Chancery Case continued. Particularly in the Southern planter's domain, an estate was a living world with crops, livestock, and human property.[55] Curators were typically appointed when a person died intestate and a significant amount of time might pass before heirs could be determined, located, and notified. A Curator might be equally important when a will was contested as in Randolph's case, and the court perceived a significant amount of time might pass before the case was resolved. In this situation, the court was very correct to appoint a Curator. Randolph's estate would remain open for at least ten years, requiring a wide variety of decisions far beyond an Administrator's duties of keeping accounts and distributing property.

Additionally, once it was determined that Mitchie B. Jefferson was pregnant, how would the birth of this new heir affect the estate? Even if she took her dower right, would the estate remain open until the youngest of Randolph Jefferson's heirs reached majority?

Surely the court could not assign Hardin Perkins or Thomas Jefferson as Curator, as they were nominated as Executors in the first will and could be seen as biased against Mitchie's interests. Besides, if Thomas Jefferson had been appointed Curator, evidence of his involvement would have survived in his private papers. Likewise, the Court could not assign one of Randolph's sons or one of Mitchie's brothers, no matter how responsible they might be. The Court must find an objective, responsible, unrelated third party, which, in the small world of Buckingham County, might not be so easy to do.

Lilburne indicates that Snowden would soon be rented . . . but to whom? He also relates that the house has burned, completely destroyed. It would follow, then, that the renter was not planning to live there, but simply to work the fields. This might indicate someone in the neighborhood or general vicinity would be most likely to rent the farm, or, perhaps there was a tenant house on the property.

The cause of the fire at Snowden is unknown; while it may have been accidental, the burning of Randolph Jefferson's papers before the next court hearing might have complicated and delayed the proceedings. Either side in "Jefferson vs. Jefferson" might have benefitted from this destruction of evidence. It is also important to note that prior to the fire, Mitchie had moved out, so her personal property was presumably safe.

Lilburne gives no indication when Mitchie moved to her mother's home and there is still no mention of Mitchie's pregnancy. If she had just left Snowden, her "confinement" at Woodlawn lasted less than a month. When John Randolph Jefferson was born on March 8, 1816, it was likely at the Pryor homeplace, four to five miles north of Buckingham Courthouse. He would spend the first two years of his life at Woodlawn.[56]

Lilburne, having decided not to rent part of Snowden, had rented the ferry from a Mr. Thomas. A common name in both northern Buckingham and southern Albemarle, no more is currently known about Mr. Thomas, though he may have a Lewis family connection. The following month, Lilburne wrote to his uncle again, this time from the village of Warren, just up the James River from Scott's Ferry. It, too, had a ferry crossing often used by Thomas Jefferson, and perhaps this is where Lilburne had rented "the ferry." In the following letter written on Friday, April 19th, Lilburne refers to the previous Monday, April 15th and the April Term in Buckingham Courthouse.

Warren April 19 1815 (sic)

Dr uncle

There has been no court on the account of the judge not coming down he sent a mesenger to the courthouse on monday stating that he was very ill but that he would try and be down on Wednesday but yesterday Mr. Booker received a letter from him stating that he was quite disabled and could not attend but that he would have a call court to try my father's will we were ready to come to a trial every essential witness on our side was at the court house but I believe the widow was not.

I am Sir your affectionate nephew

James L. Jefferson[57]

So it was that "Jefferson vs. Jefferson" was not heard and would continue on to the next court. There is no indication that William Wirt appeared, and he may have abandoned the Pryors cause by then and this is why the widow was not ready to go to court again. Who was the judge who was so ill, and what was Mr. Booker's role in the case? He is very likely Edward Booker of Prince Edward County, who represented Benjamin Bondurant in "Benjamin Bondurant (assignee) vs Jefferson" in the Prince Edward District Court. Based on Lilburne's letter, Booker may have been representing the Jefferson brothers.

Lilburne's letter is the last known reference to "Jefferson vs. Jefferson," until 1820 when Zachariah Nevil, Administrator for Randolph Jefferson, deceased, is present in the Albemarle courts attempting to collect debts owed the estate.

In 1815, Randolph's children were spread around the surrounding counties. Nancy Nevil was in Nelson County, as was her brother, Robert Lewis Jefferson. Thomas Jefferson, Jr., Randolph, Jr., and Field were all living in Fluvanna County. Lilburne was in the vicinity, at Scott's Ferry and Warren. There is no personal property tax record for Buckingham in 1816, so it is unclear if any of the boys resided in Buckingham that year. While no evidence has surfaced concerning a renter for Snowden, in 1817, Randolph Jefferson, Jr. paid personal property tax in Buckingham. That year, he was married and had at least one child, a son named Thomas Jefferson (b. abt. 1814). He did not own property in Buckingham and may be renting some or part of the farm.

<u>1817</u>   Isham R. Jefferson   1 white male, 4 slaves; 1 horse   tax: $2.90

In 1817 and 1818, Mitchie was still living in Buckingham at Woodlawn with some members of her family and her infant son. The 1818 personal property tax also reveals that undivided property remained in David Pryor's Buckingham County estate, including five slaves and four horses. Mitchie's brothers, Langston, Zach, and Zane Pryor still resided in Buckingham, while her brothers, William S. and Nicholas B. Pryor, were long gone from the county. Banister and John C. Pryor lived nearby in Prince Edward County.

In 1817, Mitchie was taxed on two slaves and a horse. A year later she was taxed on only one slave over age 16. These might have come to her from her father's estate or previously were given to her by Randolph Jefferson.

<u>1817</u>   Mickey (sic) Jefferson   2 slaves; 1 horse   tax: $1.50
<u>1818</u>   Mikey (sic) B. Jefferson   1 slave over 16   tax: $ .70

During 1818, Mitchie was taken to court for delinquent bills by William, John, and Jacob Moon, who traded under the name "William Moon & Co." Buckingham attorney, Walter L. Fontaine, represented the plaintiffs who resided in Albemarle County. Mitchie's accounts were settled in two payments. On March 12, 1818, she paid $200 toward her debt. The balance was paid on August 11, 1818, at $136.04.[58] It is very possible that she sold her horse and a slave to cover her obligation to the Moon store. Since she was sued personally, rather than the estate of Randolph Jefferson, these debts likely represent purchases she made after August of 1815.

The Widow Jefferson Leaves Buckingham County

Sometime in late 1818 or early 1819, the Pryor estate was divided and Susannah Pryor took her dower right (166 acres and 50 acres), leaving the remaining 2/3 of David Pryor's estate to be divided among his living children.

Mitchie sold her fraction to her brother, Dr. William S. Pryor. It amounted to two parcels of 28 acres and 37 acres (four miles north of the courthouse).

Concurrent with the settling of the Pryor estate, Mitchie and her brother, Zachariah Pryor, were in court in early 1819. The surviving court receipt does not reveal the nature of the complaint; however, on February 8, 1819, the Buckingham Court collected $111.14 for Guerrant Moseley, etc. vs. M. B. Jefferson & Zach: B Pryor.[59]

About this time, Mitchie left for Tennessee with her son, John Randolph Jefferson. Zach and their mother, Susannah Pryor, may have also gone west at this time.[60] Mitchie may have appealed to the Curator of Randolph Jefferson's estate for funds before moving to Tennessee. Like Lilburne, she was dependent on the estate. In theory, she could borrow up to her dower right. After she removed to Tennessee, chances of getting anything out of the estate might plummet.

On May 1, 1819, Mitchie married Josiah Johnson in Williamson County, Tennessee. Johnson was born in Virginia, and Mitchie may have known him from Buckingham, Albemarle, or the general neighborhood of Snowden. What, if any connection, he had to the Mr. Johnson mentioned in Thomas Jefferson's 1815 deposition is unknown. The Mr. Johnson of the deposition was a merchant, probably at Warren. In Tennessee, Josiah Johnson was a planter and also served as a county Judge.[61]

Did the accusation of Mitchie's possible adultery ever become public? Did she simply give up her fight to receive a portion of Randolph Jefferson's estate and decide to leave when her mother and brother, Zachariah, left for Tennessee? Was there some arrangement with Mr. Johnson before she departed? Her marriage appears to be rather immediate upon arrival in Tennessee, supporting the probability that they had previously known each other. This is believed to be Josiah Johnson's first marriage. He and Mitchie settled in Williamson County, Tennessee, where their son, James Monroe Johnson, was born on July 2, 1820. Mitchie died between her second son's birth and her husband's marriage in the autumn of 1825.

In September of 1825, Josiah Johnson married Fereby (a.k.a. Ferreby) Hyde; the marriage was recorded in Williamson County.[62] If Josiah Johnson waited at least the customary year before he married again, Mitchie died before September of 1824. The 1830 census in Williamson County enumerates Johnson (40-50), with a boy (10-15) and a woman (30-40). The boy is likely Mitchie's son, James Monroe Johnson.

Josiah Johnson's last will was recorded in the July 1858 Court in Williamson County. At the time of his death, he was a man of considerable property with holdings very comparable to Randolph Jefferson's at Snowden. His outstanding number of horses may indicate he was a breeder. His estate included about 1514 3/4 acres of land, 36 negroes, 36 horses, 67 cattle, 140 hogs, 160 sheep, etc.[63] Fereby survived him; Mitchie's son, James Madison Johnson, and Fereby's son, Josephus A. Johnson, served as Josiah Johnson's Executors.[64]

What of John Randolph Jefferson? He had just turned three years old when his mother remarried. It is not known if John stayed with his mother until her death. One Pryor family history suggests that John Randolph was raised by his uncle, Nicholas B. Pryor, who lived in Nashville.[65] Indeed, in 1830, there is a boy between ten and fifteen years old living with Nicholas B. Pryor in Davidson County, Tennessee. He certainly could be John Randolph Jefferson.

He was probably living in Davidson County before August 18, 1829, when a bill of sale was recorded there in his name. He was thirteen years old. It reads, "I, John R. Jefferson, have sold a certain mulatto girl named Ruth, aged about eighteen years. 11 Aug 1828."[66] It seems likely that Ruth had been Mitchie's property and was bequeathed to John Randolph Jefferson. Born about 1810, she may have come to Tennessee from Snowden.

When Mitchie B. (Pryor) Johnson died, sometime between July 2, 1820 and September 1, 1825, she left behind the still unsettled Randolph Jefferson estate which remained open until at least 1826. Did her death ultimately result in the closing of the estate? In 1826, John R. Jefferson was still a minor. If the estate closed that year, it is possible he was denied any inheritance. A thin series of facts suggests this possible story.

Mitchie B. (Pryor) Jefferson Johnson's death finally ended the case of "Jefferson vs. Jefferson." Mitchie died between her second son's birth on July 2, 1820, and her husband's remarriage in September of 1825. Mitchie's young son, James Monroe Johnson, was then raised by his father and stepmother. John R. Jefferson was sent to Nashville in Davidson County, where he was raised by the Nicholas B. Pryor family. John was five to ten years old at the time of his mother's death.

Martha (Jefferson) Randolph had the last word on both Mitchie and Nancy (Jefferson) Nevil when she wrote to her daughter, Ellen (Randolph) Coolidge, on September 1, 1825:

> ... Mrs. Judge Johnson (Mrs R– Jefferson) and Mrs Nevill (your old friend Nanny Dish) are both dead. the poor clergymen who pronounced their funeral orations must have been considerably embarrassed for a subject for the accustomed eulegium on such occasions.[67]

The Lingering Estate of Randolph Jefferson

Between 1820 and 1822, Albemarle County court records reveal that Randolph Jefferson's son-in-law, Zachariah Nevil, was acting as Administrator of Randolph's estate.[68] How long Nevil had served in this role is currently unknown. Nevil's appointment as Administrator is somewhat surprising. First, he was a Nelson County resident slightly removed from Snowden and the Buckingham Court. Second, his wife, Nancy, may or may not have made claims against her father's estate. She was not included in the May 1808 will, and it is unknown if she, or any of her children, were named in the will presented by Mitchie Jefferson.

On October 14, 1820, Zachariah Nevil, "administrator for Randolph Jefferson deceased," was represented by his attorney in Albemarle's Superior Court against John Scott, who claimed to have already paid a debt to the estate that Nevil was trying to collect, and the case was "set aside." A year later, on October 10, 1821, Scott's attorney acknowledged Nevil's complaint and the court conceded that Scott owed the estate $1,000. The settlement was a payment of $500, with interest, from December 25, 1819.

On May 7, 1822, Nevil (or his attorney) was back in the Albemarle Court; this time he initiated a complaint against John Scott and William Moon, claiming a debt of $1,167.88 to the Randolph Jefferson estate. Again the debt was discharged with a lesser amount: $583.94 with interest from October 17, 1821.

Were these debts created in Randolph Jefferson's lifetime, or did Scott and Moon borrow money from the estate? In any case, Nevil's persistence may have resulted in a sizable settlement in favor of the Jefferson heirs. Lacking accounting for Randolph's estate, whether or not the debts were ever collected remains unknown.

In 1830, Zachariah Nevil died in Nelson County, and the accounts for his estate reveal that his administrators paid two debts to "Rives & Brown" in October of 1830. One was for the Estate of Cornelius Nevil, deceased, who was Zachariah's brother. The amount of the account was $40.06. The other was for "the administrator of the estate of Randolph Jefferson," for $450.34. There is no other indication as to exactly what "Rives & Brown" might have supplied Nevil as administrator for Randolph Jefferson.[69]

Nevil had long been a tobacco inspector at the Tye River Warehouse in Nelson County, and, in 1826, he became associated with Robert Rives and Alexander Brown in the manufacturing of tobacco in Nelson. He may have borrowed funds from the Jefferson estate as part of this venture and died before repaying it.[70]

~

In 1820-1822, John Scott III (1796-1829) and Capt. William Moon (1770 - bet.1833-1840), who owned Stony Point and Lower Plantation, were significant Scottsville players.

John Scott formally established the "Town of Scottsville" in 1818, owned and operated the ferry, and held significant land surrounding the town. Scott's entrepreneurial talents were behind the initial development of Scottsville. Both he and Moon were fundamental in conceiving the town's future. The 1820 census confirms that "P.F. Jefferson," Randolph's son, Field, was living near both Scott and Moon, poised to become another significant resident of Scottsville.[71]

This is the same William Moon who took Randolph Jefferson to court in 1802. He is likely the same Mr. Moon who extended Mitchie Jefferson a little too much credit at the Scott's Ferry store. In 1821, Moon significantly expanded his holdings, purchasing 655 acres known as Lower Plantation, located on Totier Creek, directly across the James River from Snowden. The sellers were John Scott and Elizabeth Wood. Moon had overextended himself, however, and soon sold the property to cover debts.[72] In 1861, William Moon's son, John S. Moon, put Stony Point up for sale. It was described as one mile from Scottsville on the road to Charlottesville. Proving, once again, the interconnectedness of

Randolph Jefferson's small world, in a few years, John S. Moon would become the owner of Snowden.[73]

John Scott III was a youthful entrepreneur, close in age to Field and Lilburne Jefferson. Their distant Randolph connection may or may not have influenced Scott-Jefferson relations in this generation.[74] However, another family alliance was about to connect John Scott and the Jeffersons. On May 15, 1823, John married Susan Bathurst Bolling, the granddaughter of Mary (Jefferson) and John Bolling, the great-niece of Thomas and Randolph Jefferson. According to Virginia Moore, their wedding reception was held at Monticello.[75] A letter written from Thomas Jefferson's granddaughter, Virginia J. (Randolph) Trist, to her husband, Nicholas P. Trist, tells part of the story.

> Monticello June 5th 1823
> ...I must not omit my only piece of <u>news</u>, the marriage of Susan Bolling and John Scott and their intended removal to Alabama, in a few days. they dined with us the day before yesterday, and Susans Beauty is more resplendent than ever, and her countenance I think improves as she a[d]vances in life, which is a good sign of her intellect.[76]

Indeed, the newlyweds did not remain long in Scottsville, moving to Somerville, Alabama shortly after their marriage. Their oldest daughter, Pocahontas Bolling Scott, was born in Alabama on July 22, 1825.[77]

Scott and Moon were also connected to Capt. John Harris, who eventually purchased Snowden. William Moon was married to John Harris' niece, Charlotte Digges. The two men had business connections as well. Harris, who had no sons, was grooming the Moon boys as heirs to his vast enterprises. Scottsville historian, Richard Nicholas, further discusses the Moon-Scott connection:

> An interesting side note to the Moon family was the close relationship between Edward Harris Moon (1805-1853) and the family of John Scott III. ... The origin and circumstances of that relationship have not been discovered, but Moon, who lived at "Viewmont," the old Joshua Fry homestead in Albemarle County, appears to have assumed the

overdue tax obligation for a number of Scottsville lots owned originally by John Scott III after the latter's death in 1829. After purchasing the contested lots and paying the back taxes, Moon even went so far as to donate four and a half of those lots back to Susan B. Scott and her three daughters. The reason for his generosity is not known.[78]

Concurrent with the growing influence of Scott-Moon-Harris in the burgeoning Scottsville, Randolph Jefferson's youngest sons were getting in on the ground floor in the developing town. The July 11, 1815 survey of Scottsville by William Woods, S.A.C., shows that "Fields" Jefferson already owned Lot 6 in the heart of town, his involvement there pre-dating its incorporation.[79] In 1818, Lilburne followed his older brother's lead, investing in Lot 42 and Lot 43. The lots cost $296. In 1823, Lilburne bought Lot 44 from John Scott and his wife, Elizabeth.[80] How this real estate speculation on the part of Field and Lilburne might be connected to the Scott and Moon debts to Randolph Jefferson's estate is currently unknown.

What of the fate of Snowden? Many questions remain unanswered. Who was the court-appointed Curator who managed both the real and personal property, including the slaves? Who rented the farm in 1816? Did they also lease the slaves? Were some or all of the slaves sold/auctioned? Were they eventually divided among the legatees?

Most of Randolph's sons were living nearby, anticipating the settling of the estate. Initially after Randolph's death, Thomas, Jr., Isham Randolph, and James Lilburne all paid personal property taxes in Fluvanna County. The Nevils lived in southern Nelson County on the Rockfish, and Peter Field had settled in Scottsville, Albemarle County. Mitchie, who remarried in 1819, and John Randolph Jefferson, were living in Tennessee.

Several indicators suggest that the Chancery suit may have finally been settled about 1822. Mitchie Jefferson (or her heir) may have received a settlement, leaving Randolphs' first family with approximately 1,500 acres, perhaps a dozen slaves and other personal property. In 1822, Randolph's estate paid $29.69 tax on 1649 acres. Yet, in November, the estate advertised the farm as containing upwards of 1,500 acres. In 1826, purchaser John Harris would pay tax on just 1,454 acres.

The first indication that the situation at Snowden was changing came in 1822. After an absence from the Buckingham County personal property tax records, Randolph's sons began to reappear in the county. An advertisement in the *Richmond Enquirer* reveals that administrator, Zachariah Nevil, was preparing to sell the farm and that Peter Field Jefferson was living at Snowden. That year, Field paid personal property taxes in Buckingham, on himself, twelve slaves and four horses.

As Administrator, Zachariah Nevil scheduled the auction for November 1, 1822. If Snowden failed to sell for an acceptable price, he planned to rent it for a year.

Valuable James River Land For Sale.
Will be sold to the highest bidder in the town of Scottsville, Albemarle county, on the 1st of November next, the tract of land lying in Buckingham county, on James river, belonging to the estate of Randolph Jefferson, dec'd. containing upwards of 1500 acres, of which 180 are river bottoms; a considerable proportion of the high lands is first rate. The tract abounds with a plenty of good timber. A further description is deemed unnecessary, as it is presumed purchaser will view the land prior to the day of sale; I will only add, the advantages that attend this land entitle it to rank of any tract of its size in the upper country. The payments will be accommodating, and made known on the day of sale. Those wishing to view the land are referred to Mr. Peter F. Jefferson, who lives on the premises.
Should the land not be sold on the above mentioned day, it will be RENTED for twelve months.
ZACH. NEVEL, Adm'rer with the will annexed of Randolph Jefferson dec.[81]

The outcome of the November auction is unknown. However, during 1823, Peter Field returned to Scottsville; Robert Lewis and James Lilburne Jefferson may have taken over at the farm. It seems likely that the brothers were renting it or, at the very least, operating it for the estate. During 1824 and 1825, Lewis and Lilburne continued to pay personal property taxes in Buckingham, indicating they may have settled in at Snowden for several years. Lilburne remained unmarried; Lewis, a widower, had one son, Elbridge, age ten to twelve.[82]

The transition in the management of Snowden began in early 1823. On February 1, 1823, Robert Lewis Jefferson and another white male were listed on the personal property tax records in Fluvanna County. On April 5th, James Lilburne Jefferson was taxed in Buckingham with two slaves and one horse. Then, on May 24th, Robert Lewis Jefferson was also taxed in Buckingham, paying for thirteen slaves and five horses. This is a sudden and dramatic jump in Lewis' property and the numbers are similar to previous counts of taxable slaves and horses at Snowden.[83] In 1824 and 1825, fourteen slaves and five horses continued to be charged to Robert Lewis Jefferson, indicating that he was overseeing the farm.

Then, between 1825 and 1826, Lewis' number of taxable slaves dramatically dropped from fourteen to six. Likewise, his horses dwindled from five to one. This apparent "loss" in personal property actually reflects the final sale of the Snowden lands and the distribution or sale of property, matching the sale date of the farm to Capt. John Harris of Viewmont on September 29, 1826. That deed of trust represented the third, and final, piece of property Harris purchased in Buckingham's Horseshoe Bend.

In 1810, Randolph Jefferson was taxed on 1,827 acres. Between 1815 and 1816, the taxable land dropped to 1,649 acres. The difference is the approximate 178 acres purchased by Charles A. Scott, who promptly sold the land to John Harris. That conveyance is clear in the Buckingham County land tax records. In 1816, Harris was taxed on 178 7/8 acres, described as "conveyed from R. Jefferson to Scott."[84] In 1820, the 1,649 acres of land that now made up Snowden remained in the estate.

In 1826, the land was taxed for the last time in Randolph's estate and Snowden's 1,649 acres appeared on the Buckingham County Land Tax. The 1827 land tax record indicates that, during the previous year, the land remaining in Randolph Jefferson's estate was conveyed by the Administrator to John Harris of Albemarle, adding Snowden to his expansive network of plantations, mills, and commercial ventures. However, Harris was taxed for only 1,445 acres of land. There appear to be 205 "missing" acres. This discrepancy between the 1826 and 1827 taxable quantity is currently inexplicable.[85]

How the funds from the sale of the land were distributed and the ultimate price the farm brought is also unknown, though it was likely at least $12,000. A copy of the deed of sale has not yet been found. However, a surprising advertisement began running in the *Richmond Enquirer* on January 31, 1829. It indicated that Harris had not yet paid the full amount due the estate.

Valuable James River Land For Sale.
By virtue of a Trust Deed executed by John Harris to us, to secure to Zachariah Nevil certain sums of money dated the 29th Sept 1826, and recorded in the Clerk's Office of Buckingham county 24th December 1828; in conformity to the request of said Nevil, in order to pay the two last installments secured by the deed, viz: four thousand dollars due the 10th of June 1828, and the same sum due the 10th June 1829; and to pay the charges of sale, we shall, on the premises proceed on the 21st day of March next, to sell for cash, the tract of land in the deed of trust mentioned, called Snowden, lying in Buckingham, supposed to contain ( ) acres; the tract contains a large quantity of most valuable low grounds, and the high land is perhaps the finest tobacco land in Buckingham county; it lies on James River opposite Scottsville, and the late residence of Randolph Jefferson, dec'd. It is probable that the persons interested may make on the day of sale, some modification of the terms so as to allow some credit for a part of the purchase money, in which event the title will be retained till final payment. We believe the title good, but only convey as trustees.
JESSE JOPLIN
THOMAS DANIEL[86]

Again, the details of the March transaction are unknown; however, Harris secured the deed, apparently paying the remaining sum of $8,000. One indicator that the Jefferson sons successfully received funds from the sale of Snowden is a significant change in circumstances for Robert Lewis Jefferson. In about 1828,

Lewis completed the purchase of a 531-acre plantation on Buckingham County's Sharps Creek. After years of unsettled living, the widower established himself as a Buckingham planter. Robert Lewis Jefferson was, at last, "back home."

CHAPTER FOURTEEN

The Jefferson Brothers: A Study in Contrasts

> *The only exact testimony of a man is his actions....*
> Thomas Jefferson to Louis H. Girardin, 1815

Volumes and volumes have been written and will continue to be written about Thomas Jefferson. His multi-dimensional, often seemingly irreconcilable personality draws both scholars and the reading public. Joseph Ellis fittingly dubbed him "The American Sphinx;" indeed, Thomas Jefferson's complexities will long mystify us.

Randolph Jefferson, on the other hand, was a simple man who never would have drawn any attention, except that he happened to be the brother of an extra-ordinary person. All indications are that Randolph was a rather typical central Virginian of his social rank. Evidence strongly suggests that he was an uncomplicated, practical man. In good gentry style, he married his first cousin, Anne "Nancy" Lewis, and together they raised six children to adulthood. Also typical, a second, younger wife followed after Nancy's death.

Randolph may have lacked self-confidence. He certainly lacked ambition. It is entirely possible that when his signature accompanied Thomas Jefferson's on a document, Randolph had been swayed by Thomas' enthusiasm. Likewise, Randolph might have been influenced by his Lewis cousins, his son-in-law, Zachariah Nevil, and, later, his second wife, Mitchie, and her Pryor family. Thomas summed up Randolph's personality saying, "in all the occasions of life diffidence in his own opinions, an extreme facility and kindness of temper, and an easy pliancy to the wishes and urgency of others made him very susceptible of influence from those who had any views upon him."[1]

Randolph is not known to have had any serious impairments in his youth. As he aged, his health concerns grew, but not abnormally so. At age forty-five he required spectacles, which his brother purchased for him in Philadelphia.[2] When he was fifty-six, he suffered an attack of kidney stones and recovered.[3] He may have enjoyed a good drink; if so, Randolph had lots of company in his generation. There is no indication that alcohol got in the way of the management of Snowden, or that depression plagued Randolph as it did other members of his extended family. If he suffered debilitating headaches as did his brother, it is never mentioned in their correspondence.

Randolph's education far exceeded the average Virginian of his day, though it exemplified the planter class which expected their children to be gentlemen, but rarely scholars. Like most of his compatriots, he signed petitions, talked politics, voted, and enjoyed being part of his county's Militia, at least in times of peace. He went to court, lost some causes and won others. His most outstanding anomaly was that he did not carry heavy debt, something that could not be said for other members of his family. In short, as the proprietor of Snowden, Randolph Jefferson appears to have lived and died an upstanding, yet unremarkable, citizen of Buckingham County, Virginia.

Randolph Jefferson: A Mighty Simple Man

In the past, the vast majority of Jefferson scholars have dismissed Randolph Jefferson as historically insignificant. The few discussions concerning him have been based on very limited material, incessantly repeating the same sources without offering context or critical evaluation. Fires at Shadwell, Snowden, and Buckingham Courthouse destroyed most of the documents that might have further illuminated Randolph's life. Thus, he has been even easier to marginalize as a muddy boots farmer.

Over the years, Randolph has been increasingly maligned, his intelligence a particular target. To be sure, Randolph Jefferson was not an intellectual. Likewise, he was far from incompetent. His rustic humor has been misinterpreted as dimwittedness, notably the telling of the corn and squirrel story, in which

Randolph purportedly advised Thomas, "If you don't want the squirrels to eat your corn, don't plant the outside row!"[4]

In 1815, when Thomas stated that his brother did not possess the "skill for judicious management of his affairs," this was not meant to summarize Randolph's life-long abilities.[5] Age and/or ill-health may have compromised his competency at the end of his life, yet Thomas did not perceive this situation so disastrous that Randolph required guardianship or that his wife, Mitchie, should assume his affairs.

Currently, there is no reason to believe that Randolph did not manage Snowden independently for the entirety of his adult life. The very fact that he preserved the majority of his patrimony reveals basic competency, if not some sound business practices and judgment. Considering that the "smart" Jefferson brother died deeply in debt (the direct result of his personality traits, as well as personal and historical calamities), Randolph shines as a practical and frugal example by comparison.[6] Apparently, Randolph knew and accepted his limitations and the limitations of his plantation, content to maintain what he was given.

In 1871, Sarah Nicholas Randolph, Thomas Jefferson's great-granddaughter, noted in her book, *The Domestic Life of Thomas Jefferson*, "It is curious to remark the unequal distribution of talent in this family – each gifted member seeming to have been made so at the expense of one of the others."[7] Her delicately-stated observation had real staying power as Jefferson family legend. Mental deficiency, mental illness, and outright idiocy did plague the Jeffersons, especially those who inter-bred too often with other members of the Randolph family. The question remains – to what degree was Randolph Jefferson compromised intellectually or emotionally or both?

In 1951, Nathan Schachner, in *Thomas Jefferson: A Biography*, dedicated a few paragraphs to Randolph. Early in Thomas' story, Schachner dismisses Randolph's education in Williamsburg as ineffectual: "While he later managed to marry, have offspring and cultivate his land, all the evidence points to the fact that his grade of intelligence was barely sufficient, with his brother's aid, to grapple with the ordinary difficulties of life."[8] This gross exaggeration is incomprehensible, especially considering that Thomas personally approved the expenses to send Randolph to school at William and Mary in 1771. Would he

Martha (Jefferson) Carr (1746-1773). (THOMAS JEFFERSON FOUNDATION AT MONTICELLO)

be such a fool as to send an idiot brother to Williamsburg?[9] Given Randolph's record with Snowden, his decade-long service as an officer in the Buckingham Militia, and his legacy measured in the successful lives of his children, the man was far from incompetent. It is certain he was never an adult dependent as was his sister, Elizabeth, who spent her entire life in the care of her mother and siblings. Importantly, Thomas saw no need to prevent his brother's ascent to his property.

Randolph Jefferson's simplicity may have been, in part, misleading and was actually the expression of a straightforward, pragmatic nature. Where Thomas waxes poetic, Randolph is curt, to the point. The contrast is apparent in their exchange concerning the death of their sister, Martha Carr.

On September 6, 1811, Thomas wrote: "Without any increase of pain, or any other than her gradual decay, she expired three days ago, and was yesterday deposited here by the side of the companion who had been taken from her 38. years before. She had the happiness, and it is a great one, of seeing all her children become worthy and respectable members of society and enjoying the esteem of all."[10]

On October 6, 1811, Randolph replied: "I received yours of the twenty six of last month and am extremly sorry to hear of my sisters death and would of bin over but it was not raly in my power but it is what we may all expect to come to either later or sooner."[11]

Interestingly, his reflections on death were very fresh. In October of 1811, Randolph was still recovering from a severe attack of kidney stones and was not physically able to ride to Monticello for his sister's funeral. In the intense pain of the "attack," he feared his own death. He closed the October 6th letter, saying, "I have just got over a very severe tack of the gravil I could not of survived many [h]ours had I not got releaf from a physician immidately."[12]

Of course, the contrast in eloquence between the Jefferson brothers is yet another way in which Randolph has "come up short" in the eyes of those who have bothered to look. The Jefferson-to-Jefferson letters are among the few pieces of surviving evidence revealing Randolph's life and thoughts. His rough prose – rode for road, fare for far, and hear for here – has contributed to the conclusion that he was mentally deficient. With a yardstick like Thomas Jefferson for a brother, who wouldn't come in a very poor second? However, when Randolph's letters are juxtaposed with those of his sister, Lucy, or his brother-in-law, Charles L. Lewis, his prose is quite "normal" for the day.

Additionally, book learning is not the only measure of intelligence. It is entirely possible that Randolph possessed other kinds of intelligence, notably innate musical talent and an intuitive knowledge of agriculture. These might have been areas in which Randolph bested his brother. There is no way to know.

In materialist terms the brothers could not have been more different. Thomas had an insatiable appetite for books, good food and wine, and elegant living. He spent far beyond his means. Randolph, on the contrary, was not a spender and became extremely distressed when his second wife, Mitchie, freely piled up debt against his hard conserved property. Due to the loss of Randolph's personal ledgers and estate papers, there is no record of what he may have loaned to others. It is entirely possible that Randolph's diffidence and "kindness of temper" may have made him an "easy touch" for neighbors and relations.

> Dear Brother   Octor 6, 11
>
> I Received yours of the twenty six of last month and am extreamly sorry to hear of my sisters death and would of bin over but it was not raly in my power but it is what we may all expect to come to either later or sooner I got mr forgot to call and leave this letter for me as he was going to albemarle court and recomended it to him to make mention to cello his first days stage I intend coming over some time next month which I expect will be towards the last of the month as I shall be very busy a getting my crop of wheat down to Richmond and sowing my present crop you will not forget to take care of my puppy if you have not given him a way to any one I expect by this time he must be large I have just got over a very severe tack of the Gravil I could not of survived many ours had I not got releaf from a Physician immidately. my wife and family rents there respects to you and family I am yos Affectionately
>
> Rh: Jefferson

Randolph Jefferson to Thomas Jefferson, 6 October 1811. (Courtesy of Small Special Collections, University of Virginia)

As to other aspects of Randolph's personality, his letters reveal that he displayed great patience, such as waiting for Thomas to send him a prized sheepdog and for suffering separation from his beloved silver watch when it was sent away on lengthy repairs. He frequently showed deference for and indeed sought counsel from his much older and more accomplished brother. Notable exceptions include his rejection of Thomas' elaborate scheme for rotating crops at Snowden and his "failure" to ask Thomas' advice concerning his marriage to Mitchie.

According to Thomas, Randolph was diffident, easily influenced, and kind-tempered. Was he essentially passive? Was he weak-willed? Even late in life, he was not at all meek in his response to Mitchie's alleged excessive spending at the stores at Warren and Scott's Ferry. His solution, which included devising a secret mark for orders placed by him, is a bit convoluted and might suggest that he was not comfortable with direct confrontation, nor was his famous brother.[13]

Lastly, but no less importantly, there is no reason to believe that Randolph acted with everyone else the way he acted with his brother, who easily may have towered over him, more like a father than a sibling. Amongst his peers – neighbor Capt. Hardin Perkins, brother-in-law Charles L. Lewis, or the esteemed Rev. Martin Dawson – Randolph Jefferson may have acted and appeared very differently indeed. In the absence of eloquent depositions from any of them, however, we are left with Thomas' point-of-view.

Old Age and Harmony at Snowden

When Randolph Jefferson wrote his "last" will in May of 1808, he was fifty-two years old. Significantly, he valued a fair and equal distribution of his property among his five sons over the preservation of Snowden in his family. With so many legatees to consider, rather than cut Snowden into slivers, he requested that his land and other property be sold and the proceeds dumped into a "hotchpot," to be divided equally among the Jefferson brothers. The apparent ease with which Randolph detached himself from Snowden is a bit surprising. It is impossible to imagine Thomas Jefferson giving up his patrimony at Monticello though, ironically, it, too, was not passed on to his descendants.

Importantly, this division did not include Randolph's human property. In keeping with his strong desire to retain his slaves, he requested that they not be sold, but stay with his sons.

Lewis Miller depicts a "Negro Dance" in Lynchburg, Virginia, in his sketch dated August 18, 1853. (LEWIS MILLER, "SKETCHBOOK OF LANDSCAPES IN THE STATE OF VIRGINIA, 1853-1867," ABBY ALDRICH ROCKEFELLER FOLK ART MUSEUM, THE COLONIAL WILLIAMSBURG FOUNDATION, WILLIAMSBURG, VA., GIFT OF DR. AND MRS. RICHARD M. KAIN IN MEMORY OF GEORGE HAY KAIN)

RANDOLPH JEFFERSON, FIDDLER

Thomas Jefferson encouraged Randolph's "polite education" and particularly supported his brother's natural proclivity for the violin, arranging for expensive lessons with his own mentor, Frances Alberti. Perhaps the most oft-quoted summation of Randolph Jefferson was made by former Monticello slave, Isaac Jefferson, who dictated his memoirs to Rev. Charles Campbell in 1847. In them, Isaac remembered Monticello and various Jefferson family members, describing Randolph as follows: "Old Master's brother, Mass Randall, was a mighty simple man: used to come out among black people, play the fiddle and dance half the night; hadn't much more sense than Isaac." It is understood that "Old Master" Thomas would not have behaved as Randolph did coming out among black people to make music and "socialize;" however, the fact that Randolph "fiddled," rather than played the violin, suggests a lively and spontaneous approach to music. His late hours reveal less discipline than his "bookish" brother. If the era (and Isaac Jefferson) had not been so conscious of distinct and separate roles of "master" and slave, and the complex social lines that separated them, it might be concluded that a youthful Randolph Jefferson simply knew how to have a good time. Randolph's characteristic "naturalism" is well exemplified in his approach to music. If he indeed preferred indigenous or folk music to formal European sonatas or chamber music, the improvisational musicians of the slave quarter might have been among his favorite companions, especially in his youth. While this could appear to be unseemly behavior for a gentleman in a class-bound society such as 18th-century Virginia, a lone white member of a jazz ensemble would be accepted differently in modern times. Randolph may not have pursued the classical violin, yet he valued music and his instrument. In 1808, he still owned a violin and specifically wished his son, Lewis, to inherit it.

[Joanne L. Yeck, "A Most Valuable Citizen: A Profile of Randolph Jefferson," *Magazine of Albemarle County History* (2011), 1-37.]

There is every reason to believe that Thomas and Randolph Jefferson were very much alike in their vision of creating a harmonious old age, surrounded by their children and grandchildren. Like his brother, Thomas, Randolph looked forward to seeing Snowden populated by the next generation. While Randolph's retirement did not mean stepping down from a hectic international stage, as it did for Thomas, the passing years naturally enforced a slower pace and Randolph expected to turn the management of Snowden over to his sons.

Randolph's wife and the mother of his children, Anne Lewis, died in about 1800, leaving him a widower for most of the first decade of the 19th century; however, Randolph was never alone. His sons delayed marriage, staying on at the farm. Then, in about 1807, Zachariah and Anne Scott "Nancy" (Jefferson) Nevil came from their Nelson County plantation to join the family at Snowden. There, Zachariah went into partnership with the Jeffersons, and industry, as well as agriculture, bustled at Snowden. Nancy arrived with little James "Lilburne" Nevil and her daughter, Louisiana, likely was born at Snowden. Then, in October of 1808, Thomas, Jr. married, bringing his Albemarle bride, Polly Lewis, home to Buckingham and the expanding collective at Snowden. For several years, it seems that Randolph had achieved the domestic ideal that Thomas Jefferson so longed for – the Virginia planter at his ease, surrounded by his children and grandchildren.

Then Randolph "spoiled" it all when he married Mitchie B. Pryor, a woman young enough to be his daughter. May-December second marriages were far from unique among Virginia planters, though they were rarely appreciated by the children of the first wife. Inheritances were often threatened and sometimes completely disappeared in favor of the younger wife and her children. The younger Jeffersons were no exception in their dislike for Mitchie. Thomas, Jr. and Peter Field were particularly disrespectful towards their father's new wife.[14]

A member of an upstanding, "middling," Buckingham family with a long presence in central Virginia, Mitchie had undoubtedly lived all her life at Woodlawn, the Pryor plantation located about four to five miles from Buckingham Courthouse.[15] However, it was her personality, not her pedigree that was the issue and, one by one, her presence drove Randolph's family from Snowden. By the end of 1812, Randolph had Mitchie at his side and enjoyed new family ties with the Pryors. Randolph's own family was splintered, at best, and the quiet harmony both Jefferson brothers valued was lost.

The Jefferson Slaves

The most oft-quoted summation of Randolph Jefferson's relationship with black Virginians was not made by Thomas Jefferson, but by former Monticello slave, Isaac Granger Jefferson, who dictated his memoirs to Rev.

Isaac Jefferson, daguerreotype, 1847. (ISAAC JEFFERSON. MEMOIRS AND DAGUERREOTYPE, 1847, IN THE TRACY W. MCGREGOR LIBRARY, ACCESSION #2041, SPECIAL COLLECTIONS, UNIVERSITY OF VIRGINIA, CHARLOTTESVILLE, VIRGINIA)

Charles Campbell in 1847. In them, Isaac remembered Monticello and various Jefferson family members, describing Randolph as follows: "Old Master's brother, Mass Randall, was a mighty simple man: used to come out among black people, play the fiddle and dance half the night; hadn't much more sense than Isaac."[16]

Born about 1775, if Isaac observed these exuberant late night behaviors first-hand, Randolph was at least in his thirties when he was enjoying music "among the black people." At that time, Randolph was living twenty miles from Monticello and these would have been very inconvenient midnight frolics if they were at his brother's plantation. It seems more likely that Isaac repeated what his elders said – perhaps disapprovingly – about Randolph's unseeming behavior at Shadwell.

In the larger context of 18th-century life on the "frontier" in Albemarle County, Randolph's boyhood fraternization with slaves actually would have been quite normal. His initial years at Shadwell were remarkably different

"A Typical Negro Cabin." (SOCIAL LIFE IN OLD VIRGINIA BEFORE THE WAR, BY THOMAS NELSON PAGE, ILLUSTRATED BY GENEVIEVE COWLES AND MAUDE COWLES, 1897)

from Thomas' early life in the far more cultivated neighborhood of Tuckahoe, downstream on the James River, near the city of Richmond. Isolated in still-developing Albemarle County, with only sisters at home and his closest cousins several miles away, Randolph's first companions would have been the community immediately around him. This community happened to be enslaved and happened to be black. His first "best friend" may have been Peter, the slave given to him virtually "at birth." Unlike the years of segregation that followed, Randolph Jefferson's experience of growing up in the 1760s, fatherless and without a brother at home much of the time, meant that black men and boys were his most available male mentors and peers. Apparently, Randolph gravitated toward them and enjoyed their company.

Personality and circumstance both likely influenced the Jefferson brothers in their attitudes towards their own slaves, particularly when it came to the dilemma of selling one. Thomas, who not only inherited slaves from Peter Jefferson, but also from his father-in-law, John Wayles, struggled all his life with the buying and selling of slaves. Over many years, Thomas sold dozens of individuals; he also bought many, often expressing compassion for family groups or the wishes of slaves who had long been hired out to be sold to their employers. In 1773, Thomas purchased Ursula, her husband, George, and their two children, to please his wife, Martha. By doing so he also united Ursula and George, who had been living at separate plantations in Cumberland County. In 1792, at the request of his servant, Mary Hemings, Thomas Jefferson sold her along with her youngest children, Bob and Sally, to Thomas Bell, a Charlottesville merchant and Mary's common-law husband. In the case of Mary Hern, Thomas purchased her and her two sons from his nephew, Randolph Lewis, at the persistent request of her husband, Moses, a Monticello slave. Outstanding, of course, from Randolph Jefferson's point of view, was Thomas' failed attempt to unite Orange and Dinah in Buckingham County.[17]

The much smaller sphere of Snowden meant that every person's involvement on the farm counted; lives were tightly interwoven. Surely Cary's fate at Monticello did not encourage Randolph Jefferson to part with other slaves and his difficulty in finding a spare girl to send to Monticello to learn the spinning jenny, serves as a reminder that there were no idle bodies at Snowden. Randolph's efforts to keep Jane and perhaps her family with him, when she had been impounded to cover debts to William Moon, and his ultimate desire to "keep all my slaves as long as I live," speaks clearly of the interpersonal involvements of owner and slave at mid-sized plantations.[18]

Ironically, Thomas and Randolph Jefferson have shared many more headlines in death than they ever did in life and Randolph recently has garnered national attention as the designated fallback father of Sally Hemings' children. Did Randolph Jefferson father "slave" children at Monticello or elsewhere? This question persists into the 21st century and there is no definitive answer.

As early as 1958, Pearl M. Graham, who wrote "Thomas Jefferson and Sally Hemings," mentioned a rumor that Randolph Jefferson had fathered "mulatto" children.[19] Yet, prior to the 1998 DNA results which revealed that a Jefferson was the father of Eston Hemings, Sally's children most often were attributed to Peter or Samuel Carr, nephews to Randolph and Thomas Jefferson. When DNA ruled out a Carr as Eston's father, Randolph was dragged into the spotlight, with Isaac Granger Jefferson's brief character summary quoted as damning evidence.[20]

True, Eston Hemings' DNA matches Randolph Jefferson to the same degree it matches Thomas Jefferson, and approximately two dozen other Jefferson males alive at the time of Eston's conception. Among them, the Jeffersons at Snowden were in closest proximity to Monticello, living about twenty miles to the south, fueling the argument that Randolph or his oldest sons are excellent candidates as Eston's father.[21]

While Snowden and Monticello are not distant by modern standards, in 1800, twenty miles of uncertain roads were not considered convenient. Visiting between Snowden and Monticello took planning and was often thwarted, as is evidenced in several of the Jefferson-to-Jefferson letters. If it were possible to cross the James River at Scott's Ferry, Monticello could be reached in one day. The ferry did not always operate, the water being too low, too high, or frozen and there was the cost of ferriage to consider.[22]

Despite the lack of any conclusive evidence, Randolph Jefferson and his possible mixed-race children continue to be discussed among academics, journalists, and writers of history. If middle-aged liaisons at Monticello are unlikely, it might be remembered that Randolph did not marry until he was nearly twenty-five years old. There was plenty of time for him to father a mixed-race child before he committed himself to Anne Lewis. Everything from an encounter or two to an involved relationship is imaginable. Randolph might have simply "sowed some wild oats" or, as a young romantic, indulged in an interracial *affaire de coeur*. Absolute proof, though, will likely forever remain elusive.

Randolph Jefferson's Legacy

In sum, who was this man, the only surviving brother of Thomas Jefferson? It is certain that Randolph Jefferson lacked the intellectual curiosity which drove and inspired his brother. He lacked political ambition and was satisfied to "retire" with Captain in the Buckingham Militia as his ultimate public title. As a result, Randolph's realm and sphere of influence was infinitely smaller than that of Thomas Jefferson.

There is nothing in the record to indicate that Randolph was unhappy with his lot in life. He was likely free of the internal conflicts which many believe haunted his brother, making him far less complex and, therefore, far less intriguing than Thomas. Randolph's "cover" indeed may have been his "book." What you saw is what you got.

Indications are that Randolph was an easygoing man, especially in his youth. As he aged, he became less tolerant of his sons' criticisms. He struggled to maintain control over his finances and his second wife's behavior. An open, guileless friendliness may have attracted "users," and resulted in interpersonal disappointments. Still, Randolph was resilient.

Impressively, he was free of substantial debt. In short, he had preserved his patrimony, Peter Jefferson's "Fluvanna Lands." Ultimately, the combination of too many sons and a problematic second marriage meant that Snowden was sold out of Randolph's estate, passing into the hands of others. Rather quickly, all traces of Randolph Jefferson's reign atop Buckingham County's Horseshoe Bend faded away.

Randolph's values and sensibilities, however, were passed on to his children and to their children, many of whom remained near their homeplace, settling in Fluvanna, southern Albemarle, and southern Nelson counties. They ranged from "solid citizens" to "movers and shakers" at the local level.

His only daughter, Anna Scott "Nancy" Jefferson, married Zachariah Nevil, a successful planter, tobacco inspector, and Nelson County delegate to the Virginia General Assembly. Son, Thomas Jefferson, Jr., farmed in Fluvanna and Albemarle, married three times, lived into his 90s, but left no descendants. In the 1830s, Isham Randolph Jefferson left Fluvanna County for Kentucky where he continued life as a planter, leaving many descendants. James "Lilburne" Jefferson ventured as far as Lynchburg, where he died unmarried in 1836.

Robert "Lewis" Jefferson initially purchased a farm in Buckingham County, which he gave to his son, Baptist minister Elbridge Gerry Jefferson. Then, Lewis farmed his second wife's homeplace, Rock Castle, in southern Albemarle and purchased land nearby at Porter's Precinct, which his widow, daughter, and son-in-law developed into a service center at the crossroads.

Peter Field Jefferson, likely Randolph's most influential son, was a shaping force in the transformation of Scott's Ferry into Scottsville in Albemarle County. There he owned the ferry, which crossed the James River and connected to the landing at Snowden, a general store, a tobacco warehouse, and a grist mill. He was also instrumental in the building of the canal at Scottsville.[23]

When Randolph Jefferson died on August 7, 1815, his twin sister, Anna Scott (Jefferson) Marks, was with him. His brother, Thomas, tried and failed to reach his ailing brother. Traveling from Monticello, having reached the northern bank of the James River, in sight of Snowden, he learned that his only brother was already dead. Though no grave has been located, where else would Randolph Jefferson be laid to rest but at Snowden, high on the hill, overlooking the Horseshoe Bend?

In the absence of a eulogy for Randolph Jefferson, this line, lifted from a laudatory obituary of his grandson, Lafayette Nevil, stands as a beautiful reminder of the highest values of Virginia gentility and which characterized many members of the Jefferson family:

> He was a gentleman, who possessed, in a very distinguished degree, all those propensities of benignity, which adorn and dignify the human nature.[24]

The Jefferson Brothers: A Study in Contrasts ~ 351

"View of the West Front of Monticello" by Jane Pitford Braddick Peticolas (1791-1852), c. 1827. (THOMAS JEFFERSON FOUNDATION AT MONTICELLO)

"View from Monticello Looking Toward Charlottesville" by Jane Pitford Braddick Peticolas (1791-1852), c. 1827. (THOMAS JEFFERSON FOUNDATION AT MONTICELLO)

These watercolors by Jane Pitford Braddick Peticolas were created shortly after the death of Thomas Jefferson. They depict both Thomas' grand lifestyle at Monticello as well as the dramatic view from his little mountain. His choice to value the view was costly. Getting water and other supplies up the mountain was often challenging and expensive. Randolph Jefferson's comparatively simpler life at Snowden suited his frugal personality. A spring on Snowden's high ground also guaranteed a convenient and reliable water source.

ACKNOWLEDGMENTS

To begin, I wish to express my gratitude to the Thomas Jefferson Foundation and the Robert H. Smith International Center for Jefferson Studies for supporting my investigation of Randolph Jefferson and Snowden with a Jefferson Fellowship in 2010. Andrew O'Shaughnessy, Saunders Director at ICJS, the staff at Kenwood, and many fellow scholars created a stimulating environment nestled below Thomas Jefferson's Monticello. Particular thanks to Research Librarian, Anna Berkes, who always makes working at the beautiful Jefferson Library even more pleasurable, and to all who maintain the highly reliable and rapidly growing Thomas Jefferson Encyclopedia.

Core to this study were Bernard Mayo's collection of Jefferson-to-Jefferson letters, reprinted and expanded by James A. Bear, Jr.; the Princeton University Press collections of The Papers of Thomas Jefferson, especially the Retirement Series; and Jefferson's Memorandum Books, edited and annotated by James Bear and Cinder Stanton. My profuse thanks to them both for creating this invaluable resource for Jefferson scholars. It is just beginning to be mined and holds the seeds of many future studies. Additionally, Susan Kern's groundbreaking work concerning the Jefferson homeplace, Shadwell, made readily accessible the world of Peter and Jane Jefferson.

There are numerous colleagues and friends to thank at historical societies, libraries, and museums, all of whom tirelessly shared their expertise about the Jefferson family, as well as deep knowledge of Albemarle and Buckingham counties. Among them are Cynthia Burton, Al Chambers, Alan Pell Crawford, Steven Meeks, Richard Nicholas, and most especially, Cinder Stanton, whose ongoing support and unsurpassed knowledge of Thomas Jefferson as a planter and as a slaveholder is inestimable. Her infusion of information and inspiration into both academia and her community makes an enormous difference for all Jefferson scholars.

Research assistants and companions in the field included Judy Brown, Patt Freedman, George Hambrick, Mary Eleanor Hitselberger, Brenda Kitchen, and Ed Lay. Betty Jo Tyson and her extended family, current owners of Snowden's high ground, along with Richard and Peggy Moseley, who live where the lands of Snowden meet the James River, welcomed me again and again to explore the farm where Randolph Jefferson spent most of his life. They always made visiting today's Snowden a delightful experience.

The stellar staff at the Library of Virginia, especially Virginia Dunn who persisted to locate the records of the Prince Edward District Court, provided me with endless hours of assistance over several years of research. In between trips to Virginia, Margaret Downs Hrabe, Reference Coordinator, Albert and Shirley Small Special Collections Library, University of Virginia, provided scanned copies of original letters from the Thomas Jefferson Papers. Margaret Thomas (Historic Buckingham), Quatro Hubard (Virginia Department of Historic Resources), and the staff at the Virginia Historical Society helped with both information and images. Connie Geary, Scottsville Museum's digital archivist and webmaster, and the encyclopedic Raymon Thacker illuminated details about the Jefferson family's relationship to Scott's Ferry and Scottsville. In Ohio, Greg Estes and Jeanne Waselewski, Interlibrary Loan Services, Dayton Metro Library, and Chris May, Interlibrary Loan Services, Washington-Centerville Public Library, kept reels of microfilm flowing northward.

Many thanks to my friends, my family, and all my hospitable Virginia cousins who listened patiently during this book's gestation and to my editor, Nancey Bridston Hocking, and my graphic designer, Carole Ohl, who again contributed their skills to this project, bringing it to a tangible conclusion.

Lastly, I want to remember my very first mentor in Southern History, Jack Kirby (1938- 2008). His delightful encouragement was there at the inception of my study of Randolph Jefferson and Snowden; his presence at its fruition is sorely missed.

SELECTED BIBLIOGRAPHY

Archival Collections and Electronic Archives

Coolidge Collection of Thomas Jefferson Manuscripts, 1705-1827, Massachusetts Historical Society. Boston, Massachusetts.

Thomas Jefferson Collection, Missouri History Museum.

Thomas Jefferson Encyclopedia, http://www.monticello.org/site/research-and-collections/tje

Thomas Jefferson Papers, 1606-1827, Library of Congress, http://memory.loc.gov/ammem/collections/jefferson_papers/

Thomas Jefferson Papers, 1764-1826, Manuscripts Department, The Huntington Library. San Marino, California.

Thomas Jefferson Papers, Albert and Shirley Small Special Collections Library, University of Virginia, http://www2.lib.virginia.edu/small/collections/tj/

Thomas Jefferson Papers: An Electronic Archive, Massachusetts Historical Society, http://www.masshist.org/thomasjeffersonpapers/cfm/search.cfm

Books and Articles

Abercrombie, Janice L. and Richard Slatten. *Virginia Revolutionary Publick Claims*. Athens, GA: Iberian Publishing Company, 1992.

Bear, James A., Jr. and Lucia C. Stanton, editors. *Jefferson's Memorandum Books: Accounts with Legal Record and Miscellany, 1767-1826, Volumes I & II*. Princeton, NJ: Princeton University Press, 1997.

Betts, Edwin Morris and James Adam Bear, Jr., eds. *The Family Letters of Thomas Jefferson*. Columbia, MO: University of Missouri Press, 1966.

Burton, Cynthia H. *Jefferson Vindicated: Fallacies, Omissions, and Contradictions in the Hemings Genealogical Search*. Keswick, VA: Privately published, 2005.

Boyd, Julian P., Charles T. Cullen, John Catanzariti, Barbara B. Oberg, et al, eds. *The Papers of Thomas Jefferson*. Princeton, NJ: Princeton University Press, 1950-.

Crawford, Alan Pell. *Twilight at Monticello: The Final Years of Thomas Jefferson*. New York, NY: Random House, 2008.

Dewey, Frank L. *Thomas Jefferson, Lawyer*. Charlottesville, VA: University of Virginia Press, 1986.

Ellis, Joseph J. *American Sphinx: The Character of Thomas Jefferson*. New York, NY: Vintage Press, 1998.

Ely, Melvin Patrick. *Israel on the Appomattox: A Southern Experiment in Black Freedom from the 1790s Through the Civil War*. New York, NY: Vintage, 2005.

Gordon-Reed, Annette. *The Hemingses of Monticello: An American Family*. New York, NY: Norton, 2008.

Grundset, Eric G., editor. *Buckingham County, Virginia Surveyor's Plat Book, 1762 - 1858*. Baltimore, MD: Clearfield Company, 1998.

Grundset, Eric G., editor. *Land lying in the County of Albemarle: Albemarle County, Virginia Surveyors' Plat Books Volume I, Parts 1 and 3, and Volume 2, 1744 - 1853 [and 1892]*. Privately published, 1998.

Kern, Susan. *The Jeffersons at Shadwell*. New Haven, CT: Yale University Press, 2010.

Kimball, Marie. *Jefferson: The Road to Glory, 1743 to 1776*. New York, NY: Coward-McCann, Inc., 1943.

Kranish, Michael. *Flight From Monticello: Thomas Jefferson at War*. Oxford, England: Oxford University Press, 2010.

Lay, K. Edward. *The Architecture of Jefferson Country*. Charlottesville, VA: University of Virginia Press, 2000.

Looney, J. Jefferson, ed. *The Papers of Thomas Jefferson: Retirement Series*. Princeton, NJ: Princeton University Press, 2004- .

Malone, Dumas. *Jefferson the Virginian*, Boston, MA: Little, Brown and Company, 1948.

Maloney, Eugene A. *A History of Buckingham County*. Waynesboro, VA: Charles F. McClung, Printer, Inc., 1976.

Mayo, Bernard with additions by James A. Bear, Jr. *Thomas Jefferson and His Unknown Brother*. Charlottesville, VA: University Press of Virginia, 1981.

Merrill, Boynton, Jr. *Jefferson' Nephews: A Frontier Tragedy*. Lincoln, NE: University of Nebraska Press, 2004.

Moore, John Hammond. *Albemarle: Jefferson's County, 1727 - 1976*. Charlottesville, VA: Albemarle County Historical Society, 1976.

Moore, Virginia. *Scottsville on the James*. Richmond, VA: The Dietz Press, 1969, reprinted 1994.

Nicholas, Richard L. "The First Hundred Years of the Scottsville Community," *Magazine of Albemarle County History*. Charlottesville, VA: Albemarle Charlottesville Historical Society, 2010.

Randolph, Sarah N. *The Domestic Life of Thomas Jefferson*. New York, NY: Harper, 1871.

Rosen, Carl Coleman. *Revolutionary Patriots of Buckingham County, Virginia*. Westminster, MD: Willow Bend Books, 2002.

Stanton, Lucia. *"Those Who Labor for My Happiness": Slavery at Thomas Jefferson's Monticello*. Charlottesville, VA: University of Virginia Press, 2012.

Ward, Roger G. *Buckingham County Records: Land Tax Summaries & Implied Deeds, 1782-1814, Volume 1*. Athens, GA: Iberian Publishing Company, 1993.

Ward, Roger G. *Buckingham County Records: Land Tax Summaries & Implied Deeds, 1815-1840, Volume 2*. Athens, GA: Iberian Publishing Co., 1994.

Whitley, Edythe Rucker. *Genealogical Records of Buckingham County, Virginia*. Baltimore, MD: Clearfield, 2000.

Woods, Rev. Edgar. *Albemarle County in Virginia*. Baltimore, MD: Clearfield Co., 1997.

Yeck, Joanne L. *"At a Place Called Buckingham"... Historical Sketches of Buckingham County, Virginia*. Kettering, OH: Slate River Press, 2011.

Yeck, Joanne L. "A Most Valuable Citizen: A Profile of Randolph Jefferson," *Magazine of Albemarle County History*. Charlottesville, VA: Albemarle Charlottesville Historical Society, 2011.

NOTES

ABBREVIATIONS

<u>Repositories</u>

CSmH	Huntington Library, San Marino, CA
DLC	Library of Congress, Washington, DC
MHi	Massachusetts Historical Society, Boston, MA
ViCMRL	Thomas Jefferson Library, Thomas Jefferson Foundation, Inc., Charlottesville, VA
Vi	Library of Virginia, Richmond, VA
ViHi	Virginia Historical Society, Richmond, VA
ViU	University of Virginia Library, Charlottesville, VA
ViW	College of William and Mary Library, Williamsburg, VA

<u>Frequently Cited</u>

MB = Bear, James A. Jr., and Lucia C. Stanton, eds. *Jefferson's Memorandum Books: Accounts, with Legal Records and Miscellany, 1767-1826.* Princeton, NJ: Princeton University Press, 1997.

MACH = *Magazine of Albemarle County History*, 1940 - date.

PTJ = Boyd, Julian P., Charles T. Cullen, John Catanzariti, Barbara B. Oberg, et al, eds. *The Papers of Thomas Jefferson*. Princeton: Princeton University Press, 1950 - .

PTJ:RS = Looney, J. Jefferson, ed. *The Papers of Thomas Jefferson: Retirement Series*. Princeton: Princeton University Press, 2004 - .

TJE = *Thomas Jefferson Encyclopedia*, http://www.monticello.org/site/research-and-collections/tje.

UB = Mayo, Bernard with additions by James A. Bear, Jr. *Thomas Jefferson and His Unknown Brother.* Charlottesville, VA: The University Press of Virginia, 1981.

VMHB = *Virginia Magazine of History and Biography*, 1893 - date.

WMQ = *William and Mary Quarterly*, 1892 - date.

A Word about Transcriptions

I have used the second edition of *Thomas Jefferson and his Unknown Brother* as the primary reference for the Jefferson-to-Jefferson letters; however, when it comes to the interpretation of Randolph Jefferson's script, I do not always concur with the transcriptions of editors Mayo and Bear. Where the editors interpreted many of Randolph's "m," "w," and "n," as capital letters, I believe this was simply the way he formed these letters and that he did not intend them to be capitalized. For example, compare Mayo and Bear, p. 41 (original letter) and p. 47 (transcription). Thus, my transcriptions are not always identical to those of Mayo and Bear, nor are they identical to those included in the various volumes of *The Papers of Thomas Jefferson* and *The Papers of Thomas Jefferson: Retirement Series*.

Chapter One: Peter Jefferson, Gent.

1 Today the Jefferson homeplace lies in Chesterfield County, which was formed from Henrico County on May 25, 1749.

2 For more about early Jefferson generations, see Marie Kimball, *Jefferson: The Road to Glory, 1743-1776* (New York, NY: Coward-McCann, Inc., 1943), 2-15; Dumas Malone, *Jefferson the Virginian* (Boston, MA: Little, Brown and Company, 1948), 3-33.

3 George Carrington Mason, "The Colonial Churches of Henrico and Chesterfield Counties, Virginia: Part II," *VMHB* (April 1947), 147-158.

4 Kimball, *Jefferson: The Road to Glory*, 14.

5 Land Office Patents and Grants, Digital Collections, Vi, accessed December 2011, http://lva1.hosted.exlibrisgroup.com/F/?func=file&file_name=find-b-las30&local_base=CLAS30. Peter Jefferson's property was adjacent to the land of French patentee Peter Foure (a.k.a. Pierre Ford). See Priscilla Harriss Cabell, *Turff & Twigg: The French Lands* (Richmond, VA: Carter Printing Co., 1988), 1:143-146.

6 Ibid., 395. Martha (Jefferson) Goode left a will recorded in Powhatan County. See Powhatan County Will Book 2, p. 276.

7 Thomas Jefferson II will (written 15 March 1725, recorded in April of 1731), Henrico County Records, 1725-37, p. 293; Kimball, *Jefferson: Road to Glory*, 12. For a transcription of the will and more about Peter Jefferson's parents and siblings, see Lynn Mayes, "Thomas Jefferson (1656-1697)," *Virginia from the Southwest*,

accessed December 2011, familytreemaker.genealogy.com/users/m/a/y/Lyndall-J-Mayes/WEBSITE-0001/UHP-0115.html. Thomas Jefferson II's property was diminished before his death, first by a fire and then by the loss of a lawsuit over a debt equaling 6,480 pounds of tobacco. See Malone, *Jefferson the Virginian*, 10. The land Peter Jefferson inherited was part of a large patent in Henrico County, which later lay in Goochland, and even later in Powhatan County. Granted on July 12, 1718 to Thomas Jefferson, Thomas Turpin, John Archer, and Robert Eseley, these men received 1,500 acres "at a place called Fine Creek; beginning above the upper fall of the said creek, to a branch of the upper Manakin Town Creek; to mouth of Spring Run, for three pounds and the importation of 18 persons." See Land Office Patents and Grants, Digital Collections, Vi. The Jefferson, Turpin, and Archer families would remain close and intermarry.

8   In October of 1731, Henrico Parish records show a payment of 1,200 lbs. of tobacco to "Peter Jefferson, Son and Ex'r. of Thomas Jefferson, and Field Jefferson, brother to Peter Jefferson, for fferriages." See R.A. Brock, *The Vestry Book of Henrico Parish, Virginia, 1730-1773* (Greenville, SC: Southern Historical Press, Inc., 1995), 7. Just preceding the Jefferson entry is one for Thomas Branch, who was also paid 1,200 lbs. of tobacco for ferriages. Branch may be close kin to the Jeffersons, operating the landing on the north side at Osbornes. The marriage contract of Peter's grandmother, Mary (Branch) Jefferson, to Joseph Mattocks (16 November 1700, recorded on 1 April 1701), reveals the proximity of the Branch, Field, and Jefferson families, all of whom lived near the James River along Proctors Creek in Henrico County. Joseph Mattocks and Mary Jefferson chose as their trustees Seth Ward, Christopher Branch, and Thomas Jefferson. Witnesses were Peter Field, Ben Branch, Mary Jefferson, and Ashly [?] Branch. See Benjamin B. Weisiger, III, *Henrico County, Virginia: Deeds 1677-1705* (Athens, GA: Iberian Publishing Co., 1996), 115-116.

9   Landon C. Bell, *The Old Free State* (Baltimore, MD: Clearfield Company, Inc., 1995), 1:330.

10 John Woodson became Peter Jefferson's brother-in-law when Woodson married Jane Randolph's younger sister, Dorothea, in 1751. Within a month, Daniel Scott of Scott's Ferry in Albemarle wed their sister, Anne Randolph. John Woodson acted as security for the Scott marriage. See "Marriage Bonds in Goochland County," *WMQ* (October 1898), 98, 100.

11 Nathaniel Mason Pawlett, *Goochland Road Orders, 1728-1744* (Charlottesville, VA: June 1975, Revised 2004), 20-21; Ann K. Blomquist, *Goochland County, Virginia Court Order Book 3, 1731-1735* (Westminster, MD: Heritage Books, 2006), 307.

12 Blomquist, *Goochland County, Virginia Court Order Book 3*, p. 359

13 Ibid., 374.

14 Benjamin B. Weisiger III, *Goochland County, Virginia: Wills & Deeds 1736-1742* (Athens, GA: Iberian Publishing Co., 1997), 8. The bond for Peter Jefferson as Sheriff of Goochland County was recorded on September 30, 1737 and signed by Isham and William Randolph.

15 Benjamin B. Weisiger III, *Goochland County, Virginia: Wills & Deeds 1728-1736* (Athens, GA: Iberian Publishing Co., 1995), 75.

16 "Marriage Bonds in Goochland County," *WMQ* (October 1898), 98.

17 Jefferson family Bible, with notations (from 1752-1861), Small Special Collections, ViU.

18 For a detailed history of Shadwell, see Susan A. Kern, *The Jeffersons at Shadwell* (New Haven, CT: Yale University Press, 2010).

19 Kern, *Shadwell*, 17.

20 Albemarle County Order Book, 1744-1748, 28 February 1744; Virginia Moore, *Scottsville on the James* (Richmond, VA: The Dietz Press, 1994), 5. Joshua Fry lived at what became known as Viewmont, located nine miles northeast of Scottsville on today's Highway 20, between Keene and Carter's Bridge. For a thumbnail sketch of Joshua Fry, see Kimball, *Jefferson: The Road to Glory*, 23-24. The Bellew name is also recorded as Ballow, Ballowe, Ballew, Belew, and Bailou. Judge Bellew had extensive land holdings on the Slate, Rivanna, and James Rivers. The extended Bellow family became long-term neighbors of the Jeffersons at Snowden.

21 Richard L. Nicholas writes, "From 1745 until 1818 the hamlet and future site of Scottsville were identified variously as the 'Court House Landing,' 'Scott's Ferry,' 'Scott's Landing,' or 'Scott's Ferry Landing.' The Scott appellation was obviously derived from the dominance of the family and its ownership of extensive property at both the courthouse and ferry." See Richard Ludlam Nicholas, "The Early History of the Founding of Scottsville, Virginia and the Surrounding Area, 1732-1830" (Unpublished manuscript, 2009), 7.

22 Daniel Scott held 550 acres on the north side of the Fluvanna River, at the Horseshoe Bend. Today, the counties of Albemarle, Fluvanna, and Buckingham meet there at the historic town of Scottsville.

23 John Hammond Moore, *Albemarle: Jefferson's County, 1727-1976* (Charlottesville, VA: Albemarle County Historical Society, 1976), 1.

24 There is no absolute accounting of Peter Jefferson's land holdings. Marie Kimball concluded that, combining purchase and patents, he personally acquired over 25,000 acres. If all the land in joint ventures is totaled, Jefferson's name was associated with over 71,000 acres. See Kimball, *Jefferson: The Road to Glory*, 309-311.

25 Moore, *Scottsville on the James*, 7. In the 18th century, a hogshead fully packed with tobacco officially weighed 1,000 pounds. This immense wooden barrel measured 48 inches long and was 30 inches in diameter at the head.

26 Henry S. Randall, *The Life of Thomas Jefferson* (New York, NY: Derby & Jackson, 1858), 1:13-14. In 1874, James Parton imagined an indefatigable Peter Jefferson, conquering the western wilderness. "When surveying in the wilderness," wrote Parton, "he could tire out his assistants, and tire out his mules; then eat his mules, and still press on, sleeping alone by night in a hollow tree, to the howling of the wolves." See James Parton, *Life of Thomas Jefferson, Third President of the United States* (Boston, MA: James R. Osgood, 1874), 9. Fawn Brodie retold the shed story: "... it is said, [he] once directed three slaves to pull down a ruined shed which had been girdled with a rope. When after repeated straining they failed, he 'bade them stand aside, seized the rope, and dragged down the structure in an instant." See Fawn M. Brodie, *Thomas Jefferson: An Intimate History* (New York, NY: Norton, 1974), 34.

27 Frederick Clifton Pierce, *Field Genealogy* (Chicago, IL: Hammond Press, W.B. Conkey Company, 1901), 1060.

28 Sarah N. Randolph, *The Domestic Life of Thomas Jefferson* (New York, NY: Harper, 1871), 19-20.

29 "Tuckahoe," *TJE*, accessed December 2011, http://www.monticello.org:8081/site/research-and-collections/tuckahoe.

30 Peter Jefferson took the oath as Vestryman on April 12, 1748. See William Lindsay Hopkins, *St. James Northam Parish Vestry Book, 1744-1850* (Athens, GA: Iberian Publishing Company, 1993), 7-8.

31 Moore, *Scottsville on the James*, 2.

32 Albemarle records, abstracted by Eric Grundset, are as follows: "19 Feb 1746: Peter Jefferson 400 acres lying on and adjoyning South side Fluvanna; surveyed for Benjamin Harris transferred to Peter Jefferson; courthouse road shown; by Thomas Turpin; adjoining Lad's line, David Watkins. 17 Feb 1746: Peter Jefferson 225 acres, lying on South side Fluvanna; surveyed for David Watkins. NB: Certificate for the above land is made out for Pet. Jefferson ye sher. having returned the said Watkins insolvent June 30, 1747 (sic); courthouse road shown; adjoining John Lad." See Eric Grundset, *Land lying in the County of Albemarle: Albemarle County, Virginia Surveyors' Plat Books Volume I, Parts 1 and 3, and Volume 2, 1744-1853 [and 1892]* (Privately published, 1998), 2. During this period of rapid western expansion, a patent or "new" land was granted by the King of England and was subject to the law that the patentee must build a structure on the granted land within three years and clear three out of every fifty acres in order to secure the patent. If the land was not thus improved, it reverted to the Crown and could be re-surveyed by a new potential patentee. Before 1754, Peter Jefferson was able to assemble 1,550 acres of

"new land" in his name running along the Fluvanna River, on the west side of the Horseshoe Bend. The precise source of all of the 1,550 acres is unclear.

33 William Walter Hening, *Statutes at Large*, 5:364-365, accessed January 2011, http://vagenweb.org/hening/vol07-20.htm.

34 Albemarle County Order Book 1744-1748, p. 33.

35 In October of 1748, Scott's permission to operate a ferry was described as "On the Fluvanna, from Scott's to Battersby's, or Noble Ladd's, in Albemarle, for a man three pence, for a horse the same." See Hening, *Statutes*, 6:21.

36 Amos Ladd's patents were the first held in the immediate area, claiming the prime land at the tip of the river's Horseshoe Bend and making him the first owner of a significant portion of the land eventually incorporated in Peter Jefferson's Snowdon. In 1723, "Amos Lad" patented "400 acres on the south side of the Fluvanna" in what was then Goochland County. In 1728, he patented another 400 acres on the south side of the Fluvanna, described as opposite "Totero" Town, on the river at the tip of the Horseshoe Bend. This was virgin wilderness, where only the trace of a long abandoned Native American village remained. See Land Office Patents and Grants, Digital Collections, Vi; Benjamin B. Weisiger, III, *Goochland County, Virginia: Wills & Deeds 1728-1736* (Athens, GA: Iberian Publishing Co., 1995); Benjamin B. Weisiger, III, *Goochland County, Virginia: Wills & Deeds 1736-1742* (Athens, GA: Iberian Publishing Co., 1997).

37 Weisiger, *Goochland County, Virginia Wills and Deeds, 1736-1742*, p. 52.

38 Amos Ladd's recorded will is now lost, but is referred to in the surviving Albemarle court orders. See Albemarle County Order Book, 1744-1748, p. 84.

39 Albemarle County Deed Book 1, pp. 231-233.

40 There is a gap in the existing Albemarle County deeds between 1752 and 1758. John Ladd's transfer of 400 acres to Peter Jefferson has not been located; however, it is referenced on the survey recorded in November of 1754.

41 Albemarle County Surveyor's Plat Book No. 1, Plat 1, p. 293.

42 Born in Virginia to Virginian parents, Peter Jefferson had to reach back several generations to connect to Wales, Mt. Snowdon, and memories of a Welsh homeland preserved through family tradition. His son, Thomas, wrote, "The tradition in my father's family was that their ancestor came to this country from Wales, and from near the mountain of Snowdon, the highest in Gr. Br." See Thomas Jefferson, *Autobiography of Thomas Jefferson, 1743-1790* (New York, NY: 1914), 3. In 2007, "WalesOnline" reported that, "tests on two British men, who have the Jefferson surname, have shown they have an identical 'K2' section of DNA, which shows the US president does indeed have links with the UK." See "DNA tests prove Jefferson's Welsh lineage," accessed December 2011, http://www.walesonline.co.uk/news/

wales-news/tm_headline=dna-tests-prove-jefferson-s-welsh-lineage&method=full &objectid=18592199&siteid=50082-name_page.html.

43 "Last Crossing of the Scottsville Ferry, 1907," accessed December 2011, http://scottsvillemuseum.com/transportation/homeB62cdB16.html.

44 R. B. Moon's 1853 map of Scottsville shows the "Snowdon Ferry House." See Nicholas, "The Founding of Scottsville," 3. The Albemarle County Order Book, 1747-1748 does not contain an ordinary license for Ladd or for Jefferson; however, the order books for 1749-1783 are destroyed. These likely contained more than one ordinary license for a tavern at Snowdon. Also, after 1761, any ordinary license for the Snowdon Ferry House would have been approved by the Court of Buckingham County and those order books, 1761-1869, are also lost.

45 Moore, *Albemarle*, 39-40.

46 Albemarle County Order Book, 1774-1748, pp. 233-234.

47 Moore, *Scottsville on the James*, 10.

48 Rev. William Douglas, *Douglas Register* (Baltimore, MD: Clearfield Co., 2007), 84. John Woodson and Valentine Wood served as witnesses. See Goochland County Marriage Register, 1730-1853, p. 264.

49 Goochland County Will and Deed Book 4, p. 168. Isham Randolph signed his last will on April 6, 1741; it was recorded on December 21, 1742. He named William Randolph, Esq., Col. Richard Randolph, William Randolph, Jr., Beverly Randolph, and Peter Jefferson as guardians for his children. He made his wife, Jane, his sole executor.

50 Goochland County Will and Deed Book 8, pp.168-169. Jane (Rogers) Randolph, born c. 1692, signed her will in Goochland County on December 5, 1760, which was probated on July 21, 1761. Jane had received all of her husband's land in Goochland and Amelia Counties and all of his slaves. In her will, she did not name her daughter, Jane, nor her daughter, Mary, who married Charles Lewis. She included Anne, Daniel Scott's widow, who, by then, had remarried Pleasants, to whom she gave a slave, Dianna, and her gold watch. She also left bequests to daughters, Elizabeth Raley (a.k.a. Railey), Dorothea Woodson, and Susanna Randolph. Jane also remembered Rev. William Douglas with five guineas. She left Dungeness to her son, William, but it was ultimately inherited by his brother, Thomas Randolph (born abt. 1745), who married Jane Cary in 1768. The reason for her exception of Jane Jefferson and Mary Lewis is unknown. See Robert Isham Randolph, "The Sons of Isham Randolph of Dungeness," *VMHB* (October 1937), 383-386; http://freepages.genealogy.rootsweb.ancestry.com/~warejamesbakercalder/randolphisham0001.htm#id602.

51 "Marriage Bonds in Goochland County," *WMQ* (October 1898), 98. Peter Jefferson's friends and neighbors in Goochland joined him in the development of southern

Albemarle and Scott's Ferry. For example, Dr. Arthur Hopkins (d. 1766/69) of Goochland County, who signed the Jeffersons' marriage bond, patented land in Albemarle as early as 1732, amassing extensive property in the area including a ninety-nine-acre tract adjoining John Scott of Scott's Ferry. In Goochland during the 1740s, Hopkins was a Vestryman and a Justice of the Peace. Later, between 1761 and 1766, he served as a Justice in Albemarle. See Nicholas, "The Founding of Scottsville," 166.

52 Upon Daniel Scott's death, Peter Jefferson not only lost a brother-in-law, but also a business partner. It is interesting to note that Jefferson's brother, Field, operated a ferry in Lunenburg County, which he continued until his death on February 10, 1765. Likewise, their sister, Martha, recreated her childhood life by the river when she married Bennet Goode, who operated a ferry (later known as Jude's Ferry) in Goochland. Concurrent with Peter Jefferson's acquisition of the landing at Snowdon, Jefferson was involved with the development of a town at Beverley Randolph's plantation, Westham, located downriver in Henrico County. Near Richmond, "Beverley Town" was laid out at the point where Westham Creek entered the James River. When Beverley Randolph died in 1751, he instructed that the area of the town be conveyed in fee simple to the subscribers. Significantly, Peter Jefferson purchased lot # 57 – "the ferry lot." See Thomas Jefferson's "Farm Book," 32, 127, accessed December 2011, http://www.masshist.org/thomasjeffersonpapers/farm/. In a letter to Thomas Taylor, Thomas Jefferson describes the laying out of Beverley Town by his father: "I was a boy of about 8. years age, living with my father at Tuckahoe, when this transaction took place, and well remember his going to Westham to lay off the town, and his mentioning the price at which the lots were sold, which was a doubloon (£4-6) apiece the purchasers drawing their numbers (if I recollect) by lot, and not chusing them." See Thomas Jefferson to Thomas Tylor, 28 December 1814, *PTJ:RS*, 8:169-171; Susan A. Kern, "The Jeffersons at Shadwell: The Social and Material World of a Virginia Family," dissertation (College of William and Mary, 2005), 312-314.

53 "Randolph Family," W. G. Stanard, *WMQ* (October 1899), 119-122. In April of 1754, Anne was summoned to appear in Cumberland County Court to declare whether she would take on the administration of the estate of her late husband. See Cumberland County Order Book 3, p. 166. The following month, she was granted a certificate to administer the estate. See Cumberland County Order Book 3, p. 168.

54 It is clear from Peter Jefferson's estate accounts that the four acres opposite the county seat came complete with houses – plural. Certainly until Buckingham County was created in 1761, and the courthouse moved to Charlottesville c. 1761-1762, traffic was steady across the river to Scott's Ferry. On November 30, 1757, just months after Peter Jefferson's death, Mr. Richard Murray, Ordinary Keeper, paid the estate for "Rent of the houses opposite to Albemarle Court house with 4 acres of Land at £4 pr. annum from this time." Murray continued to rent the four acres of

land, complete with houses, through June of 1762 when he paid £3 and 5 shillings to the estate. He may have given up the lease when the courthouse moved. See John Harvie, Peter Jefferson Estate Account Book I, p. 36, CSmH.

55 Sources disagree as to how long the Jeffersons resided at Tuckahoe. According to Jane Jefferson's Booke, Lucy Jefferson was born at Tuckahoe on October 10, 1752. Thomas Mann Randolph, born in 1741, reached his majority in 1762.

56 For more about John Biswell, see Kern, dissertation, 157-159; Fiske Kimball, "In Search of Jefferson's Birthplace," VMHB (October 1943), 314; Kern, *Shadwell*, 40.

57 Kern, dissertation, 32, 35. The mill at Shadwell provided a service center for the local community. Among the individuals using the mill were Benjamin Sneed, tutor to the Jefferson children, and Manus Burger, a local blacksmith.

58 Kern, *Shadwell*, 27.

59 Ibid., 26-40; Peter Jefferson Estate Inventory, Albemarle County Will Book 2, pp. 41-47. For a transcription of the Jefferson inventory, see Kern, dissertation, "Appendix II.a: "Inventory of the Estate of Peter Jefferson, 1757," pp. 448-455.

60 Randall, *Jefferson*, 1:2.

61 Kern, *Shadwell*, 29, 33-38. Peter Jefferson's office may have been an addition, not part of the heated main house; however, in Kern's suggested floor plan, the office is included as one of the four main rooms of the ground floor.

62 Kern, dissertation, 44.

63 Kern, *Shadwell*, 29.

64 Peter Jefferson Estate Inventory, Albemarle County Will Book 2, p. 42.

65 Kern, *Shadwell*, 29-32; Kern, dissertation, 47.

66 Ralph E. Fall, *The Diary of Robert Rose* (Falls Church, VA: McClure Press, 1977), 1, 33. Rev. Rose knew both the Tidewater and the Piedmont, intimately. A successful planter, he lived in Albemarle and at Bear Garden, on the Tye River (today in Nelson County). His diary, which places him at the homes of various pioneers in Albemarle, Buckingham, and Amherst Counties, is an extraordinarily valuable record of the period, 1746-1751.

67 "Fry-Jefferson Map," *TJE*, accessed December 2011, http://www.monticello.org/site/house-and-gardens/fry-jefferson-map-virginia; "Joshua Fry," *TJE*, accessed December 2011, http://www.monticello.org/site/jefferson/joshua-fry; "Viewmont," accessed December 2011, http://scottsvillemuseum.com/homes/homeLOC160638pr.html. Peter Jefferson's professional and political colleagues were likely his core associates. Intermarriages of the Fry-Maury-Nicholas-Scott-Walker families in the late 18th century established kinship webs which formed the next generation of Albemarle's elite. Following their fathers who had founded the

English-based culture in Albemarle, the next generation was born in the Piedmont. To what degree the wives and children of these men were friendly with each other is less known; however, both Thomas and Randolph Jefferson would study and live in their homes – Thomas at Rev. James Maury's Classical School for Boys and Randolph at Col. Charles Lewis' Buck Island and Col. John Nicholas' Seven Islands. These marriages reflect the Tidewater mergers of the previous generations. Martha Fry married John Nicholas; Margaret Fry married John Scott; Joshua Fry married Peachy Walker; Henry Fry married Mildred Maury; Rev. Matthew Maury married Elizabeth Walker, etc. See Rev. Edgar Woods, *Albemarle County in Virginia* (Baltimore, MD: Clearfield Co., 1997): Fry (197-198), Maury (268-269), Nicholas (289-291), Scott (312-313), and Walker (334-336).

68 Woods, *Albemarle County*, 235-238. Peter Jefferson also served as Joshua Fry's executor upon Fry's death in 1754.

69 Hayes, *The Road to Monticello*, 25; Kern, dissertation, 79.

70 Peter Jefferson Will, Albemarle County Will Book 2, p. 32-34.

71 Albemarle's new courthouse in Charlottesville was completed in 1763, indicating that the transition took some time, but which ultimately left Scott's Ferry with considerably less traffic and immediate possibilities for growth.

Chapter Two: Jane Jefferson's Shadwell

1   In 1816, Randolph Jefferson's home in Buckingham County burned, destroying his personal papers. In 1869, the Buckingham Courthouse burned, eliminating the vast majority of public documents concerning his affairs and contribution at the local level. The few letters and other miscellaneous documents that survive in Thomas Jefferson's papers constitute the largest single source of material concerning Randolph Jefferson. For a concise biographical sketch, see Joanne L. Yeck, "A Most Valuable Citizen: A Profile of Randolph Jefferson," *MACH* (2011), 1-37.

2   Brodie, *Thomas Jefferson: An Intimate History*, 36, 41.

3   Kern, dissertation, "Appendix III: '"Jane Jefferson – Her Booke:' Jefferson Family History and Bibliography," 459-485; Jefferson family Bible, with notations (from 1752-1861), Small Special Collections, ViU.

4   Brodie, *Thomas Jefferson: An Intimate History*, 43; "Isham Randolph," Virginia Historical Society website, accessed December 2011, http://www.vahistorical.org/dynasties/ishamrandolph.htm; "Jane Randolph," *TJE*, accessed December 2011, http://www.monticello.org/site/jefferson/jane-randolph-jefferson.

5   "Journals of the Council of Virginia in Executive Sessions, 1737-1763," *VMHB* (January 1907), 225-227. Isham Randolph is buried at the Randolph seat, Turkey Island.

6   Kern, *Shadwell*, 70. Goochland County court records bear out Jane (Rogers) Randolph's strength and determination. During the December Court of 1742, the fifty-year-old Jane presented Isham Randolph's last will and testament and was declared Executrix of her husband's estate. By the November Court of 1743, Jane had initiated multiple chancery cases for monies owned the estate. See Blomquist, *Goochland County, Virginia Court Order Book 5, 1741-1745*, pp. 177, 301, 365-366.

7   Randall, *Jefferson*, 1:16-17.

8   Kern, *Shadwell*, 19; "Randolph Family Portraits," Virginia's Colonial Dynasties, accessed December 2011, http://www.vahistorical.org/dynasties/randolphfamily.htm.

9   Peter Jefferson Estate Account Books, CSmH; Peter Jefferson Estate Papers, MHi.

10 Peter Jefferson Will, Albemarle County Will Book 2, p. 32.

11 Thomas Walker to Thomas Jefferson, 19 January 1790, *PTJ*, 16:114-115. Cate was bequeathed to Elizabeth by her father. See Will of Peter Jefferson, Albemarle County Will Book 2, p. 32.

12 "The Carr Family," *TJE*, accessed December 2011, http://www.monticello.org/site/jefferson/carr-family. Thomas Mann Randolph identified Martha's son, Samuel Carr, as the father of Betty Brown's children. A Monticello slave, Betty Brown was a member of the Hemings family. If Randolph was correct, Carr may have fathered Betty's last two children. See Annette Gordon-Reed, *The Hemingses of Monticello: An American Family* (New York, NY: Norton, 2008), 678.

13 Randolph, *Domestic Life*, 38-39.

14 Randall, *Jefferson*, 1:41.

15 Ibid.

16 Jane Jefferson, Jr. Estate Inventory, Albemarle County Will Book 2, pp. 227, 233. For a transcription of the inventory of Jane Jefferson, Jr, see Kern, dissertation, "Appendix II. c.: "Inventory of the Estate of Jane Jefferson Jr., 1768," 457-458; Kern, dissertation, 62.

17 Kern, dissertation, 53.

18 In a letter dated January 10, 1790, Thomas Jefferson mentions "the youngest child, my sister Nancey." See Thomas Jefferson to Thomas Walker, 18 January 1790, *PTJ*, 16:113; "Mary Jefferson Bolling," *TJE*, accessed December 2011, http://www.monticello.org/site/jefferson/mary-jefferson-bolling.

19 *PTJ*, 15: 632-35. In 1787, this laying on of fat likely would be associated with the good health they were enjoying, not seen as counter to it.

20 Randolph, *Domestic Life*, 104.

Chapter Three: The Early Education of Randolph Jefferson

1 Randall, *Jefferson*, 1:14.

2 Goochland County Will Book 4, pp. 110-112.

3 Merrow Egerton Sorley, *Lewis of Warner Hall: The History of a Family* (Baltimore, MD: Genealogical Publishing Co., Inc., 1991), 345-349.

4 The staggered generations made Charles Lewis, Jr, born in 1721, a relatively young uncle to Thomas Jefferson, born in 1743.

5 Land Office Patents and Grants, Digital Collections, Vi.

6 Boynton Merrill, Jr., *Jefferson's Nephews: A Frontier Tragedy* (Lincoln, NE: University of Nebraska Press, 2004), 4.

7 Woods, *Albemarle County*, 21.

8 Merrill, *Jefferson's Nephews*, 8; Albemarle County Will Book 2, pp. 403-405; Sorley, *Lewis of Warner Hall*, 347.

9 Sorley, *Lewis of Warner Hall*, 345-349.

10 Charles Lewis Estate Inventory, Albemarle County Will Book 2, pp. 403-405.

11 Charles Lewis Will, Albemarle County Will Book 2, pp. 399-400.

12 Merrill, *Jefferson's Nephews*, 32.

13 Col. Charles Lewis Will, Albemarle County Will Book 2, p. 403. The Lewises may have been producing cloth while Randolph Jefferson boarded with them; they certainly did in the 1780s. The Jefferson children were accustomed to their mother and older sisters carding and spinning. Jane Jefferson worked a cotton wheel and was skilled with hackle and brushes for "dirty wool" and raw cotton. Young Jane also spun and had "a pair of cards" for carding wool. At age sixteen, Martha received her own wheel. See Kern, dissertation, 49-50, 133.

14 Albemarle County Will Book 2, p. 404.

15 Kern, dissertation, 46-47.

16 Albemarle County Will Book 2, pp. 399-400.

17 The extreme closeness of the Lewis family is reflected as early as 1776, in the will of Charles Lewis, Jr.'s son-in-law, Charles Lewis of St. Anne's Parish (North Garden), who married Mary Randolph Lewis. His will is witnessed by three brothers-in-law: John Thomas, Bennett Henderson, and Charles L. Lewis. He names his beloved wife, Mary, as Executrix, along with his father-in-law, Col. Charles Lewis of Buck Island, Bennett Henderson, another brother-in-law, Charles Hudson, and his own brother, Nicholas Lewis, as Executors. His trusted friends were his extended Lewis family, without exception. See Albemarle County Will Book 2, pp. 376-377.

18   *MB*, 140. An Albemarle road order dated 1792, which includes the Sneed property, reveals its proximity to both Shadwell and the Randolph property at Edgehill, mentioning "Richard Johnson is appointed Surveyor of the road from Benjamin Sneeds to the Shop Branch above Shadwell with the following Male Labouring hands Viz Col⁰ John Harvies, Thos M Randolph's Tho's Jefferson's at Shadwell and his own." See Nathaniel Mason Pawlett, *Albemarle County Road Orders 1783-1816* (Charlottesville, VA: December 1975, Revised 2004), 26. Richard Johnston was Benjamin Sneed's son-in-law. See Dick Baldauf, "The Family of Benjamin Sneed," *The Sneed Reports* (April 2008), 5, accessed December 2011, homepages.rootsweb. ancestry.com/~lksstarr/html/snead.html.

19   Edwin Morris Betts and James Adam Bear, Jr., editors, *The Family Letters of Thomas Jefferson* (Columbia, MO: University of Missouri Press, 1966), 161-162. When Martha wrote the letter, she was at Belmont plantation, which sat just north of Sneed's school.

20   The Jefferson records favor the spelling "Snead," whereas the Albemarle County and Sneed family records prefer "Sneed." Benjamin Sneed's obituary in Danville, Kentucky, printed his name as "Sneed."

21   The initial £6 for 1762, paid on May 24, 1762, might be in advance of Randolph's stay. He was six and one-half years old at the time. The £6 paid on July 28, 1764, may be in arrears for the year of 1763. The two installments of £3/each on July 29, 1764 and on January 4, 1765 would then cover his stay during 1764-1765. See Peter Jefferson Estate Account Books, CSmH; Peter Jefferson Estate Papers, MHi.

22   Frank L. Dewey, *Thomas Jefferson Lawyer* (Charlottesville, VA: University of Virginia Press, 1986), 10-11. Thomas Jefferson was at Shadwell for Christmas of 1762 and did not return to Williamsburg until October of 1763. An outbreak of smallpox delayed his return. See Ibid., 11.

23   Other authors have drawn different conclusions. In 1942, Bernard Mayo wrote that Sneed taught at the Lewis home: ". . . the younger children were being instructed by a Mr. Benjamin Snead. In the years 1764 and 1765, Mr. Snead held school a few miles southeast of Shadwell at the Buck Island plantation of Charles Lewis, whose wife was a sister of Mrs. Peter Jefferson. During these years Harvie's account books show payments to Lewis for Master Randolph's boarding, and to Snead for his schooling." See Bernard Mayo, *Thomas Jefferson and His Unknown Brother Randolph* (Charlottesville, VA: The Tracy W. McGregor Library, University of Virginia, 1942), 8. Author Boyton Merrill elaborated Mayo's opinion, while criticizing Sneed's lack of rigor: "As the younger Jefferson children reached school age they were sent to Buck Island, where Mr. Benjamin Snead was employed as a tutor by Charles Lewis, Jr. During the years of 1764 and 1765, Randolph Jefferson, Peter's youngest son, boarded with this aunt and uncle at Buck Island during school sessions. Snead's demands upon his students were probably not too rigorous, if the

later correspondence of some of his pupils is any indication." See Merrill, *Jefferson's Nephews*, 10-11.

24 Alexander Brown, *The Cabells and Their Kin: A Memorial Volume of History, Biography, and Genealogy* (Boston, MA: Houghton, Mifflin and Company, 1895), 190-191.

25 *MB*, xlv. There is no additional evidence of hired tutors for Randolph Jefferson until 1771, when he resides in Williamsburg.

26 Dewey, *Thomas Jefferson Lawyer*, 10-11. Thomas Jefferson was at home through most of the spring of 1766, and in the fall he was appointed as a Justice in Albemarle. Ibid., 25.

27 *Virginia Gazette* (Purdie and Dixon), 22 February 1770.

28 *MB*, 158-159.

29 "Shadwell," *TJE*, accessed December 2011, http://www.monticello.org/site/research-and-collections/shadwell.

30 Kimball, "In Search of Jefferson's Birthplace," 315; Kern, *Shadwell*, 64-65. When Fiske did his work in the 1940s, it was believed that the house was not rebuilt; however, in 1991, archaeological work uncovered two cellar foundations. The smaller of the two clearly post-dated the main house.

31 John Harvie's death in 1767 also stripped Thomas Jefferson of a potential advisor, the man who had been most intimately familiar with the estate.

32 Kern, dissertation, 226.

33 Ibid., 229.

34 Kern, *Shadwell*, 141.

35 *MB*, 177.

36 Ibid.; Case Book, No. 433; Thomas Jefferson Accounts, "Randolph Jefferson, 1771-1779," Small Special Collections, ViU.

37 "700 years of Bates Family History," accessed December 2011, http://www.garrettfamily.info/family-history/bates/Bates-Family-History-1.pdf, 14-16; Grundset, *Albemarle County, Virginia Surveyors' Plat Books*, 12; Pawlett, *Albemarle Road Orders 1744-1744*, p. 16. On a surviving, fragmented Buckingham County tithe list for 1764, Elizabeth Bates appears with her overseer, William Burnitt, indicating her husband's death prior to 1764. See Mary Bondurant Warren, *Buckingham County, Virginia Church and Marriage Records* (Athens, GA: Heritage Papers, 1993), 5.

38 *MB*, 65. "Earlier in 1768, Buckingham client Charles Patteson owed Thomas Jefferson £2/4/3." See Ibid., 58.

39 "Isaac Bates, *Forney and Clark Genealogy*, accessed December 2011, http://forneyclarkgenealogy.com/getperson.php?personID=I3788&tree=forneyclark;

"Adkins/Spence/Perry/Smith Connections," accessed December 2011, http://wc.rootsweb.ancestry.com/cgi-bin/igm.cgi?op=REG&db=poki00&id=I079001; Edythe Rucker Whitley, *Genealogical Records of Buckingham County, Virginia* (Baltimore, MD: Clearfield, 2000), 7-30.

40   Buckingham County Personal Property Tax, 1782.

41   Ibid.; Kern, dissertation, 228.

Chapter Four: A World beyond the Piedmont: Williamsburg and the College of William and Mary

1   Rev. James Maury's school was about twelve miles from Shadwell, in Albemarle's Fredericksville Parish. Thomas Jefferson boarded with the Maury family during 1758-1760.

2   The College of William and Mary was founded in 1693 by royal charter of King William III and Queen Mary II.

3   *PTJ*, 1:3. John Harvie (1706-1767) was one of several executors of Peter Jefferson's estate. He actively handled the accounts for many years, and Thomas Jefferson needed Harvie's approval for college expenses. Col. Peter Randolph (1717-1767), the son of William Randolph Jr. and Elizabeth Peyton Beverley, was the first cousin of Jane (Randolph) Jefferson. He lived at Chatsworth, near Richmond, north of the James River, a plantation which ultimately consisted of 20,000 acres and was worked by 250 slaves. In 1738, he married Lucy Bolling and, in 1751, built his impressive mansion. He died on July 8, 1767; his sons, Robert and Beverley, attended William and Mary after his death. For more about Col. Peter Randolph and Chatsworth, see Marc R. Matrana, *Lost Plantations of the South* (Jackson, MS: University of Mississippi, 2009), 26-27.

4   Jefferson, *Autobiography*, 6. William Small was born in England, came to Virginia in 1758, returned to England in 1764, and died in Birmingham in 1775. He introduced the lecture system at William and Mary and was an "intimate friend" of James Watt, the Scottish inventor of the steam engine, and of Erasmus Darwin, the English physician, eminent scientist, and grandfather of Charles Darwin. According to Dumas Malone, Small arrived in Virginia as "Mr. William Small," and upon return to England called himself William Small, M.D. See Malone, *Jefferson the Virginian*, 53. For more about Small see "Early Courses and Professors at William and Mary College," *WMQ* 14 (1905), 71-83; "Papers Relating to the College," *WMQ* 16 (1908), 162-173; J. A. C. Chandler, "Jefferson and William and Mary," *WMQ*, Second Series, 14 (1934), 304-307; Herbert L. Ganter, "William Small, Jefferson's Beloved Teacher," *WMQ*, Third Series, 4 (1947), 505-511.

5   Thomas Jefferson to L.H. Girardin, 15 January 1815, Andrew A. Lipsomb and Albert Ellery Bergh, eds., *The Writings of Thomas Jefferson* (Washington D.C., 1903), 14:231. Louis Hue Girardin was appointed Professor of Modern Languages at William and Mary in 1803.

6   Thomas Jefferson, *Notes on the State of Virginia* (www.forgottenbooks.org), 157-162; Glenn Patton, "The College of William and Mary, Williamsburg, and the Enlightenment," *Journal of the Society of Architectural Historians* (1970), 24-32; Ludwell H. Johnson III, "Sharper than a Serpent's Tooth: Thomas Jefferson and His Alma Mater," *VMHB* (1991), 145-162.

7   There were four components to William and Mary: the Grammar School, the School of Philosophy, the Divinity School, and Brafferton, a school for the education of Native Americans, which was completed in 1723. The School of Philosophy was divided into Natural Philosophy and Moral Philosophy.

8   *MB*, 261.

9   In 1934, J.A.C. Chandler wrote, "However, after a year at college Jefferson found out that he had wasted a lot of money on dress and horses and extravagant living. It is difficult to tell where Jefferson would have gone if he had not come under the influence of Dr. William Small, of the College faculty." See Chandler, "Jefferson and William and Mary," 305. In a lengthy letter written in 1808, Thomas Jefferson counseled his favorite grandson, Thomas Jefferson Randolph, who was sixteen years old and heading for school in Philadelphia. In it, Thomas shared wisdom gained from his own college experience and expressed concern for his grandson's moral as well as academic success. See Thomas Jefferson to Thomas Jefferson Randolph, 24 November 1808, Betts and Bear, *Family Letters*, 362-365.

10 *MB*, 262.

11 Ibid., 261-262.

12 Thomas Jefferson was first elected to the Virginia House of Burgesses in 1769, at the age of twenty-six.

13 *MB*, xlv-xlvi. There is no written record of interaction between the Jefferson brothers while Randolph was in school at William and Mary; however, Thomas' protracted presence in the capital was likely reassuring for Randolph and may have broadened his social experience. Additionally, during 1771 and 1772, Thomas was particularly mobile and active as an attorney; during that time he handled 137 cases in 1771 and 153 cases in 1772. For more concerning Thomas Jefferson's career as an attorney see Frank L. Dewey, "Thomas Jefferson and a Williamsburg Scandal: The Case of Blair V. Blair," *VMHB* (1981), 44-63; Dewey, *Thomas Jefferson, Lawyer*.

14 *MB*, 30.

15 Ibid., 289.

16   The Bursar's Books for William and Mary charged Randolph for his board beginning October 14, 1771. For more about Anderson's Tavern see Helen Bullock, "Wetherburn's Tavern Historical Report, Block 9 Building 31 Lot 20 & 21," accessed December 2011, http://research.history.org/DigitalLibrary/View/index.cfm?doc=ResearchReports\RR1165.xml.

17   *Virginia Gazette* (Purdie and Dixon), 28 April 1774, p. 3.

18   *Virginia Gazette* (Purdie and Dixon), 7 March 1771, p. 3; *The Wigmaker in Eighteenth-Century Williamsburg: An Account of his Barbering, Hair-Dressing, & Peruke-Making Services, & some Remarks on Wigs of Various Styles* (Williamsburg, VA: Colonial Williamsburg Foundation, 2004), 31-32. Randolph Jefferson may also have made use of Hubbard's wig shop. As a young gentleman entering Williamsburg society, Randolph certainly needed his hair cut and possibly dressed. He may or may not have sported a periwig. Hubbard died in 1779 and left a last will, mentioning wig-making tools.

19   Mark R. Wenger, "Thomas Jefferson, the College of William and Mary, and the University of Virginia," *VMHB* (1995), 339-374. Originally, the building was simply known as "the College." Much later, it was named for the famous English architect, Sir Christopher Wren, to whom an 18th-century author attributed the building's design. See "The Wren Building," accessed December 2011, http://www.wm.edu/about/history/historiccampus/wrenbuilding/index.php; "Thomas Jefferson's College," accessed December 2011, http://www.wm.edu/about/history/tjcollege/?svr=web.

20   During Thomas Jefferson's time at William and Mary, Martha Bryan held the position of housekeeper, retired in 1761, and was replaced by Isabella Cocke, who was let go in 1763 for behaving "much amiss in her Office of Housekeeper, not only in Contempt of the unanimous Resolves of this Society . . . but likewise in other respects: therefore they think proper to desire her to Finish her Year, and provide her self with some other Place. Resol That an Advertisement be inserted in the *Gazette* to desire a Man capable of managing the Housekeeper's Business in the College to apply to the President & Master." See Wenger, "Thomas Jefferson, the College of William and Mary, and the University of Virginia," 356; "Papers Relating to the College," 171; "Journal of the Meetings of the President and Masters of William and Mary College,"1763-1764, *WMQ* (1895), 43-46; Journal of the Meetings of the President and Masters of William and Mary College," 1769-1771, *WMQ* (1905), 148-157.

21   The Bursar's records indicate that Margaret Garret died on the job on January 25, 1773. See "Bursar's Books, 1770-1777," 60. She was immediately replaced by Mrs. Maria Digges. The remainder of Margaret's salary was paid to her Executors. In 1775, Maria Digges was accused of misconduct by a group of ushers and students,

but was retained by the college. See "Journal of the Meetings of the President and Masters of William and Mary College," 27 May 1775, *WMQ* (1906), 1-14.

22 "Journal of the Meetings of the President and Masters of William and Mary College," 1771-1772, *WMQ* (1905), 230-235. In 1775, Phoebe Dwit gave a deposition in the controversy concerning housekeeper, Maria Digges. Phoebe was accused of neglecting the sick to attend to Mrs. Digges. She signed her deposition with her mark. See "Journal of the Meetings of the President and Masters of William and Mary College," 27 May 1775, *WMQ* (1906), 1-14.

23 "Journal of the Meetings of the President and Masters of William and Mary College," 1771-1772, 233. In May of 1772, Mr. James Innis replaced Mr. Marshall as Assistant Usher of the College. In 1775, Innis was one of the individuals who brought the complaint against Maria Digges.

24 "Accounts of Salaries in 1770," *WMQ* (1898), 188-189. One of James Nicholson's several jobs was the "Cutting and Carting of the College wood," described here as "Surveyor of the Woodcutters."

25 According to Robert Polk Thomson, Winkfield was well-known around Williamsburg as one of the African-American slaves at William and Mary. See Robert Polk Thomson, "The Reform of the College of William and Mary, 1763-1780," *Proceedings of the American Philosophical Society* (1971), 187-213. For the reference to Winkfield, see Ibid., 201.

26 *MB*, 295. In 1775, Winkfield's name was appropriated by William and Mary faculty members. That year, student disorder was rising while enrollment and revenues fell. When the faculty defended themselves in the *Virginia Gazette*, they signed their letter, "Winkfield." See *Virginia Gazette* (Pinkney), 26 October 1775.

27 "Papers Relating to the College," 170; "List of slaves owned by the College of William and Mary, circa 1780," SCRC Database, accessed December 2011, http://scrc.swem.wm.edu/index.php?p=collections/controlcard&id=8030&q=slaves+owned+by+the+. William and Mary is conducting a long term project concerning the history of African Americans at the college. Called the "Lemon Project," it is named in memory of one of the College's slaves. See "The Lemon Project: A Journey of Reconciliation," accessed December 2011, http://www.wm.edu/sites/lemonproject/index.php.

28 Wenger, "Thomas Jefferson, the College of William and Mary, and the University of Virginia," 344.

29 Robert Polk Thomson, "Reform of the College of William and Mary, 1763-1780," 203.

30 The cost of room and board at the College was fairly consistent with previous years; in 1754, it was £13 per session, with an additional £2/10 for the board of personal servants. Although a variety of house servants attended to the various needs of

the students, a few scholars brought personal valets with them to William and Mary. In 1754, only eight out of fifty-two tuition-paying students brought slaves with them, including George and Carter Braxton, who each had a "negro boy" to serve them. See "Students in 1754 at William and Mary College," *WMQ* (1898), 187-188; "Notes Relative to Some of the Students Who Attended the College of William and Mary, 1770-1778," *WMQ* (1921), 116-130.

31 "Notes Relative to Some of the Students," 125-126; "Randolph Family," accessed December 2011, http://homepages.rootsweb.ancestry.com/~marshall/esmd18.htm#id29.

32 Ibid., 129; Phillip Hamilton, "Education in the St. George Tucker Household: Change and Continuity in Jeffersonian Virginia," *VMHB* (1994), 167-192; "St. George Tucker," accessed December 2011, http://www.history.org/Almanack/people/bios/biotuck.cfm; "St. George Tucker House," accessed December 2011, http://research.history.org/Architectural_Research/Research_Articles/ThemeBldgs/Tucker.cfm.

33 "Notes Relative to Some of the Students...," 117, 123.

34 *Virginia Gazette* (Purdie and Dixon), 27 June 1777, p. 3.

35 Hugh A. Garland, "School Days of John Randolph," *WMQ* (1915), 1-10.

36 Hamilton, "Education in the St. George Tucker Household," 169.

37 "Notes Relative to Some of the Students...," 125.

38 Richard Channing Moore Page, M.D., *Genealogy of the Page Family in Virginia* (New York, NY, Publisher's Print Co., 1893), 108.

39 Wenger, "Thomas Jefferson, the College of William and Mary, and the University of Virginia," 342.

40 Ibid., 342-350. Joining the William and Mary students for meals and for chapel were the Indian students, who lodged at Brafferton. By 1771, this component of education at William and Mary was coming to an end. The Master of Brafferton was Rev. Emanuel Jones, who, in 1770, also functioned as the Librarian for William and Mary and as Clerk to the [Faculty] Society. In 1775, the last Native American attended under this program, and Thomas Jefferson, in his capacity as Governor of Virginia, officially closed the school in December of 1779. See "Accounts of Salaries in 1770," *WMQ* (1898), 188; "The Brafferton," accessed December 2011, http://www.history.org/almanack/places/hb/hbbraff.cfm; "The Indian School at William and Mary," accessed December 2011, http://www.wm.edu/about/history/historiccampus/indianschool/index.php.

41 Wenger, "Thomas Jefferson, the College of William and Mary, and the University of Virginia," 351.

42 Wenger, "Thomas Jefferson, the College of William and Mary, and the University of Virginia,", 359-361; Fiske Kimball, "Jefferson and the Public Buildings of Virginia: I. Williamsburg, 1770-1776, *Huntington Library Quarterly* (February 1949), 115-120. Thomas Jefferson idealistically imagined a space that significantly expanded the existing building, despite the reality of diminishing enrollment at the time. Due to various delays, Jefferson's proposed fourth wing was suspended and never completed.

43 Thomson, "Reform of the College of William and Mary," 196.

44 "Journal of the Meetings of the President and Masters of William and Mary College," 1771-1772, p. 235.

45 Wenger, "Thomas Jefferson, the College of William and Mary, and the University of Virginia," 340; "A Grammar School Boy at William and Mary College," *WMQ* (1915), 278-280; "Early Courses and Professors at William and Mary College." In 1779, the Grammar School was discontinued, and the study of Latin and Greek did not resume at William and Mary until 1791, when Rev. John Bracken re-established the Grammar School. *Virginia Gazette* (Purdie and Dixon), 30 July 1772, p. 2; Thomson, "Reform of the College of William and Mary," 193. In 1770, Rev. Josiah Johnson received a salary of £150 as professor of Humanities, and in 1772, he was elected rector of Bruton Parish. See "Accounts of Salaries in 1770," *WMQ* (January 1898), 188.

46 Thomson, "Reform of the College of William and Mary," 202.

47 "A Grammar School Boy at William and Mary College," 280. Addressed to St. George Tucker, Esqr. Williamsburg, young Harry wrote his letter in June of 1793.

48 *MB*, 296.

49 Kevin P. Kelly, "The White Loyalists of Williamsburg," *Colonial Williamsburg Interpreter* 17 (1996); E. Alfred Jones, "Two Professors of William and Mary College," *WMQ* (1918), 221-231; Mark R. Wenger, "Thomas Jefferson, the College of William and Mary, and the University of Virginia," *VMHB* (1995), 339-374.

50 Jones, "Two Professors of William and Mary College," 225; *UB*, 8. George Murray, 5th Earl of Dunmore (30 April 1762 – 11 November 1836), was known as Viscount of Fincastle until his father's death in 1809. In the first edition of *Thomas Jefferson and His Unknown Brother Randolph*, author Bernard Mayo wrote, "One of Randolph's teachers at William and Mary, who also seems to have been his special tutor, was the Reverend Thomas Gwatkin, professor of natural philosophy and mathematics, a young man not long out of Jesus College, Oxford. An able teacher, Gwatkin in 1774 became the tutor of Lord Fincastle, eldest son of the Earl of Dunmore." Mayo went on to speculate, "Doubtless he found teaching Governor Dunmore's son much pleasanter, and more rewarding, than teaching the brother of Thomas Jefferson, then a radical member of the House of Burgesses from Albemarle."

51 Thomson, "Reform of the College of William and Mary," 202.

52 *Virginia Gazette* (Rind), 15 August 1771, pp. 2-3.

53 George W. Pilcher, "Virginia Newspapers and the Dispute over the Proposed Colonial Episcopate, 1771–1772," *Historian* (November 1960), 98-113; Jones, "Two Professors of William and Mary College," 223, 226-227; Thomson, "Reform of the College of William and Mary," 198-201. Among the four men was Rev. Samuel Henley, professor of Moral Philosophy and College Chaplain. Rev. Henley (1740-1815) was a commentator, poet, clergyman, schoolmaster, and, eventually, a friend to Thomas Jefferson. He was nominated to his post at the College in 1769 and probably came to Virginia with Rev. Gwatkin in 1770.

54 Jones, "Two Professors of William and Mary College," 222, 226; Thomson, "Reform of the College of William and Mary," 201. Gwatkin not only tutored Lord Dunmore's son, but also served as the family's Chaplain. Once in England, Rev. Gwatkin placed a claim for £100, covering his lost property in Virginia, which the British government paid in full. A second claim for £300 addressed the loss of his annual income. He was awarded £200 towards that claim, plus a yearly pension of £100.

55 Jones, "Two Professors of William and Mary College," 222-223; Kelly, "White Loyalists." Eventually, Dunmore had Gwatkin appointed vicar in Chousley, Berkshire; the position earned him £80 a year. Rev. Samuel Henley also reestablished himself in England, married in 1776, became assistant master of Harrow School, and was appointed curate in a parish in Northall, Middlesex.

56 "The Flat Hat Club," *WMQ* (1917), 161-164. The archives at William and Mary's Earl Gregg Swem Library do not contain a list of members for the Flat Hat Club nor anything further concerning Thomas Gwakins' affiliation with it. Many of the earliest records of William and Mary have been lost to fire, military occupation, and deterioration, leaving significant gaps in the history of the college.

57 *MB*, 293.

58 Ibid., xlvi.

59 Ibid., 295-296.

60 "Bursar's Book, 1770-1777," William and Mary, Special Collections, Earl Gregg Swem Library.

61 Ibid., 80. Thomas Jefferson's accounts include this credit from Robert Miller: "November 2: By cash from Rob. Millar (sic) credited by Jos. Hornsby as board for R. Jefferson 0.10.10." Bear and Stanton, editors of *Jefferson's Memorandum Books*, note: "Credit the estate of P. Jefferson 10/10 rec. of Rob. Millar wch. was overpd. by Mr. Hornsby for my brother's board at college." See "The Estate of Peter Jefferson, 1765-1778," CSmH; *MB*, 296. Robert Miller, a Scot and a Loyalist, was

Treasurer of the College (1770), Bursar of the College (1772), and Comptroller of the port of Williamsburg (1773). As the political storm heated up in Williamsburg, he eventually received "daily threats and insult for being outspoken and a revenue officer." Miller was a merchant in Williamsburg. His partner, William Maitland, also a Loyalist, took over as Assistant Treasurer of the College when Miller left the Colonies in June of 1775. Miller announced his departure from the Colonies in 1775. See *Virginia Gazette* (Purdie), 7 April 1775, p. 3; Kelly, "White Loyalists."

62 "Bursar's Books, 1770-1777," 80.

63 "Bursar's Books, 1770-1777," 80 (contra).

64 "Bursar's Books, 1770-1777," 83.

65 "Notes Relative to Some of the Students...," 117, 121.

66 *UB*, 2-3.

67 *Virginia Gazette* (Purdie and Dixon), 19 December 1771, p. 3. In April of 1772, Bursar Miller repeated the announcement in the *Gazette*. See *Virginia Gazette* (Rinds), 23 April 1772, p. 4; Robert Miller, "William and Mary College, 1771," *WMQ* (1924), 277-278.

68 Randolph Jefferson does not appear in the surviving "Account Book of Doctors John Minson Galt and William Pasteur, 1771-1780." The accounts apply primarily to the doctors' apothecary shop and include items for sale from candles to butter to sundries, although some "visits" (house calls?) are included. Customers range from Rev. Thomas Gwatkin to Patrick Henry, Esq.; see "Account Book of Doctors John Minson Galt and William Pasteur, 1771-1780," microfilm reels M-120 and M-1079, John D. Rockefeller, Jr. Library, Williamsburg, Virginia. This would seem to indicate that Randolph's account was for medical services, not an account at the apothecary.

69 *MB*, 446. In 1776, Thomas Jefferson used Galt's service for his servant, Martin, paying 13/9, and, in 1779, Thomas purchased "oil" from Galt, paying £14-8. See *MB*, 429, 480. For more about Dr. John M. Galt's career and life in Williamsburg see Mary A. Stephenson, "Nicolson Store Historical Report, Block 17 Building 4 Lot 56," accessed October 27, 2010, http://research.history.org/DigitalLibrary/View/index.cfm?doc =ResearchReports\ RR1337.xml; A. Lawrence Kocher, "Pasteur & Galt Apothecary Shop Architectural Report, Block 17 Building 32 Lot 56," accessed November 2011, http://research.history.org/DigitalLibrary/View/index.cfm?doc=ResearchReports\RR1375.xml; "Pasture & Galt Apothecary Shop," accessed November 2011, http://www.history.org/almanack/places/hb/hbpast.cfm; "John Minson Galt," accessed November 2011, http://scrc.swem.wm.edu/wiki/index.php/John_Minson_Galt.

70 *Virginia Gazette* (Purdie and Dixon), 2 February 1769, p. 3. Galt had competition in town in the person of Dr. William Pasteur, who had established his first Williamsburg shop in 1759. Pasteur had apprenticed under Dr. George Gilmer, Sr. of Williamsburg and had studied for about a year at St. Thomas' Hospital, London. Galt and Pastuer eventually became partners. Together the doctors provided surgical, midwifery, dental, and general medical services. The partnership lasted until 1778, at which time Pasteur retired from medicine and Galt continued in the business. During the Revolution, Galt became a military surgeon and served as director of the state apothecary until 1780. In 1795, he became a visiting physician to the Public Hospital and, later, was appointed to its board of directors, serving in both capacities until his death in 1808. See "Pasture & Galt Apothecary Shop."

71 *Virginia Gazette* (Purdie and Dixon), 21 September 1768, p. 4.

72 *Virginia Gazette* (Purdie and Dixon), 24 October 1771, p. 3.

73 In 1777, Thomas Jefferson recorded another medically related entry in Randolph Jefferson accounts within the Peter Jefferson Estate. On October 2, 1777, Thomas paid Dr. Gilmer £2/7/6 toward Randolph's account. If this is another delayed payment, it could relate back to Randolph's return from Williamsburg. Randolph Jefferson's interest in Peter Jefferson's estate did not close until September 5, 1779, and, until that time, Thomas continued to pay miscellaneous bills for Randolph.

74 While Randolph Jefferson did not return to William and Mary, his accounts contain a currently inexplicable notation dated 1774 regarding his schooling: "cash to Gordon the schoolmaster 14/6 ...." The location of Gordon's school or the nature of Randolph's studies is currently unknown. This, too, may have been a delayed payment and the instruction could have preceded his time in Williamsburg. See "The Estate of Peter Jefferson, 1765 - 1778," CSmH.

75 *MB*, 334-335. This February payment to William Page may have further colored Bernard Mayo's interpretation that Randolph stayed at school until early 1773.

76 *MB*, 334-335. During January and February of 1773, Thomas traveled through Goochland, Henrico, Chesterfield, and Fluvanna counties. In January, he purchased slaves, Ursula, George and Bagwell, from the Fleming estate in Cumberland County; they would become a core family at Monticello. He also noted that he sold the previous runaway slave, Sandy, to Col. Charles Lewis in Albemarle, relieving himself of a troublesome servant. Thomas was also at the Randolph seat Tuckahoe, where he paid the waterman and stayed several times at Byrd's ordinary in Louisa County.

77 *A Provisional List of Alumni, Grammar School Students, Members of the Faculty, and Members of the Board of Visitors of the College of William and Mary in Virginia. From 1793 to 1888 ... Issued as an Appeal for Additional Information* (Richmond: Division of Purchase and Printing, 1941), 23; accessed April 2012, http://swem.wm.edu/departments/special-collections/exhibits/exhibits/provlist/frame.htm.

78 *MB*, 295-296.

79 One historian unjustly dismissed Randolph Jefferson's letters to his brother, Thomas, as "illiterate" scrawling. While Randolph's simply worded and often phonetically spelled writings make it eminently apparent that he was not a man of letters, his communications are clear, often covering a wide range of concerns. See Nathan Schachner, *Thomas Jefferson: A Biography* (New York, NY: Appleton-Century-Crofts, Inc., 1951), 923.

Chapter Five: Randolph Jefferson, Patriot

1 Following Peter Jefferson's death, the spelling of Snowdon rather quickly drifted to Snowden, which likely reflected pronunciation and this was the spelling consistently used by Randolph Jefferson. During this same period, the westward extension of the James River was less and less referred to as the Fluvanna River and was simply referred to as the James.

2 Whitley, *Genealogical Records of Buckingham County*, 18, 22. The Buckingham records for these two years, housed at the Library of Virginia, were transcribed by Edyth Rucker Whitley. According to Research Notes distributed by the Library of Virginia, "In seventeenth-and eighteenth-century Virginia, the term tithable referred to a person who paid (or for whom someone else paid) one of the taxes imposed by the General Assembly for the support of civil government in the colony. In colonial Virginia, a poll tax or capitation tax was assessed on free white males, African American slaves, and Native American servants (both male and female), all age sixteen or older. Owners and masters paid the taxes levied on their slaves and servants." See "Colonial Tithables (Research Notes Number 17)," Vi.

3 Jess "Bootwright" is probably Jess Boatwright, a common surname in Buckingham.

4 Constantine Perkins Will, Goochland County Deed Book 9, p. 206; William Kearney Hall, *Descendants of Nicholas Perkins of Virginia* (Ann Arbor, MI: University Microfilms International, 1980); "Marriage Bonds in Goochland County," *WMQ* (January 1899), 197.

5 Following John Johnston's death, William Fontaine took over the school at Seven Islands. The son of Col. Peter Fontaine, he taught first at William Cabell's Union Hill (Amherst County) in 1770-1771, then at his father's home in 1772-1773, then at John Nicholas' in 1774-1775. See Brown, *Cabells and Their Kin*, 190-191.

6 Despite the fact that John Nicholas' residence fell out of Albemarle and into Buckingham County in 1761, he remained Clerk of Court in Albemarle County, serving from 1749-1792. See Whitley, *Genealogical Records of Buckingham*, 1.

7 For more about the Nicholas family, see Richard Ludlum Nicholas, "The Nicholas Family of England and Virginia," (Houston, TX: A.V. Ammott Sons Bookbinders,

Inc., 1988, 1989); Richard Ludlum Nicholas, "Land Ownership in the Horseshoe Bend of Northern Buckingham County and Southern Albemarle County and the History of the Nicholas Family in That Area," (Charlottesville, VA: R. L. Nicholas, 2002).

8   *Virginia Gazette* (Purdie & Dixon), 17 June 1773, p. 4. Oddly, the location of the plantation is not stated, while the previous ad placed by Francis Moseley begins, "Taken up, in *Buckingham* . . ." and the next ad begins, "Taken up, in *Goochland*. . . ."

9   Thomas Jefferson Accounts, "Peter Jefferson Estate, 1763-1790," CSmH.

10 *MB*, 202, 441, 470. Agents, lawyers, doctors, and merchants often waited years to settle their bills. Thomas Jefferson paid this account in the fall of 1778. Carter and Trent had dissolved their partnership in early 1774, indicating these expenditures were made long before 1778. See *Virginia Gazette* (Purdie & Dixon), 17 February 1774, p. 3. In the summer of 1778, Thomas settled several other miscellaneous bills charged to their father's estate, including £3 which Randolph owed William Rice, who was a stone and brick mason. See Ibid., 470.

11 "Elizabeth Jefferson," *TJE*, accessed December 2011, http://www.monticello.org/site/jefferson/elizabeth-jefferson; Kern, dissertation, 385.

12 *MB*, 369-370. Thomas Jefferson noted that the river was 18 l. higher than when Nicholas Lewis' bridge was washed out. See Ibid., 370.

13 There was minimal damage beyond the epicenter in Petersburg, Virginia. Still, Virginians reacted with understandable alarm. See "Major Earthquakes in Virginia," accessed August 2010, http://www.dmme.virginia.gov/DMR3/majorvaearthquakes.shtm.

14 *MB*, 370.

15 Ibid.

16 Wilson Miles Cary to Miss S. N. Randolph, n.d. privately owned. See "Elizabeth Jefferson," *TJE*.

17 Ibid.

18 Thomas Jefferson's "Farm Book" listed them as follows: "Slaves conveyed by my mother to me under the power given her in my father's will as an indemnification for the debts I had paid for her. [At] Lego. Caesar. [At] Shadwell. Sall. Lucinda. 1761. Simon. 1765. Cyrus. Nov. 1772. Squire. Belinda. Val. 1760. Charlotte. Mar. 1768. Minerva. Sep. 1771. Sarah. Dec. 1772. See Jefferson, "Farm Book," 10.

19 Jane Jefferson Will, Albemarle County Will Book 2, p. 367.

20 Thomas Jefferson Accounts, "Estate of Jane Jefferson dec[d], 1776 - 1778," CSmH.

21 Jane Jefferson Estate Inventory, Albemarle County Will Book 2, p. 356.

22 Moore, *Scottsville on the James*, 32.

23 Ibid., 33.

24 *MB*, 437.

25 Ibid., 438. James MacPherson (1736-1796) was a Scottish poet.

26 Ibid., 450. Thomas Jefferson owned a copy of the play.

27 Peter Jefferson Estate Inventory, Albemarle County Will Book 2, pp. 41-47; Kern, dissertation, 448-455.

28 Ibid. Additionally, there was "A Cart and Whels" and two pair of old cart wheels listed in the inventory.

29 Harold B. Gill, Jr., "A Sport Only for Gentleman," *Colonial Williamsburg Journal* (Autumn 1997), accessed December 2011, http://www.history.org/history/teaching/enewsletter/volume4/march06/sport.cfm.

30 Lynn Mayes, "Notes for Thomas Jefferson, II," *From Virginia Through The Southwest*.

31 Isaac Jefferson, "Memoirs of a Monticello Slave," in James A. Bear, Jr., ed., *Jefferson at Monticello: Recollections of a Monticello slave and a Monticello Overseer* (Charlottesville, VA: University Press of Virginia, 1976), 20.

32 Buckingham County Personal Property Tax, 1797.

33 *MB*, 458-459.

34 Lucia Stanton, "Free Some Day," in *"Those Who Labor for My Happiness": Slavery at Thomas Jefferson's Monticello* (Charlottesville, VA: University of Virginia Press, 2012), 118.

35 *MB*, 440-441, 446-447, 449, 451. It is unclear where Randolph acquired these blankets or if there was weaving being produced at or near Snowden. They may have come from Buck Island, where the Lewis family operated looms.

36 Thomas Jefferson Accounts, "Randolph Jefferson, 1771-1779," Small Special Collections, ViU . Beginning on February 2, 1774, Thomas Jefferson hired Phill from Randolph at £11/year, eventually purchasing him. See Thomas Jefferson Accounts, "The Estate of Peter Jefferson, 1765-1778," CSmH.

37 "Continental Association of 20 October 1774," *Thomas Jefferson Papers*, Printed Broadside, American Memory, DLC, accessed November 2011, http://hdl.loc.gov/loc.mss/mtj.mtjbib000093.

38 *PTJ*, 1:666.

39 *Virginia Gazette* (Prudie), 22 August 1777, pp. 2-3; "Thomas Nelson, Jr.," *Signers of the Declaration*, accessed December 2011, http://www.nps.gov/history/history/online_books/declaration/bio35.htm.

40   Charles H. Cureton, "Nelson's Corps of Light Horse, 1778," *Military Collector & Historian, Journal of the Company of Military Historians* (Winter 1983), 180-181.

41   Ibid., 181.

42   *Virginia Gazette* (Prudie), 1 May 1778, p. 2.

43   Despite the fact that Randolph Jefferson had reached his majority, his father's estate was not settled, and Thomas Jefferson still kept scant accounts for some of Randolph's expenses. It should not be concluded that Thomas was controlling Randolph's money in 1778, but rather that he was assigning, at Randolph's request, monies remaining in the estate.

44   *MB*, 465.

45   Thomas Jefferson Accounts, "Randolph Jefferson, 1771-1779," ViU.

46   *MB*, 481.

47   Cureton, "Nelson's Corps of Light Horse, 1778," 181. *The Company of Military Historians* offers a plate showing the uniform of Nelson's Corps of Light Horse. See *The Company of Military Historians*, accessed December 2011, http://www.military-historians.org/.

48   *Virginia Gazette* (Prudie), 5 June 1778, p. 3.

49   *Virginia Gazette* (Prudie), 12 June 1778, p. 2.

50   "General Nelson's Heirs," Document No. 9, *Journal of the House of Delegates of the Commonwealth of Virginia* (Richmond, VA: Thomas Ritchie, 1831), 1.

51   Ibid. Proof was presented to the Virginia House of Delegates from the auditor's and treasurer's offices, "that general Nelson fully settled his account with the public for all the expenses incurred in equipping and marching the troop of cavalry to Philadelphia, and that on the 15th January, 1779, he received a warrant in full for the balance due to him, amounting to £636/19/11." See Ibid.

52   Cureton, "Nelson's Corps of Light Horse, 1778," 181. Cureton concludes, "The Council subsequently turned the equipage of the corps over to the 1st Regiment of Continental Light Dragoons."

53   Green Clay to Thomas Jefferson, 4 May 1823, Thomas Jefferson Papers, DLC, accessed December 2011, http://memory.loc.gov/ammem/collections/jefferson_papers/. Clay's letter concerns the situation with the Cherokee. Thomas Jefferson responded on May 28, 1823.

54   Nathaniel Mason Pawlett and Howard H. Newlon, Jr., "The Route of the Three Notch'd Road: A Preliminary Report," accessed December 2011, http://www.virginiadot.org/VTRC/main/online_reports/pdf/76-r32.pdf, 12, 31.

55 "Prince/Printz Family History," accessed December 2011, http://www.princefamilyhistory.com/brusheswithuspresidents.htm.

56 J. T. McAllister notes, "A search of the records of Buckingham County failed to disclose any record of Militia Officers qualifying in it during the Revolution." See J. T. McAllister, *Virginia Militia in the Revolutionary War* (Westminster, MD: Heritage Books, copyright 1913), 26.

57 "About Virginia Revolutionary War Public Service Claims," accessed December 2010, http://www.lva.virginia.gov/public/guides/opac/revpscabout.htm. The Publick Claims for Buckingham County are preserved in the Court Booklets and Lists, the Certificates, and the Commissioners Books I and II.

58 Janice L. Abercrombie and Richard Slatten, *Virginia Revolutionary Publick Claims* (Athens, GA: Iberian Publishing Company, 1992), xii.

59 *UB*, 4.

60 For more about Buckingham County's contribution to the Revolutionary War, see Abercrombie and Slatten, *Buckingham County, Virginia Publick Claims* (Athens, GA: Iberian Publishing Co., n.d.); Whitley, *Genealogical Records of Buckingham County, Virginia*, 50-70; Carl Rosen, *Revolutionary Patriots of Buckingham County, Virginia* (Westminster, MD: Willow Bend Books, 2002); Joanne L. Yeck, *"At a Place Called Buckingham"* . . . *Sketches of Buckingham County, Virginia* (Slate River Press, 2011), 33-46.

61 Michael Kranish, *Flight from Monticello: Thomas Jefferson at War* (Oxford: Oxford University Press, 2010), 105-113.

62 H.J. Eckernrode, *A Calendar of Legislative Petitions, Arranged by Counties: Accomac-Bedford* (Richmond, VA: D. Bottom, 1908), 28; Howell Lewis, "Charles Lewis," *WMQ* (July 1922), 194.

63 Sorley, *Lewis of Warner Hall*, 346, 604-605; Woods, *Albemarle County*, 265. Col. Nicholas Lewis (1734-1808) was a close associate of Thomas Jefferson. The younger Randolph Jefferson may have been friendly with his children. Nicholas and his wife, Mary Walker, had seven children; Jane was one of four girls. See "Nicholas Lewis," *TJE*, accessed December 2011, http://www.monticello.org/site/research-and-collections/nicholas-lewis.

64 Peter Nicolaisen, "Thomas Jefferson and Friedrich Wilhelm von Geismar: A Transatlantic Friendship," *MACH* (2006), 1-27.

65 John Moore, *Albemarle*, 64; Thomas Anburey, *Anburey's Travels, Through the Interior Parts of America* (Carlisle, MA: Applewood Books, 2007).

66 Woods, *Albemarle County*, 37-38.

67 Yeck, *"At a Place Called Buckingham,"* 33-46.

68 Public Service Claims Certificates, Buckingham County, Reel 10, Vi; William Kearney Hall, Descendants of Nicholas Perkins of Virginia (Ann Arbor, MI: University Microfilms International, 1980), 27; Abercrombie and Slatten, *Buckingham County, Virginia Publick Claims*, 11.

69 *PTJ* 4:414-415.

70 Missing records obscure Randolph Jefferson's initial involvement with the Buckingham County Militia. By 1787, he was recommended to the rank of Lieutenant, though may have served prior to that date.

71 Public claims presented by John Cox make it clear that prisoners were held at the tavern at Buckingham Courthouse. Cox's claims are also undated. See Abercrombie and Slatten, *Buckingham County, Virginia Publick Claims*, 9.

72 *PTJ*, 4:645-647.

73 Abercrombie and Slatten, *Buckingham County, Virginia Publick Claims*, 6, 14.

74 "Nathanael Greene," *New World Encyclopedia*, accessed December 2010, http://www.newworldencyclopedia.org/entry/Nathanael_Greene.

75 Yeck, *"At a Place Called Buckingham,"* 36-39.

76 *MB*, 508-509.

77 Ibid. During the war, Thomas Jefferson kept track of inflation, which was steady. By 1781, inflation was rampant, dramatically depreciating paper money. In January, the value of one pound specie equaled £75 in paper money. By April, that same pound was equal to £100 paper. In December, Jefferson recorded that a pound equaled £1000 paper. See Ibid., 514-515.

78 Ibid., 391; Kranish, *Flight from Monticello*, 194-196. When Arnold and his second in command, Col. Simcoe, approached Richmond, Thomas Jefferson quickly rode the "six miles up the James River to Westham, the site of an arms factory. He had earlier ordered the transfer of important state papers to Westham; now he feared the place would be attacked, and he intended to supervise the movement of arms and documents across the James River. Once there, he ordered Daniel Hylton, one of the wealthiest men in Richmond, to take charge of the operation. Seven wagons were filled with arms and gunpowder and driven to a courthouse and church far from Richmond." See Ibid., 189.

79 "Revolutionary War Raids and Skirmishes in 1781," accessed December 2010, http://www.myrevolutionarywar.com/battles/1781s.htm. In January, Arnold and Simcoe had plundered the town when Governor Jefferson declined to surrender. See Kranish, *Flight from Monticello*, 191-194.

80 Ibid., 248; *PTJ*, 5:644-645; Kranish, *Flight from Monticello*, 248-249, 256-257. A quorum was not present until May 28th. See Moore, *Albemarle*, 64.

81 Militia Commission Papers, 1777-1858, Virginia Governors' Records, Vi.

82 *PTJ*, 5:554-555. On April 13, 1781, George Gilmer wrote to Thomas Jefferson concerning changes in Albemarle: "Yesterday George Twyman, C. L. Lewis, James Marks, and Isaac Davis mounted the rostrum.... Reuben Lindsay, John Marks and James Minor Commissioners." Change abounded in Albemarle politics and the local Militia. Gilmer continued, "Twyman and C.L.L. would have been returned but they could not hang together. Party business seems to be propagating very rapidly." See Ibid., 430-431. On April 17, Thomas Jefferson wrote to David Jameson about the appointments, saying, "The court of Albemarle on the resignation of John Coles County Lieutenant and Nicholas Lewis Colo. have passed by Reuben Lindsay who was Lt. Colo. and a man of as much worth as any in the county . . . and have recommended (as report said) John Marks to be County Lieutt. who was formerly a very junior captain and retired, not possessing an inch of property in the county or other means of obtaining influence over the people, and of a temper so ungovernable that instead of reconciling he will by his manner of executing revolt the minds of the people against the calls of government. . . ." See Ibid., 468-469. Lindsay had recommended Marks; however, in the end, it was Lindsay and Lewis who got the commissions.

83 "Reuben Lindsay," accessed December 2011, http://www.monticello.org/site/research-and-collections/reuben-lindsay; Lay, *Architecture of Jefferson Country*, 142.

84 *PTJ*, 6:30.

85 Ibid., 31.

86 Abercrombie and Slatten, *Buckingham County, Virginia Publick Claims*, 6.

87 *PTJ*, 6:35-36.

88 Ibid., 38-39. Ultimately, Governor Jefferson did not empower Col. White, but asked Lafayette to pass a pre-existing warrant onto White. This action guaranteed that the warrant already had the approval of the Virginia House of Representatives.

89 Moore, *Scottsville on the James*, 36.

90 Kranish, *Flight from Monticello*, 278-287. Michael Kranish reports in great detail the steps leading up to and following Thomas Jefferson's journey from Monticello to join his family.

91 Moore, *Scottsville on the James*, 37-38. Moore's explanation of the Jefferson family lineage is confused. Thomas Jefferson's nephew, Peter Field Jefferson, had a daughter, Frances. She married Valentine Foland. Their son was Peter V. Foland.

92 *MB*, 510-511.

93 *PTJ*, 4:261.

94 Ibid., 265.

95   *PTJ*, 4, 265.

96   Robert R. Howison, *A History of Virginia* (Richmond, VA: Drinker and Morris, 1848), II:270-271. John Hammond Moore explains how the stores came to be at Albemarle Old Courthouse: "Military supplies followed the government to the Piedmont, although Governor Jefferson discouraged a concentration of matériel at Charlottesville. As a result, these goods were divided among a variety of sites – Point of Fork, Henderson's Warehouse on the Rivanna where Milton would soon appear, the old county courthouse, Scott's Ferry on the James, and Charlottesville – and many wagon loads were stranded somewhere along the way from Richmond." See Moore, *Albemarle*, 64. Buckingham citizens contributed manpower to moving the stores from Point of Fork to Albemarle Old Courthouse, including Joseph Cabell and Francis Spencer, who lost a canoe in public service and lent twelve Negro fellows for eleven days. See Abercombie and Slatten, *Buckingham County, Virginia Publick Claims*, 21. Col. Joseph Cabell (1732-1798) was a Justice for Albemarle County. When Buckingham and Amherst were cut out of Albemarle, Cabell owned land in all three counties. He served Buckingham as Burgess from 1768-1771, continuing in statewide politics until 1790. Cabell's figure loomed large in the northern section of Buckingham, where he and Randolph Jefferson undoubtedly had contact.

97   Brown, *Cabells and Their Kin*, 193-195.

98   Ibid., 193.

99   Rosen, Revolutionary Patriots, 65; Abercrombie and Slatten, *Buckingham County, Virginia Publick Claims*, 6, 7. John Watkins may be the husband of Dorothea Ballow, sister of Snowden neighbor, Thomas Ballow. Ware Oglesby is a kinsman of neighbor, Hardin Perkins.

100 Abercrombie and Slatten, *Buckingham County, Virginia Publick Claims*, 2, 6-7.

101 Ibid., 6-7.

102 Zachariah Nevil, Randolph Jefferson's future son-in-law, was a private with the 3rd Regiment of the Light Dragoons. Nevil served during 1781-1783 for twenty-nine months, twenty-seven days, beginning on May 20, 1781. It seems likely he may have camped at Snowden in June and July of 1781.

103 Abercrombie and Slatten, *Buckingham County, Virginia Publick Claims*, 6. Presley Peter Thornton (d. 1811) served in the 3rd Regiment of Light Dragoons under Col. George Baylor to the end of the war, and was honorably discharged. See Francis B. Heitman, *Historical Register and Dictionary of the United States Army* (Washington, D.C.: Government Printing Office, 1903), 959.

104 Stephen Perkins, overseer, was likely the nephew of Hardin Perkins, Sr., who lived adjacent to Snowden at Perkins Falls.

105 Abercrombie and Slatten, *Buckingham County, Virginia Publick Claims*, 1.

106 Abercrombie and Slatten, *Buckingham County, Virginia Publick Claims*, 6-7.

107 Ibid., 16; Public Service Claims, Certificates, Reel 10, Buckingham County, #0505, Vi.

108 Public Service Claims, Commissioner's Book I, Reel 5, p. 202-203, Vi; Public Service Claims Commissioner's Book II, 166, Vi.

109 It is clear from the surviving letters and notations in Thomas Jefferson's memos that correspondence existed beyond the surviving sample. Mayo and Bear estimate there were fifty letters written between the Jefferson brothers. Additionally, informal modes of communication, both written and oral, undoubtedly took place. Randolph Jefferson definitely "dropped by" Monticello and likely sent servants there with or without a hasty note. Servants, who were known to each other, may have met in public places and exchanged news about Snowden and Monticello, bringing messages home to their respective owners.

110 It is currently unknown if Randolph Jefferson came up through the ranks, first serving as a Private and then as an Ensign.

111 Buckingham County, "Militia commission papers, 1777-1858," Virginia Governor's Records, Vi.

Chapter Six: Randolph Jefferson, Planter

1 Woods, *Albemarle County*, 237; *Albemarle County Marriages, 1780-1853*, 1:175; Albemarle County Marriage Bonds, loose papers, Albemarle County Courthouse, Charlottesville, VA. Anne Lewis has been referred to as Anne Jefferson Lewis, notably in *Thomas Jefferson and his Unknown Brother* and *Jefferson's Nephews*. It is doubtful that this was her name. To date, no primary document has surfaced indicating that her full name was Anne Jefferson Lewis or Anne J. Lewis. She is Anne Lewis on her marriage bond and "Anna" Jefferson in her father's will. When it was written in 1782, she was already married to Randolph Jefferson. Apparently, at some point, it was assumed that her new surname was her "middle" name.

2 Sorley, *Lewis of Warner Hall*, 371, 384. Anne may have been the Lewis' fourth daughter. She is named sixth in her father's will. His two sons, Charles Lilburne and John Lewis, are mentioned first. Ahead of Anne is Elizabeth Lewis born in 1752. Following Anne is Frances Lewis, born January 24, 1759. If the girls are in age order in the will, Anne's birth falls somewhere between 1752 and 1758.

3 Woods, *Albemarle County*, 237, 265.

4 Grace McLean Moses, *The Welsh Lineage of John Lewis (1592-1657), Emigrant to Gloucester, Virginia* (Baltimore, MD, USA: Genealogical Publishing Co., 2002); John Meriwether McAllister and Lura Boulton Tandy, eds., *Genealogies of the Lewis*

  and Kindred Families (Columbia, MO: E.W. Stephens Publishing Co., 1906); *Legend of Lewis*, accessed December 2011, http://www.legendoflewis.com/.

5 Sorley, *Lewis of Warner Hall*, 349-391. Two Lewis sons died without apparent issue. John Lewis likely died young and Isham Lewis died unmarried. Both were named in their father's 1782 will. Isham left his 1,000 acres of land to his nephew, John Moore. See Woods, *Albemarle County*, 284.

6 There is no direct mention of Lucy Jefferson's wedding in Thomas Jefferson's memorandum books. However, on June 18, 1769, he gave his sister two shillings and six pence, perhaps in anticipation of the September event. Years later, in 1783, Thomas shopped for Lucy in Philadelphia, which cost her £3.12.6. In October of 1783, he bought a doll for her. See *MB*, 144, 524, 536.

7 Lucy Lewis to Thomas Jefferson, 19 November 1807, Edgehill-Randolph Papers, #1397, Small Special Collections, ViU. While Anne and Charles Lilburne Lewis grew up with many advantages, by 1802, Charles Lilburne and Lucy Lewis were destitute and, for several years, depended on their son-in-law, Craven Peyton, before leaving for Kentucky. See Merrill, *Jefferson's Nephews*, 95-96. Lucy, like her brother, Randolph, was somewhat of a phonetic speller. Both educated by Benjamin Sneed, these Jefferson siblings possessed basic writing skills, enough to communicate clearly, if not eloquently. As a bonus, Lucy gives us insight into the sound of early 19th century speech in central Virginia.

8 *PTJ*, 3:90-91.

9 Mr. Samuel Carr (1771-1855), the middle Carr brother, is conspicuous by his absence.

10 Merrill, *Jefferson's Nephews*, 18-19.

11 Presumably, Jane Jefferson supported her daughter's marriage to her nephew, and, if she had lived to see it, would have been in favor of Randolph's marriage to her niece, Nancy.

12 Thomas Jefferson to David Jameson, 16 April 1788, *PTJ*, 5:468-469.

13 Thomas Jefferson to Philip Mazzei, 7 January 1792, Marraro, "Unpublished Correspondence," 119-20, Maruzzi Archives, Pisa, Italy.

14 Thomas Jefferson to Gideon Fitch, 23 May 1809, Thomas Jefferson Papers, MHi; Merrill, *Jefferson's Nephews*, 32, 44, 74-75.

15 Woods, *Albemarle County*, 227-228; "Craven Peyton," accessed December 2011, http://www.monticello.org/site/research-and-collections/craven-peyton; Robert Haggard, "Thomas Jefferson v. The Heirs of Bennett Henderson, 1795-1818: A Case Study in Caveat Emptor," *MACH* (2005), 1-29.

16 Woods, *Albemarle County*, 284. The Moore-Lewis children, John, Ann, and Charles, were bound as apprentices for four years to William Watson when their father died. Further complicating these Lewis interconnections, Frances Moore (the sister of Edward Moore) married John Henderson, Jr. (brother of Bennett Henderson), thus tightening the relations between the Moore and Henderson clans. See Ibid., 227-228.

17 Thomas Jefferson to Charles Lilburne Lewis, 22 February 1790, *PTJ*, 16:191-192.

18 Thomas Jefferson to Mrs. Lucy Lewis, 26 May 1806, MHi.

19 Lucy Lewis to Thomas Jefferson, 26 May 1806, Edgehill-Randolph Papers, #1397, Jefferson Papers, Small Special Collections, ViU.

20 *UB*, 13-17. Referring to Nancy as "sister," rather than "sister-in-law," was the convention of the era.

21 Maria (Jefferson) Eppes' son, Francis W. Eppes, suffered from seizures. In a letter from Martha (Jefferson) Randolph to Thomas Jefferson, she mentions that Francis' seizures are similar to convulsions that "My aunt's children have been subject to." Which aunt is not stated, and it could have been Lucy (Jefferson) Lewis. Currently, it is not known if any of Randolph Jefferson's children experienced seizures. See Betts and Bear, *Family Letters*, 252-53.

22 Merrill, *Jefferson's Nephews*, 182-186; Isham Lewis to Thomas Jefferson, 27 April 1809, *PTJ:RS*, 1:167-168.

23 Thomas Jefferson to Isham Lewis, 1 May 1809, *PTJ:RS*, 1:181-182.

24 For the complete story of the Lewis murder, see Merrill, *Jefferson's Nephews: A Frontier Tragedy*.

25 Albemarle County Will Book 2, pp. 399-401. The last will of Charles Lewis, Jr. (14 March 1721 - 14 May 1782) was proved in Albemarle County on July 12, 1782. Anne (Lewis) Jefferson's servant, Lucy, may have come to Snowden at the time of her marriage or the following year after her father's death. A slave named Lucy appears on both 1782 and 1783 lists of slaves at Snowden. A slave named "Luce" also appears on the 1783 list. See Buckingham County Personal Property Tax, 1782, 1783.

26 "October 22, 1776, Albemarle, Amherst, Buckingham, Dissenters, against established churches, and for religious equality," Early Virginia Religious Petitions, *American Memory*, DCL, accessed December 2011, http://memory.loc.gov/ammem/collections/petitions/. Sons-in-law of Charles Lewis, Jr., Charles Hudson, Bennett Henderson, and Charles Lewis of North Garden, all joined the dissenters, as did future son-in-law, Charles Wingfield. A week later, another nearly identical petition was signed by the Lewis family. Again, Charles Lewis, Jr. headed the list. Signatures also included Jefferson family friends and neighbors such as Dr.

George Gilmer and Philip Mazzei. See "November 1, 1776, Albemarle, Amherst, Dissenters, for disestablishment and religious equality," Early Virginia Religious Petitions, American Memory, DCL, accessed December 2011, http://memory.loc.gov/ammem/collections/petitions/.

27 Merrill, *Jefferson's Nephews*, 12-19.

28 Woods, *Albemarle County*, 130-131.

29 Estate Inventory of Charles Lewis, Albemarle County Will Book 2, p. 403.

30 Tillotson Parish was established in 1757, when land in Albemarle located below the James River was cut out of St. Annes' Parish. In 1761, it became the sole parish for Buckingham County. See Warren, *Buckingham County, Virginia Church and Marriage Records*, 8-13; Yeck, *"At a Place Called Buckingham,"* 9-22.

31 A survey of 150 acres, dated December 14, 1790, includes "Goodlings" Church surveyed for the Church Wardens of Tillotson Parish. The parcel adjoins Josiah Chambers and Col. Joseph Cabell. The survey illustrates the location of the church and a creek. See Grundset, *Buckingham County, Virginia Surveyor's Plat Book 1762-1858*, p. 20. Lineaus Bolling held land near "Goodwins" Church on the Scottsville Road or Howards Road, eleven miles north of the courthouse. See Roger G. Ward, *Buckingham County Records: Land Tax Summaries & Implied Deeds, Volume 2* (Athens, GA: Iberian Publishing Co., 1994), 36.

32 *PTJ*, 9:165.

33 Hening, *Statutes*, 9:165.

34 Woods, *Albemarle County*, 126; "Lynchburg VA - Church - English Church," accessed December 2011, http://files.usgwarchives.net/va/lynchburg/church/english01.txt.

35 *MB*, 460, 470, 478, 483.

36 "October 27, 1785, Buckingham, Against assessment bill," *American Memory*, DCL, accessed December 2011, http://memory.loc.gov/ammem/collections/petitions/.

37 Woods, *Albemarle County*, 134.

38 Vogt and Kethley, *Albemarle County Marriages*, 175, 238.

39 Albemarle County Deed Book 51, p. 19.

40 Gordon G. Ragland, Jr., *The Tie That Binds: The Stories of Sharon Baptist Church Buckingham County, Virginia* (Farmville, VA: Farmville Printing, 2008), 27-29.

41 Grundset, *Albemarle County, Virginia Surveyor's Plat Book*, 65. The undated survey was recorded in November of 1754.

42 Another early patentee in the neighborhood, John Bolling of Goochland and Henrico, did not live on his land and gifted some of it to Thomas Ballow. On March

18, 1734, Bolling gave the 400-acre tract above Seven Islands to Thomas Ballow, Sr. of Goochland County, "in consideration of the natural love and affection" which he bore for him. See Goochland County Deed Book 2, p. 64. This was one of two parcels Bolling patented in the Horseshoe Bend on September 28, 1732: one a little below the Seven Islands and one a little above. See Land Office Grants, Vi. Col. Bolling (1700-1757) was a prominent Judge, Burgess, and property owner who acquired thousands of acres up and down the James River. Subsequent members of the Bolling family would occupy land adjacent to the James, near Seven Islands in Buckingham County. See Nicholas, "Land ownership in the Horseshoe Bend."

43   McLaughlin, *Jefferson and Monticello*, 146-147.

44   Buckingham County Personal Property Tax, 1815.

45   Albemarle County Personal Property Tax, 1815.

46   Buckingham County Personal Property Tax, 1815. Randolph Jefferson's gig was demolished by 1814; therefore, in 1815, he was not taxed on a carriage of any kind.

47   Ibid.

48   Interestingly, the items taxed in 1815 are not consistent from county to county. The tax lists for Buckingham and Albemarle, for example, vary dramatically. Many more items are listed in Albemarle, including mills. Albemarle, then and today a more affluent county, may have been home to much more "luxury" property to tax. See Albemarle County Personal Property Tax, 1815.

49   Mutual Assurance Society, policy #339, 4 May 1799. Col. Lewis insured Monteagle with the Mutual Assurance Society. Two dwellings, connected by a "wooden-covered way," were valued at $400 for a one-story building with a shed and $800 for a two-story 24 x22 feet dwelling.

50   *MB*, 957. On March 26, 1797, Thomas Jefferson executed a bond to Donald, Scott & Co. for £80.13.10½, with interest at 5% from October 1, 1796, payable July 1, 1798. Thomas assumed Randolph's debt as part of his payment for two of his brother's slaves, Ben and Cary. There is no way to know from Thomas' entry when this account with the Scottish merchants originated or how much time it covered.

51   Special Collections, Perkins Library, Duke University. This letter was included with a purchase of Chaffin pieces. These items belonging to Randolph might have been sold before the house at Snowden burned in early 1816. Perhaps an auction of household furnishings took place as part of the liquidation of Randolph's estate. It is also possible that some of his personal property was handed down to his children prior to his death in 1815.

52   *UB*, 20.

53   John Jefferson to Thomas Jefferson, January 7, 1790, *PTJ*, 16:87-88; Buckingham County Personal Property Tax, 1800, Randolph Jefferson with John Jefferson.

54 "Craven Peyton," *TJE*, accessed December 2011, http://www.monticello.org/site/research-and-collections/craven-peyton.

55 *UB*, 57.

56 As a young man, during his time in Williamsburg, Randolph saw more of Virginia. Except for his ride to Baltimore, there is no direct evidence that he ventured out of the state. In his mature years, his experience occasionally took him beyond counties adjacent to Buckingham. In December 1809, he and his second wife, Mitchie, took their gig to Charlotte County to visit her ailing brother. See Ibid., 22.

57 Maj. Thomas Bolling Diary, 1787-1804, accessed September 2011, http://lefeberleaguecity.ieasysite.com/MajorThomasBollingDiary's1787-1804.pdf.

58 Col. William Bolling Diary, 1793-1837, accessed September 2011, http://lefeberleaguecity.ieasysite.com/ColonelWilliamBollingDiary's1793-1837.pdf; Col. William Bolling Diary, 1838-1842, accessed September 2011, http://lefeberleaguecity.ieasysite.com/ColonelWilliamBollingDiary's1838-1842.pdf. During the 1930s, William Bolling's diary was transcribed and printed in multiple parts in the *Virginia Magazine of History and Biography*. See William Bolling, "Diary of Col. William Bolling of Bolling Hall," *VMHB* (July 1935), 237- 250; (January 1937), 29-39; (January 1939), 27-31.

59 " Col. William Bolling Diary, 1793-1837," 1.

60 Ibid.

61 Ibid., 61-63.

62 Ibid., 34.

63 Ibid., 58.

64 Ibid., 64-65.

65 Ibid., 67.

66 "Register of Magistrates and Other Civil Officers, 1775-1818," Microfilm reel 458a, Vi; Jeanne Stinson, *Early Buckingham County, Virginia Legal Papers, Volume 1* (Athens, GA: Iberian Publishing Co., 1993), 21-23.

67 E. Lee Shepard, "'THIS BEING COURT DAY,' Courthouses and Community Life in Rural Virginia," *VMHB* (1995), 459-470. As a member of the Buckingham Militia, Randolph Jefferson may have mustered at the Courthouse and participated in "keeping the peace" during elections, which could get contentious.

68 William Diuguid, a Buckingham pioneer, was the founder of Diuguidsville. See "Diuguidsville,"accessed December 2011, http://www.hmdb.org/marker.asp?marker=29954.

69 After 1810, when Randolph spent extended time at his second wife's home, he was only about four miles from the courthouse, making it an easy day trip.

70 Hening, *Statutes*, 7:419-420; Yeck, "At a Place Called Buckingham," 23-31.

71 Maloney, *A History of Buckingham County*, 7.

72 *MB*, 1,161, 1,313.

73 Yeck, "At a Place Called Buckingham," 23-31.

74 Ibid., 98-110.

Chapter Seven: Matters of Money

1 *MB*, 460. In 1778, Snowden is mentioned for the first time by name in Thomas Jefferson's memoranda. Bob may be Robert "Bob" Hemings (1762-1819), son of Betty Hemings.

2 Ibid., 438, 444, 483, 509, 945, 952. For more about Ben and Cary, see Chapter Ten.

3 *UB*, 49.

4 "Phillip Mazzei," *TJE*, accessed December 2011, http://wiki.monticello.org/mediawiki/index.php/Philip_Mazzei.

5 *MB*, 474. Thomas Jefferson did considerable business with Mazzei and lent him money. See Ibid., 470. On September 2, 1779, Thomas Jefferson collected £193.4 for Philip Mazzei from Hudson Martin, who was Randolph's friend. See Ibid., 483.

6 *PTJ*, 16:128. On February 28, 1790, Thomas wrote to Randolph, "I have the pleasure further to inform you that Doctr. Walker has no right to call on us for any interest. ... [T]he balance is in his favor, the result is that he would owe us a small balance of interest not worth notice." See Ibid., 127-29. For further exchange concerning the Peter Jefferson estate, see Ibid., TJ to Walker, 18 January 1790 (pp. 112-114); Walker to TJ 19 January 1790 (pp. 114-115); TJ to John Nicholas 20 January 1790 (pp. 115-116).

7 Ibid., 207-208. An execution was a court order issued to the Sheriff to bring a defendant before the court to satisfy a debt or damages of a judgment against him or an order to seize and sell a defendant's property to satisfy a judgment. See Gregory Crawford and J. Christian Kolbe, compilers, "Judgments" *Research Notes Number 29* (Library of Virginia, June 2007) accessed November 2011, http://www.lva.virginia.gov/public/guides/rn29_judgments.pdf.

8 Ibid., 194-195.

9 Buckingham County Personal Property Tax, 1789. In 1787, all of Randolph Jefferson's slaves were subject to tax, giving a fuller picture of the community. That year he owned fourteen slaves over age sixteen and seventeen slaves under age

sixteen. He also paid tax on six horses and thirty head of cattle. See Buckingham County Personal Property Tax, 1787.

10 *PTJ*, 16:87-88, 181.

11 John Jefferson to Thomas Jefferson, 7 January 1790, Ibid., 87-88. According to a footnote in the Jefferson papers, "John Jefferson's suit against Col. James was not settled for a quarter of a century." Multiple concurrent John Jeffersons living in the vicinity make it difficult to identify this cousin precisely. He may be the father of John Jefferson who is living at Snowden in 1800.

12 Ibid., 181. Another letter from Thomas Jefferson to Nicholas Lewis, dated March 7, 1790, lists the loan to John Jefferson for £10 or £15, as well as cash out to Randolph Jefferson coming to about £20. This probably refers to the division of the estate of their sister, Elizabeth; from it, Randolph was to receive £19.19.0 from Thomas. See Ibid., 210-212; *UB*, 15. That February was a busy month at Monticello. Martha "Patsy" Jefferson wed Thomas Mann Randolph on February 23rd.

13 *MB*, 945, 952, 970, 261, 991. Richard Anderson was the Albemarle representative for Donald, Scott & Co., making this "order" an additional debt to the Glasgow merchants.

14 Ibid., 85, 161, 275.

15 "Jack Jouetts Ride," *TJE*, accessed December 2011, http://www.monticello.org/site/research-and-collections/jack-jouetts-ride. In 1819, several years after Randolph Jefferson's death, Christopher Hudson and Thomas Jefferson brought suit against two other parties, defending their portions of the limestone tract on the Hardware River. Thomas still owned the 1/6 share he had purchased from Randolph; Hudson owned 4/6 of the track. See *MB*, 1,359.

16 Woods, *Albemarle County*, 230-231.

17 In April of 1792, James Brown and Robert Rives formed a mercantile partnership, operating under the name of Robert Rives & Co. In 1794, they changed the name to Brown, Rives & Co. See "Burton v. Brown's Ex'ors & als.," Cases Decided in the Supreme Court of Appeals, *Virginia Reports, 1730-1880* (Charlottesville, VA: Michie Company, Law Publishers), 1901.

18 Woods, *Albemarle County*, 57-58.

19 Ibid., 138-139. Christopher Hudson married Sarah Anderson. There was more than one David Anderson residing in Albemarle County in 1797-1798. One was Christopher Hudson's brother-in-law; another was the son of his brother-in-law, Richard Anderson.

20 Additionally, Randolph may occasionally have attended court in neighboring counties, as he did in March of 1815 when he appeared in Albemarle for "Randolph v. Craven Peyton." See *UB*, 53-54.

21 Stinson, *Early Buckingham County, Virginia Legal Papers*, 80. No details of the case survive; only two receipts are in Archibald Austin's files.

22 There are signs that Randolph Jefferson may have spent more freely as a young man, as indicated by his undisclosed financial "circumstances" in early 1790. See *PTJ*, 16:127-29.

23 While the Buckingham court records prior to 1869 are destroyed, many Prince Edward records survive. As a result, something is known about the case as it was argued in Buckingham. Randolph Jefferson's complex relationship with William Moon lasted many years. Both Randolph and his sons shopped at Moon's Stony Point store in southern Albemarle County. See Journals and ledgers of a Stony Point, Va. store, 1809-1829, MSS 13188, Small Special Collections, ViU. About ten years after Moon took Jefferson to court, Randolph's second wife, Mitchie (Pryor) Jefferson, ran up debt at Moon's store, disturbing her husband. See Thomas Jefferson deposition, 15 September 1815, Carr-Cary Papers, #1231, Small Special Collections, ViU. After Randolph's death, Moon and Scottsville developer, John Scott, owed significant funds to the Jefferson estate. See Albemarle Law Orders Book 1809-1821 and 1821-1831.

24 Prince Edward County District Court Record Book, 1802-1804, p. 56. A memo accompanies the entry in the Prince Edward County Order Book: "This Execution to be discharged by the payment of thirty pounds with lawfull interest thereon from the 10th day of February, 1800." See Ibid; Prince Edward County District Court Order Book 1799-1805, p. 221.

25 Defendant Benjamin Morris, a Gentleman Justice for Buckingham who also served as sheriff, probably provided security for Randolph Jefferson. See "Registers of Justices and Other County Officials, 1775-1865," Buckingham County 1822, Miscellaneous Reel 458a, Vi. Interestingly, Morris' land holdings were in the southern part of the county, near Paynes Creek, where he purchased 170 acres from the Revolutionary War hero, Peter Francisco, in 1813. See Roger G. Ward, *Buckingham County Records: Land Tax Summaries & Implied Deeds, Volume 1* (Athens, GA: Iberian Publishing Co., 1993), 209.

26 Prince Edward County District Court Record Book 1802-1804, p. 56. The original papers for Prince Edward County District Court currently housed at the Library of Virginia do not include this case.

27 Nathan Wash's connection to Randolph Jefferson, the Jefferson family, or Snowden is currently unknown. Albemarle County records reveal that on January 13, 1791, a man named Nathan Wash married Mercey Wood, spinster. Samuel Wood was the bondsman and witness. Her father, Henry Wood, gave his consent. In 1810, Nathan Wash, married with children, still lived in Albemarle County.

28 Prince Edward County District Court Record Book 1802-1804, pp. 56-57.

29 Prince Edward County District Court Record Book 1802-1804, pp. 57-58.

30 Ibid., 59; Prince Edward County District Order Book, 1799-1805, 1802, p. 221.

31 Jane's age is unknown; however, her value must have been sufficient to cover Randolph Jefferson's debt. In 1793, Thomas Jefferson sold Dinah, with two of her young daughters, for £139.15.0. See Jefferson, "Farm Book," 28. Later, in 1807, Thomas paid £150 for Mary, who was twenty-seven years old, with her two young sons. See "Mary Hern," *TJE*, accessed December 2011, http://www.monticello.org/site/plantation-and-slavery/mary-hern.

32 Buckingham County Land Tax, 1802, 1803. Due to the loss of Buckingham deeds, the purchaser is unknown.

33 *UB*, 57.

34 *MB*, 1,162.

35 Prince Edward County District Court Executions Book 1797-1809, Vi. The Executions Book contains the following notation in conjunction with the April 10, 1805 judgment for "Bondurant, assignee vs Randolph Jefferson:" "1804 June 4th, 6.69, Bgm." This refers to the prior case heard in Buckingham on June 4, 1805, for which court costs were $6.69. These costs were eventually charged to Randolph Jefferson, who lost the case in the District Court.

36 Randolph Jefferson, promissory note, 4 June 1804, Prince Edward County District Court Records, 1804-1807, Vi.

37 Ibid.

38 E. Booker, P.E., bill of complaint by Benjamin Bondurant, n.d., Prince Edward County District Court Records, 1804-1807, Vi. While the statement is not dated, the back side reads: "Bondurant (assignee) vs Jefferson, D. Court P. Edward, 1804 Oct CC Conss[d] [?]." Edward Booker resided primarily in Prince Edward County, was a representative to the Virginia Assembly, and had business dealings in Buckingham. See Charles Edward Burrell, *A History of Prince Edward County Virginia* (Richmond, VA: The Williams Printing Co., 1922), 31; Ward, *Land Tax Summaries & Implied Deeds*, 1:40; Ward, *Land Tax Summaries & Implied Deeds*, 2:40.

39 Summons, 14 August 1804, Prince Edward County District Court Records, 1804-1807, Vi.

40 Buckingham County Personal Property Tax, 1815; Military Commissions Papers, Acc 42222, Buckingham files, Vi; Buckingham County Personal Property Tax, 1804; "Registers of Justices and Other County Officials, 1775-1865," Buckingham County 1801, Miscellaneous Reel 458a, Vi.

41 Benjamin Bondurant (1773-1845) was born in Buckingham County and removed to Weakly County, Tennessee, where he died on October 23, 1845, and was buried in the Bondurant Family Cemetery. See Ruby Talley Smith, Bondurant Family Genealogy, Private Collection.

42 Buckingham County Personal Property Tax, 1803-1805. In 1800, Benjamin Bondurant did not pay personal property tax in Buckingham, although he held land in the county from 1796-1814. In 1805, Benjamin Bondurant purchased 125 acres from David Bondurant, presumably his brother. In 1813, Benjamin and David were living jointly on 120 acres, adjacent to his father, Darby Bondurant, among others. Benjamin's residence was in Nelson County. See Ward, *Land Tax Summaries & Implied Deeds*, 1:38-39.

43 Summons, 14 August 1804, Prince Edward County District Court Records, 1804-1807, Vi.

44 Randolph Jefferson and David Oglesby, Bond, 16 August 1804, Prince Edward County District Court Records, 1804-1807. Vi.

45 Ann Perkins, the sister of Hardin Perkins, Sr., married William Oglesby and eventually settled in Buckingham County. David Oglesby may be their son, but the connection is unproven. See Constantine Perkins Will, Goochland Deed Book 1, p. 148; Whitley, *Genealogical Records of Buckingham*, 23, 37, 40. David Oglesby was also involved in the failed town of Plantersville in southern Buckingham. See Ibid., 133-136. He remained in Buckingham until about 1807, when he sold 84 ½ acres to David Bondurant. See Ward, *Land Tax Summaries & Implied Deeds*, 1:228. He died in Campbell County at the age of 77; Lynchburg newspapers carried his obituary. See Whitley, *Genealogical Records of Buckingham*, 103-104.

46 On the reverse side of the 14 August 1804 summons it is noted: 1804 Spt. CO. See Summons, 14 August 1804, Prince Edward County District Court Records, 1804-1807, Vi.

47 M. Branch, memorandum, 12 March 1805, Prince Edward County District Court Records, 1804-1807, Vi. On the reverse side of Branch's memorandum it reads: "D.C. Jefferson ag [?] Bondurant, asse, Bail piece."

48 Ibid. On the back of the memorandum is written: "D.C., Jefferson ag (?) Bondurant ass(ignee), Bail Piece, 1805 Apr 1st." Mathew Branch served Buckingham County as a Gentleman Justice until his death on June 30, 1821. See "Registers of Justices and Other County Officials, 1775-1865," Buckingham County 1822, Miscellaneous Reel 458a, Vi.

49 *MB*, 768.

50 Ward, *Land Tax Summaries & Implied Deeds*, 1:294-295. In 1800, Capt. John Thomas paid personal property tax in Buckingham County for six horses and four slaves. There is also a John Thomas, son of James, who paid tax that year on one

horse and one slave. It seems more likely that Capt. Thomas is Randolph Jefferson's friend, Capt. John Thomas, who considerably outlived Randolph, is mentioned in Abraham Jones' pension application, dated September 17, 1828. Jones stated that he enlisted in 1779, under Thomas, and that Thomas should be able to attest to Jones' evidence. See Whitley, *Genealogical Records of Buckingham*, 46, 67.

51 Prince Edward County District Court Executions Book 1797-1809, B, 1805, Vi. Next to the note concerning the April 10, 1805 decision are the initials "EB," which may indicate that Edward Booker represented the case.

52 Ibid. The Executions Book contains this notation in conjunction with the September 11, 1805 judgment for Bondurant, assignee, vs. Randolph Jefferson: "1805 Aug 9th, 4.42, Bgm." This refers to the prior case heard in Buckingham on August 9, 1805, for which court costs were $4.42. These costs may have been charged to Jefferson.

53 Randolph Jefferson and Thomas Jefferson, Jr., Bond, 9 August 1805, Prince Edward County District Court Records, 1804-1807, Vi. Bernard (a.k.a. Barnett) Hatcher lived on Slate River, about halfway between Buckingham Courthouse and Snowden, twelve miles northeast of the courthouse. Ward, *Land Tax Summaries & Implied Deeds*, 1:148. In 1777, when Randolph purchased Ryno from his brother, Thomas, he paid £35. If 1805 prices were similar, the sale of two quality horses could cover the debt.

54 Ibid.

55 This is probably Darby Bondurant (1749-1828), the father of Benjamin Bondurant, who lived his entire life in Buckingham County and owned land on Sharp's Creek. See Ruby Talley Smith, Bondurant Family Genealogy, Private Collection. In 1794, Darby Bondurant was an Ensign in Capt. Randolph Jefferson's Sixth Company, Buckingham County Militia. Military Commissions Papers, Accession 42222, Buckingham files, Vi.

56 Benjamin Bondurant, assignee of Hardin Perkins, notice, 24 August 1805, Prince Edward County District Court Records, 1804-1807, Vi.

57 Prince Edward County District Court Order Book 1802-1804, pp. 506-507.

58 Prince Edward County District Court Executions Book 1797-1809, Vi.

59 *MB*, 1,160 - 1,161. The following entries are also posted on August 1, 1805: Warren ferrge. .33 and vales .50. These expenses indicate that Thomas Jefferson crossed the James River at Warren from the Albemarle side to the Buckingham side. The landing at Buckingham was also known as Warren.

60 Ibid., 1,163.

61 Buckingham County Personal Property Tax, 1804, 1805, 1806.

Chapter Eight: Snowden: A Plantation in Buckingham County

1   *MB*, 750.

2   "Limestone Land," *TJE*, accessed December 2011, http://www.monticello.org/site/research-and-collections/limestone-land.

3   *PTJ*, 29:66-67; Deed, Randolph Jefferson to Thomas Jefferson, 17 April 1796, Carr-Cary Papers, #1231, ViU; *MB*, 1,052.

4   Woods, *Albemarle County*, 329. John Slaughter died the following year, in 1797. *MB*, 936, 941-942, 948.

5   Grundset, *Buckingham County, Virginia Surveyor's Plat Book 1762-1858*, p. 34. This survey represents only part of Randolph Jefferson's holdings. For many years, Randolph was taxed on 2,000 acres. On rare occasions, he sold some land to satisfy debts. Circa 1802, he sold 123 acres and his taxable land dropped to 1,877 acres. By 1810, the farm was reduced another 50 acres, to 1,827 acres. The farm remained that size until 1815. In 1816, the estate was taxed for 1,649 acres, which likely reflects the 70 acres sold to Charles A. Scott, which Randolph mentioned in a letter to Thomas Jefferson, dated April 2, 1815. In 1816, Scott sold 178 7/8 acres to Capt. John Harris, conveyed to him by Randolph Jefferson. See *UB*, 57; Buckingham County Land Tax, 1800, 1802, 1810, 1815, 1816.

6   Grundset, *Buckingham County, Virginia Surveyor's Plat Book 1762-1858*: Price Perkins, April 1797, 666 acres (p. 28); Robert Craig, 21 April 1798, 160 acres (p. 32); Price Perkins, 23 November 1798, 355 acres (p. 33); John Patteson , asst. surveyor, 18 June 1799, 1,000 acres each side Big Georges Creek and Marrs Road joining Randolph Jefferson, Price Perkins, Mr. Murrey and Robert Craig (p. 35); Robert Craig, 27 January 1801, 7 acres (p. 37); William C. Murrey, 4 December 1802, 169 acres (p. 43).

7   Ann Richardson Sclather and Dr. John Kirk Richardson, "John Richardson of Cumberland County, Virginia and Some of His Descendants," *VMHB* (April 1937), 210.

8   Cumberland County Will Book 2, pp. 465-466; Martin Richardson v. John Jefferson and Wife, Cumberland County, Virginia Chancery Records, Vi, accessed December 2011, http://www.lva.virginia.gov/chancery/.

9   Pension application, Thomas Ware, accessed December 2011, http://southerncampaign.org/pen/r11138.pdf.

10 A son, Stephen, is named in Constantine Perkins' 1769 will. See Constantine Perkins Will, Goochland County Deed Book 9, p. 206; Douglas, *Douglas Register*, 271; William Kearney Hall, *Descendants of Nicholas Perkins of Virginia* (Ann Arbor, MI: University Microfilms International, 1980), 27; Abercrombie and Slatten, *Buckingham County, Virginia Publick Claims*, 6. He is likely the Stephen Perkins

(d. 14 March 1824), who married Sarah "Sallie" Bowles and is buried in Fluvanna County at Lockland. See Susie V. Shepard, "Lockland," *Virginia Historical Inventory*, 6 July 1937, Vi.

11 Pension application, Thomas Ware. John Bryant was Randolph Jefferson's neighbor on Rock Island Creek and, in 1787, held a license for an ordinary. He may be the John Bryant who married Richard Murray's daughter. It is possible Bryant took over the Snowden Ferry House from Murray. There are two men named John Bryant on Buckingham's 1800 tax list, so absolute identification is difficult. One of the John Bryants married Judy Winfrey; the Winfrey place is also not far from Snowden.

12 In 1788, at least two men named Thomas Hall paid personal property tax in Buckingham County. On April 10th, Thomas Hall was recorded with Edward Hall, eight slaves, and two horses. On March 14th, Thomas Hall, Jr. was recorded with one horse. This is likely the man who became Randolph Jefferson's overseer.

13 In 1798 and 1799, there is no overseer listed with Randolph Jefferson on the Buckingham County Personal Property Tax records; he simply paid tax for one white male – himself.

14 *UB*, 50.

15 Ibid., 56.

16 *MB*, 1,049, 1,209; *UB*, 19-21.

17 For example, in 1797, Thomas Jefferson advised his son-in-law, John Wayles Eppes, to occupy his blacksmith in part by making coal. See Thomas Jefferson to John Wayles Eppes, 21 December 1797, *PTJ*, 29:586.

18 *UB*, 36-37.

19 Ibid., 42.

20 "Crops at Monticello," *TJE*, accessed December 2011, http://www.monticello.org/site/plantation-and-slavery/crops-monticello.

21 "Valuable James River Land For Sale," *Richmond Enquirer*, 15 October 1822, p. 4.

22 Buckingham County Agricultural Census, 1860.

23 "Peaches," *TJE*, accessed December 2011, http://www.monticello.org/site/research-and-collections/peaches.

24 *UB*, 15.

25 "The Site of the Fruit Gardens," *TJE*, accessed December 2011, http://www.monticello.org/site/house-and-gardens/site-fruit-gardens.

26 *UB*, 53.

27 Ibid., 55.

28 The spring floods of 1771 were extraordinary in Virginia, known as the "Great Fresh." See Elizabeth Dabney Coleman, "The Great Fresh of 1771," *Virginia Cavalcade* (1951), 20-22.

29 *Virginia Gazette* (Purdie and Dixon), 12 September 1771, p. 3; "Ballews in Virginia," accessed December 2011, http://www.myplanet.net/gedmnds1/ballewva.htm.

30 "RATTLE-SKULL, STONEWALL, BOGUS, BLACKSTRAP, BOMBO, MIMBO, WHISTLE BELLY, SYLLABUB, SLING, TODDY, AND FLIP: Drinking in Colonial America," accessed December 2011, http://www.history.org/foundation/journal/holiday07/drink.cfm.

31 Kern, *Shadwell*, 60-61. While no information has survived concerning the making of alcohol at Snowden, occasional purchases of whiskey were made at Moon's Stony Point store and the Nicholas' distillery by the Jeffersons. See Journals and Ledgers of a Stony Point, Va. store, Journal and Ledger for 1808-1809, Small Special Collections, ViU; Papers of Wilson Cary Nicholas, Ledger, 1796-1797, Small Special Collections, ViU.

32 "Jefferson: the Scientist and the Gardener," *TJE*, accessed December 2011, http://www.monticello.org/site/house-and-gardens/jefferson-scientist-and-gardener.

33 *UB*, 33.

34 Ibid., 33-34.

35 Ibid.

36 Ibid., 44-45.

37 Ibid., 55.

38 Buckingham County Personal Property Tax, 1797.

39 Broadside 1807, no. 2, ViHi.

40 Broadside 1796, no. 2, ViHi; Broadside 1809, no 2, ViHi.

41 *UB*, 46.

42 "Sheep," *TJE*, accessed December 2011, http://www.monticello.org/site/plantation-and-slavery/sheep. In 1818, Thomas Jefferson was still sending full-blooded Merinos to friends such as Archibald Stuart. See Paul Leicester Ford, *The Writings of Thomas Jefferson, 1816-1826* (New York, NY: G.P. Putnam's Sons, 1899), 10:109-110.

43 "Sheep," *TJE*.

44 "Textile Factory," *TJE*, accessed December 2011, http://www.monticello.org/site/plantation-and-slavery/textile-factory.

45 For more about Fanny, see Chapter Ten.

46 Buckingham County Personal Property Tax, 1813. There were undoubtedly a few younger children who did not appear on the tax record.

47 *UB*, 35-37.

48 Journals and ledgers of a Stony Point, Va. store, Journal and Ledger for 1808-1809, Small Special Collections, ViU.

49 *MB*, 1,079. Jesse Moore, Sr., who removed to North Carolina before the Revolutionary War, is mentioned in a Fluvanna County deed in 1803, placing the Moore family on the banks of the Hardware River. In 1806, his brother, Warren Moore, still lived in Albemarle when he sold 200 acres on the south side of Hardware River to William Moore, Jr. See "Ancestors of John Lee Sharp," accessed December 2011, http://familytreemaker.genealogy.com/users/s/h/a/John-L-Sharp/GENE5-0091.html.

50 *MB*, 1,201.

51 Journals and ledgers of a Stony Point, Va., store, 1808-1829, MSS 13188, Small Special Collections, ViU. Randolph Jefferson's complex relationship with William Moon, Jr. lasted many years. Beyond their dealings at the store, they wrangled in at least two court proceedings. After Randolph's death, Moon and Scottsville developer, John Scott, owed significant funds to the Jefferson estate. See Albemarle Law Orders Book, 1809-1821, 1821-1831.

52 Moore, *Scottsville on the James*, 45.

53 Permission for David Ross' Rivanna Warehouse came in 1785, while the town of Columbia was not established until 1788. See Hening, *Statutes*, 12:66-67, 682. Permission for an inspection point/tobacco warehouse and for the incorporation of the town of Warminster came simultaneously in the fall of 1788. See Hening, *Statutes*, 12:665-668.

54 Nicholas, "The Early History of the Founding of Scottsville," 13; Shepherd, *Statutes at Large of Virginia*, 1:155.

55 Richard L. Nicholas, "The First Hundred Years of the Scottsville Community," *MACH* (2010), 2-23. According to Nicholas, "the two communities competed for commercial supremacy through access to the burgeoning agricultural market of the region in such things as turnpikes, inspection stations, and town status." See Ibid., 10. Additionally, John Scott's death in 1792 contributed to the slow development of what would eventually become Scottsville.

56 After the days of the ferry, a bridge was never built crossing from Warren to the ferry landing in Buckingham County, which was also referred to as Warren. No town ever developed there. Today, Warren Ferry Road (Route 627) still leads down to the river on both banks. See Richard L. Nicholas to Joanne L. Yeck, email correspondence, 3 November 2010.

57 Wilson Cary Nicholas had a long and complicated relationship with Thomas Jefferson. Family relations were tightened when grandson, Thomas Jefferson Randolph, married Wilson Cary Nicholas' daughter, Jane Hollins Nicholas. Late in life, Jefferson acted as security for one of Nicholas' obligations; as a consequence, Nicholas' subsequent bankruptcy added greatly to Jefferson's already significant debt. See "Wilson Cary Nicholas," *TJE*, accessed December 2011, http://www.monticello.org/site/research-and-collections/wilson-cary-nicholas; Alan Pell Crawford, *Twilight at Monticello: The Final Years of Thomas Jefferson* (New York, NY: Random House, 2008), 170-172, 177-180.

58 Randolph Jefferson was on friendly terms with Nicholas and his family; Wilson Cary Nicholas, Jr. called at Snowden with apparent regularity. See *UB*, 57.

59 Lay, *The Architecture of Jefferson County*, 115.

60 Mutual Assurance Society Policy #18, Mount Warren, 16 July 1805, Vi.

61 Moore, *Albemarle*, 89. The closest grist mill to Snowden, located on the south side of the James River, is not known; though, during Randolph Jefferson's lifetime, Richard Murray operated a mill at Buckingham's Horseshoe Bend.

62 R.E. Hannum, "Brick Mill at Warren," Virginia Historical Inventory, 29 November 1937. Surviving ledgers and journals from Nicholas' mill begin on September 15, 1796. See Papers of Wilson Cary Nicholas including material on William and Mary College, 1751-1850, MSS 2343, Small Special Collections, ViU. The mill at Warren was burned during the Civil War, in March of 1865, during Gen. Philip Sheridan's raid, but was eventually rebuilt. See Richard L. Nicholas, *Sheridan's James River Campaign of 1865 Through Central Virginia* (Charlottesville, VA: Historic Albemarle, 2012). In 2010, the surviving mill functioned as an inn. See "The Carriage House at Warren Mill," accessed December 2011, www.virginia.org/Listings/PlacesToStay/TheCarriageHouseatWarrenMill/.

63 Papers of Wilson Cary Nicholas, Ledger, 1796-1797, 10 December 1796 and 13 June 1797, Small Special Collections, ViU.

64 In 1790, Anthony Murray paid tax for a William Murray. See Buckingham County Personal Property Tax, 1790.

65 Papers of Wilson Cary Nicholas, Ledger, 1798-1801, entries: 16 April 1799, 1 June 1799 and 10 June 1799; *UB*, 42.

66 K. Edward Lay and Nathaniel Mason Pawlett, "Architectural Surveys Associated with Early Road Systems," *Bulletin of the Association for Preservation Technology*, Vol. 12, No. 2 (1980), 3-36. For quote, see p. 7.

67 Hening, *Statutes*, 13:42-43.

68 Hening, *Statutes*, 12:66-67. In November of 1793, the town of New Canton was established on the south side of the James; it, too, was downriver from Snowden,

sitting in the northeast corner of Buckingham County. The ferry landing there was on William Cannon's land. New Canton was slow to develop; a tobacco warehouse was constructed there by January of 1804, making it the most convenient to Snowden. See Samuel Shepherd, *The Statues at Large of Virginia, Volume 1* (Richmond, VA: Printed by Samuel Shepherd, 1835), 269.

69   Hening, *Statutes*, 13:48. The Virginia Historical Inventory survey for Warren Ferry states that it "is one of the oldest crossings in this section. Just below the ferry site there was a ford which left the south bank of the river at about the same place where the present ferry is now. It went down the creek and came out about the mouth of the creek." See R.E. Hannum, "Old Ferry," Virginia Historical Inventory, 30 November 1936.

70   Shepherd, *Statues at Large of Virginia, Volume 1*, p. 425.

71   Along with his neighbor, Price Perkins, Randolph may also have availed himself of the Warren blacksmith shop. See Papers of Wilson Cary Nicholas, Blacksmith Ledger, 1802-1806, Small Special Collections, ViU.

72   Woods, *Albemarle County*, 58-59.

73   *UB*, 24, 42, 46. Capt. Brown was likely William Brown, who purchased Kinney's tavern in 1812, as well as the "Mr. Brown" to whom Randolph wrote in 1813 concerning the catching of the carp in the Warren mill race. In 1815, William Brown paid tax on a residence on James River and on a lot in the town of Warren. See Ward, *1815 Directory of Virginia Landowners (and Gazetteer)*, 4. In 1815, there was a John Patterson who lived on Ballinger's Creek, adjacent to Warren; this is no doubt Wilson Cary Nicholas' son-in-law. There is also a John "Patteson" of Warren. It is unclear if they are the same man. The Patteson name is very common in northern Buckingham County; however, Nicholas' son-in-law spelled his name with an "r." Ibid., 14.

74   Thomas Jefferson deposition, 1815.

75   Woods, *Albemarle County*, 304-305; 313-314; Lay, *Architecture of Jefferson Country*, 312. In 1937, R.E. Hannum wrote, "The 'still house creek' starts from a spring in the yard that was then Wilson C. Nicholas' home." See R.E. Hannum, "Distillery at Warren," Virginia Historical Inventory, 21 October 1937. Samuel Shelton died in 1826. See Woods, *Albemarle County*, 314.

76   Lay, *Architecture of Jefferson Country*, 168-169.

77   *MB*, 1,327; R.E. Hannum, "J.C. Anderson's Rock House," Virginia Historical Inventory, 30 November 1936; Lay, *Architecture of Jefferson Country*, 29. Jacob Kinney later removed to Staunton, Virginia, selling the tavern to William Brown. See Woods, *Albemarle County*, 58-59.

Chapter Nine: The Children of Randolph Jefferson

1. In this letter, Thomas Jefferson refers to his "sister," Anne (Lewis) Jefferson, who is his sister-in-law, to his daughter Martha Jefferson (b. 1772), and Mary "Polly" Jefferson (b. 1778). The day before he penned this letter to Randolph, on January 10, 1789, Thomas wrote a very similar message to his brother-in-law, Col. Charles L. Lewis, and his sister, Lucy (Jefferson) Lewis. After a probable long silence, he was reconnecting with the family before returning to Virginia. See Merrill, *Jefferson's Nephews*, 72-73. This letter written in Paris is the earliest known existing correspondence between the Jefferson brothers. The 1981 edition of *Thomas Jefferson's Unknown Brother* contains all known existing letters. Thomas Jefferson's "Summary Journal of Letters" indicates some are lost. They include letters written on 24 August 1795; 9 May, 4 September, and 3 November 1796; 27 October 1797; 8 September, 17 October, and 28 October 1798, 4 April and 6 November 1799; and 17 August 1800. See *PTJ*, 29:66-67.

2. In 1782, Randolph Jefferson was taxed in Buckingham County on thirty slaves, six horses, and forty-two head of cattle; however, he was not taxed on himself. The following year, Randolph was taxed in Buckingham on himself and thirty-two slaves, indicating the return of the Jeffersons and their servants to Snowden. See Albemarle County Personal Property Tax, 1782; Buckingham County Personal Property Tax, 1782, 1783.

3. It is possible there was a set of twins in the group; however, no direct evidence has been found to support that idea. The suggested range in possible birth dates is the result of several indicators, including the census records, some of which are given here.

4. Anne (Lewis) Jefferson's uncle was Robert Lewis of the Byrd (29 May 1739 - 10 January 1803); he married Jane Woodson, daughter of Tucker Woodson, on February 20, 1760. Anne also had a cousin, Robert Lewis, son of Col. Robert Lewis of Belvoir and Jane Meriwether, who married her aunt, Frances Lewis. It is interesting to note that in the Jefferson family Bible, one of the children (likely Randolph), while practicing penmanship, wrote "Tucker Woodson." Tucker Woodson, who became Deputy Clerk of Albemarle in 1769, had a son, Tucker Moore Woodson, whom Randolph likely knew. Albemarle Court Orders indicate a connection between the Moores and Tucker Woodson's family, mentioning "Edward Moore he being first appointed guardian for that particular purpose to Samuel Woodson orphan child of Tucker Woodson Dec[d]." See Albemarle Order Book 1791-93, p. 10. Edward Moore married Anne (Lewis) Jefferson's sister, Mildred Lewis. Currently, nothing more is known about a relationship between Randolph Jefferson and Tucker Woodson. Tucker M. Woodson owned Viewmont, the former home of Joshua Fry. When Woodson removed to Kentucky, the plantation was purchased by Capt. John Harris, who would eventually purchase Snowden. See Woods, *Albemarle County*, 356-357.

The Woodson and the Lewis families continued to intermarry. Randolph Jefferson's son, Peter Field, would eventually marry Jane Woodson Lewis.

5   Woods, *Albemarle County*, 250.

6   The Lilburne name came down through the Randolphs. Jane (Randolph) Jefferson's mother was Jane (Rogers) Randolph. Her parents were Charles and Jane (Lilburne) Rogers, who married in England. Jane Lilburne's father was William Lilburne, born 1636, of Newcastle-upon-Tyne. See "Lilburne, Randolph, Jefferson," *VMHB* (July 1918), 321-324.

7   Anne (Lewis) Jefferson had an uncle, James Lewis, who lived in Lunenburg County and died in 1764. It seems unlikely that this James influenced the Jefferson's choice of the name James Lilburne for their youngest son. The will of Col. Charles Lewis of the Byrd names: John Lewis, Charles Lewis, Howell Lewis, Robert Lewis, Elizabeth Kennon, Anne Taylor & Frances Lewis, and the other eighth part or portion thereof to the sons and daughters of my son James Lewis, deceased. See Sorley, *Lewis of Warner Hall*, 299-300.

8   "Charles L. Lewis, etc. vs. Executors of Isham Lewis," Fluvanna County, Virginia, # 1800-002, Vi, accessed October 2011, http://www.virginiamemory.com/collections/chancery/.

9   Betts and Bear, *Family Letters*, 174.

10 *MB*, 1,028-1,029. There is not another known Jefferson family death during October or November of 1800.

11 Woods, *Albemarle County*, 178, 401.

12 Information concerning the Jefferson children is gleaned from a variety of primary sources, including Chancery Cases, death and marriage records, census material, personal property taxes, the Nevil family estate records, etc. A few published sources discuss Randolph's descendants, including Moore, *Scottsville on the James*; Burton, *Jefferson Vindicated*; Merrill, *Jefferson's Nephews*; Ragland, *The Tie That Binds*; and Scottsville Museum, accessed December 2011, http://smuseum.avenue.org/.

13 *MB*, 1,010. Despite the fact that the payment went to Mrs. Sneed, Thomas, Jr.'s teacher was probably Benjamin Sneed.

14 In 1801, the Albemarle Road Orders note that Benjamin Sneed still adjoined Thomas M. Randolph and Thomas Jefferson.

15 *MB*, 1,007, 1,010, 1,021, 1,022, 1,036.

16 "Inoculation," *TJE*, accessed December 2011, http://www.monticello.org/site/research-and-collections/inoculation; B.S. Leavell, "Thomas Jefferson and smallpox vaccination," *Transactions of American Clinical and Climatological Association* (1977), 119–127.

17 *MB*, 1,038.

18 Thomas Jefferson to John Wayles Eppes, 8 April 1801, Photostat, Original owned by Luther Ely Smith, Small Special Collections, ViU.

19 Merrill, *Jefferson's Nephews*, 84-87, 351-352.

20 Vogt and Kethley, *Albemarle County Marriages*, 1:238.

21 Ibid., 175. Craven Peyton was married to Polly Lewis' sister, Jane. His relationship to John Peyton is currently unknown. Ibid., 252.

22 *UB*, 35-36.

23 Ibid., 42.

24 Ibid., 46.

25 National Archives and Records Administration, Pay roll records, 7th Virginia Regiment.

26 The 7th Virginia Regiment was stationed at Camp Carter, 28 August 1814 - 24 February 1815, under Lt. Col. William Gray. Capt. Boaz Ford, Capt. William Freeland, and Capt. William M. Holman were from Buckingham County. See Stuart Lee Butler, *A Guide to Virginia Militia Units in the War of 1812* (Athens, GA: Iberian Publishing Co., 1988), 59-60.

27 *Virginia Militia in the War of 1812*, 1:329-330, accessed December 2011, http://search.ancestry.com/search/db.aspx?dbid=48441.

28 For more about James Lilburne Jefferson and his relationship with Thomas Jefferson, see Chapter Twelve and Chapter Thirteen.

Chapter Ten: The Jefferson Servants

1 Buckingham County Personal Property Tax, 1782; Albemarle County Personal Property Tax, 1782.

2 Buckingham County Personal Property Tax, 1782.

3 Buckingham County Personal Property Tax, 1783.

4 Lucia Stanton, "Perfecting Slavery: Rational Plantation Management a Monticello," in *"Those Who Labor for My Happiness:" Slavery at Thomas Jefferson's Monticello* (Charlottesville, VA: University of Virginia Press, 2012), 71-89.

5 *MB*, 945. Donald, Scott & Co. was one of numerous tobacco importers in Glasgow, Scotland. In cases where the planter could not meet his debts, the merchants ended up acquiring land in Virginia. See Louisa Deed Book D, p.118; Donald Scott & Co. to David Terry, 26 June 1773.

6   Myrtilla was sold in 1770 to Benjamin Moore. See "The Slave Families of Thomas Jefferson," database, accessed December 2011, http://www.sylvest-sarah.com/.

7   Stanton, "Perfecting Slavery," in *"Those Who Labor for My Happiness,"* 71-89.

8   *MB*, 952.

9   This information conflicts slightly with records at *Monticello* indicating that Cary was born in 1785. Ibid., 952.

10 Interestingly, at precisely the same time in 1797, Thomas Jefferson removed two boys, Judy Hix's sons, Ben and Kit, from the nailery, making them part of the dowry for his daughter, Maria. See Stanton, "Perfecting Slavery," *"Those Who Labor for My Happiness,"* 321n29.

11 *MB*, 970, 957, 996.

12 Ibid., 970.

13 Ibid., 991. For more about the Randolph Jefferson's debt to Christopher Hudson, see Chapter Three.

14 Ibid., 1,005, 1,007.

15 Ibid., 990. Ben and Cary appear on several pages of the "Farm Book." See Jefferson, "Farm Book," 39, 52, 53, 55.

16 "Nailmaking," *TJE*, accessed December 2011, http://www.monticello.org/site/plantation-and-slavery/nailery.

17 Stanton, "Perfecting Slavery: Rational Plantation Management a Monticello," in *"Those Who Labor for My Happiness,"* 71-89.

18 Jame Hubbard proved to be a problematic personality, at least from Thomas Jefferson's point of view. Jame may or may not have influenced Cary's rebellious behavior or attitude. When Jame Hubbard successfully ran away from Monticello, Thomas Jefferson sold him to Reuben Perry, who advertised for the runaway, "James" Hubbard, in the Richmond Enquirer on April 12, 1811. See Stanton, "Free Some Day," in *"Those Who Labor for My Happiness,"* 147-148. Brown Colbert was the son of Betty Brown and the grandson of Elizabeth Hemings. Brown survived the attack, was eventually sold away from Monticello at his own request, gained his freedom, and immigrated to Liberia. See "Slaves Who Gained Freedom," TJE, accesses December 2011, http://www.monticello.org/site/plantation-and-slavery/slaves-who-gained-freedom.

19 *MB*, 1,106, 1,110; Stanton, "Free Some Day," in *"Those Who Labor for My Happiness,"* 148.

20 Thomas Mann Randolph to Thomas Jefferson, 30 May 1803. ViU.

21 Thomas Jefferson to Thomas Mann Randolph, 8 June 1803, DLC; Edwin Morris Betts, *Thomas Jefferson's Farm Book* (Chapel Hill, NC: University of North Carolina Press, 2001), 19.

22 *MB*, 1,181.

23 In 1827, Ben, valued at $300, and Lilly, valued at $100, were living at Tufton with thirty-one other slaves. They appear on the Inventory and Appraisal of the Estate of Thomas Jefferson at Tufton and Lego, dated January 11, 1827. See Bernetiae Reed, *The Slave Families of Thomas Jefferson, Volume 1* (Greensboro, NC: Sylvest-Sarah, Inc., 2007), 385. In addition to Ben and Lilly, Prof. Blaettermann bought most of the Wormley Hughes' family, including Wormley's wife, Ursula, the daughter of Bagwell & Minerva, and their children, Burwell, Critta, George, and Robert Hughes, who was a blacksmith. Blaettermann also purchased Marshall, son of Lazaria [Maria]. Most were later reunited at Edgehill, the plantation of Jefferson's grandson, Thomas Jefferson Randolph. Many years later, Randolph Jefferson's son, Thomas Jefferson, Jr., would marry Blaettermann's niece, Elizabeth Wilhelmina Siegfried. See "An Account of Sales of negroes of the Est. of Thomas Jefferson," January 1, 1829, accession no. 8937, Small Special Collections, ViU.

24 Betts and Bear, *Family Letters*, 182-183.

25 Stanton, "Free Some Day," in *"Those Who Labor for My Happiness,"* 107-113, 117-131.

26 *MB*, 992.

27 Gibson vs. Key, Chancery Case, April 1856 term, Fluvanna County, Virginia; "Hardin Perkins & Sarah Price," accessed December 2011, http://webpages.charter.net/blankenstein/hardin_perkins_&_sarah_price.htm.

28 The searchable *Virginia Gazette* ends in 1780, while Daniel E. Meaders' collection of runaway advertisement begins in 1801, leaving a significant gap in indexed and/or culled Virginia newspapers. See Daniel E. Meaders, ed., *Advertisements for Runaways Slave in Virginia, 1801-1820* (New York, NY: Routledge, 1997).

29 *MB*, 1,276.

30 Ibid., 1,288. On May 17, 1813, Thomas Jefferson noted "foreman James" returned $2.25 of the $5.00, indicating he may not have gotten as many fish as Thomas expected. See Ibid., 1,289.

31 Ibid., 1,288-1,289. He also spent $.75 on ferriage.

32 *UB*, 35-37.

33 Ibid., 42-43. The 1782 and 1783 lists of slaves at Snowden include a man named James. He may or may not be the same James who was Randolph Jefferson's property in 1813, when the carp episode took place. A common slave name, there

are at minimum four men named James at Monticello, at least two of whom may be involved in acquiring carp for Thomas Jefferson.

34 *MB*, 1,289. The "missing" $.50 covered Ned's James' ferriage.
35 *UB*, 46.
36 Ibid., 35.
37 Ibid., 42.
38 Ibid., 44-45.
39 Ibid., 46.
40 Ibid., 47.
41 Ibid., 48.
42 Ibid., 49.
43 Ibid., 50.
44 Ibid., 51.
45 In 1756, Peter, Randolph's manservant and age peer, was valued at £17.10.
46 *UB*, 49.
47 Ibid.
48 Ibid., 50. The other reference to this loan is a memo which Thomas Jefferson made on February 3, 1814. "Exchanged a watch with my brother & gave him the 40. D. ante Aug. 8. in boot." See *MB*, 1,297.
49 *MB*, 1,292.
50 *UB*, 55, 57.
51 Orange County Will Book 2, pp. 181-183, 193.
52 Orange was valued at £13 in Peter Jefferson's 1757 inventory, the lowest value put on a male child at Shadwell. At Snowdon, the child, Bellow, was valued at £8. See Kern dissertation, 453, 455.
53 Kern, dissertation, 191-192; 231-232.
54 After Culpeper County was cut out to the west, the Orange County seat moved from the location on the Rapidan River to Orange Court House in November of 1749. For more about the founding and development of Orange County, Virginia, see William W. Scott, *A History of Orange County* (Richmond, VA: Everett Waddey Co.,1907); Ann Brush Miller, "Orange County Road Orders: 1734-1749" (Orange County, VA: Orange County Historical Society), accessed September 2011, www.virginiadot.org/vtrc/main/online_reports/pdf/85-r2.pdf; Ann Brush Miller, "Orange County Road Orders: 1750-1800" (Orange County, VA: Orange County

Historical Society), accessed September 2011, www.virginiadot.org/vtrc/main/online_reports/pdf/90-r6.pdf.

55 Miller, "Orange County Road Orders: 1734-1749," 64 (rank), 94 (ordinary), 104 (bridge), 122 (mill). By 1751-1752, the construction of a permanent courthouse and prison were underway. Prior to that, the house of John Branham (a.k.a. Bramham) was used for court, "it being near where the courthouse is, with all expedition, going to be built." See William W. Scott, *A History of Orange County*, 23, 35-37. This is likely the house on Edward Spencer's property. Though Spencer did not personally operate the ordinary (just as Peter Jefferson did not operate his), Spencer remained involved. On May 23, 1743 (OS), Robert Slaughter and Edward Spencer, Gent. entered their bond for Thomas Fox's ordinary license. "On the Petition of Thomas Fox he is allowed to keep ordinary at the Court House of this County for one whole year from this Time Giving Security whereupon he with Edward Spencer Gent. his Security entered into and ack^d their bond for his keeping the said Ordinary according to Law and it is Ordered that the Clerk of this Court do prepare a License for him accordingly." See Miller, "Orange County Road Orders: 1734-1749," 108. Spencer provided security for a series of ordinary licenses: Thomas Fox (1743-1745), James Rucker (1746), and Timothy Crosthwait (1747). See Miller, "Orange County Road Orders: 1734-1749," 94, 108, 125, 134.

56 In 1743, John Branham took a nine-year mortgage on Spencer's 104 acres. See Orange County Deed Book 7, pp. 272-275; Miller, "Orange County Road Orders: 1734-1749," 127.

57 Miller, "Orange County Road Orders: 1734-1749," 100.

58 Ibid., 127.

59 Orange County Will Book 2, pp. 181-183, 193.

60 Kern, dissertation, 191-192.

61 *MB*, 380. Eventually known as Poplar Forest, Jefferson did not start building his retreat home in Bedford County until 1806. In the fall of 1774, Randolph Jefferson had not yet settled down at Snowden and it is likely that Randolph had not moved all of his slaves to Buckingham County, hence Orange's errand for Thomas Jefferson.

62 Buckingham County Personal Property Tax, 1782, 1783. Orange appears on both the tax records of 1782 and 1783: however, his sister, Nanney, does not. She may not have survived until 1782 or possibly was sold prior to that date.

63 *UB*, 17-18. Mayo and Bear conclude that Orange is Dinah's husband. Though direct evidence has not been found, she definitely had a husband at Snowden and her first surviving child was called Orange.

64 Jefferson, "Farm Book," 9, 15. Dinah was a popular slave name. By 1774, Thomas Jefferson owned multiple women named Dinah, hence the use of her birth year to

distinguish her from the others. Dinah's birth coincided with the death of John Wayles' third wife. His daughter, Martha Wayles (b. 5 November 1748), the future Mrs. Thomas Jefferson, was twelve years old at the time.

65   *UB*, 16; Jefferson, "Farm Book," 24. Thomas Jefferson's list of "Births and Deaths in 1776" includes an "anonymous" death attributed to a Dinah at Monticello. This may be a record of the death of Dinah's first unnamed infant, who was either stillborn or died very quickly. Dinah would have been about fifteen years old at the time. See *MB*, 26; David Library of the American Revolution, loose page from Thomas Jefferson's Farm Book, accessed December 2011, http://www.dlar.org/.

66   Jefferson, "Farm Book," 15.

67   *UB*, 17.

68   Woods, *Albemarle County*, 247; Thomas Mann Randolph, Jr. to Thomas Jefferson, 1793 January 24, *PTJ*, 25:91. In Thomas Mann Randolph, Jr.'s letter he states that Dinah and her children were valued by Col. Lewis and Col. Bell at 139/17/6. See Ibid.

69   Kern, dissertation, 231-232; Jefferson, "Farm Book," 28; "The Slave Families of Thomas Jefferson," accesses December 2011, http://www.sylvest-sarah.com/; *UB*, 17. Mayo may have misread "Lizzy" for "Lucy" in his transcription of the September 25, 1792 letter from Thomas Jefferson to Randolph Jefferson. She is clearly written as "Lizzy" in Jefferson's notation of the sale to James Kinsolving in the "Farm Book." Jefferson's "Farm Book," page twenty-five, which lists "Negroes Alienated from 1784-1794 inclusive," was separated from the rest of the "Farm Book" and is not part of the collection at the Massachusetts Historical Society. It can be viewed at "Classroom Monticello," accessed December 2011, http://classroom.monticello.org/kids/gallery/image/159/Jeffersons-Farm-Book-Page-25-Negroes-alienated-from-1784-1794-inclusive/.

70   "Descendants of William Leigh of Amherst County," accessed December 2011, http://leigh.editme.com/files/GenealogyReportsII/Descendants%20of%20William%20Leigh%20of%20Amherst.pdf.

71   James Kinsolving Estate Inventory, Albemarle County Will Book 9, p. 403.

72   Broadside 1807: 2, printed by W.C. Lyford, 1807, ViHi. The stud horse was imported by Wm. Lightfoot, Esq. of Charles City County and was advertised on February 24, 1807 to "stand the ensuing season."

73   *MB*, 925, 1,026.

74   Jesse Moore, Sr., who removed to North Carolina before the Revolutionary War, is mentioned in a Fluvanna County deed in 1803, placing the Moore family on the banks of the Hardware River. In 1806, his brother, Warren Moore, still lived in Albemarle when he sold 200 acres of land, south side of Hardware River to William

Moore, Jr. See "Ancestors of John Lee Sharp," accessed December 2011, http://familytreemaker.genealogy.com/users/s/h/a/John-L-Sharp/GENE5-0091.html.

75 James Kinsolving grave stone, Temple Hill, Albemarle County, Virginia.

76 Albemarle County Will Book 9, p. 403.

77 Albemarle County Will Book 10, pp. 32-33. According to the 1810 Federal Census, James Kinsolving owned twenty-one slaves, including Dinah, Lizzie and Sallie, and Dinah's already numerous grandchildren.

78 The values stated on December 2, 1829 vary somewhat from the June 1, 1829 inventory which listed the following: Bob ($120), Ned ($150), Hannah ($150), Mary ($200), Maria ($250), Phill ($550), Harry ($300), Sarah and three children, Rachel, Nelson and Pheby ($525), and Lucinda ($150). See Albemarle County Will Book 10, pp. 32-33.

79 A marriage bond for William Wood and Martha "Patsy" Kinsolving was issued on December 22, 1817. See John Vogt and T. William Kethley, Jr., *Albemarle County Marriages, 1780-1853*, 1:351. William Wood also purchased Kinsolving slaves: a man named Jesse ($350) and a boy named Ross ($162). See Albemarle County Will Book 10, p. 274.

80 Ibid. On the Kinsolving inventory, the other Elizabeth is called "Liza." She and her two children, Toby and Coy, were valued at $500, the same value that was applied to "Eliza" and her two children, Melinda and Preston. Significantly, Toby's name suggests a connection to Orange's family, Toby being his father's name. See Albemarle County Will Book 9, p. 403. Eventually, Toby sold separately for $80 to G.W. Kinsolving, indicating that Coy stayed with his mother. See Albemarle County Will Book 10, p. 274.

81 Albemarle County Will Book 13, p. 82.

82 Albemarle County Will Book 10, pp. 32-33, 273.

83 Federal Population Census, Albemarle County, Virginia, 1830.

84 G.W. Kinsolving operated the Central Hotel in Charlottesville near the courthouse. See Woods, Albemarle County, 247-248.

85 "Albemarle County Area, Charlottesville," Garden Week 2010, accessed February 2011, http://cgc.avenue.org/GuideCAGW2010.html.

86 "Albemarle: Cemetery Records – Temple Hill," accessed February 2011, http://files.usgwarchives.net/va/albemarle/cemeteries/templehill.txt. The transcript gives the family name as "Kingsolving."

Chapter Eleven: The Patient Randolph Jefferson

1. "Dogs," *TJE*, accessed December 2011, http://www.monticello.org/site/house-and-gardens/dogs. Stanton, *Mad Dogs and Faithful Servants* (Thomas Jefferson Memorial Foundation, 1989).

2. James Breig, "The Eighteenth Century Goes to the Dogs," *Colonial Williamsburg* (Autumn 2004), accessed December 20011, http://www.history.org/foundation/journal/autumn04/dogs.cfm.

3. Hening, *Statutes*, 6:488-489 accessed December 2011, http://vagenweb.org/hening/vol06-20.htm#bottom.

4. Ibid., 489. The counties covered in the statute included Goochland, though not its newly formed and sparsely populated neighbor, Albemarle County.

5. "Dogs," *TJE*.

6. Breig, "The Eighteenth Century Goes to the Dogs."

7. Bear, *Jefferson at Monticello*, 21-22.

8. *PTJ*, 15:553.

9. Ibid., 552.

10. Ibid., 16:18.

11. *MB*, 749-750.

12. *UB*, 15.

13. "Dogs," *TJE*. "Armandy" may have been Norman, one of the puppies that stayed at Monticello and was still alive in 1796. See Ibid.

14. *PTJ:RS*, 2:413.

15. "Dogs," *TJE*. By house dog, Thomas Jefferson likely did not mean a pet dog, but rather a dog comfortable interacting with people around the dwelling house and guarding the house, not simply living out in pasture land.

16. Ibid.

17. Ibid.

18. Papers of Thomas Jefferson, MHi.

19. *UB*, 27.

20. Ibid., 30.

21. Ibid., 32.

22. Ibid., 46.

23. Ibid., 52.

24 *UB*, 53.

25 Ibid., 54.

26 Ibid., 55.

27 "Dogs," *TJE*.

28 *UB*, 55.

29 *MB*, 1,192. Henry Voigt (1738-1814), clock and watchmaker, was appointed the first Chief Coiner at the U.S. Mint in Philadelphia in 1793 and held the office until his death. Appreciated for both his mechanical and design skills, Voigt designed and engraved the first U.S. coins. See http://www.usmint.gov/.

30 In 1810, Randolph Jefferson's taxes were $34.11 (land) + $8.30 (personal property), making the cost the watch equivalent to about two years' worth of taxes at Snowden.

31 The feminization of the watch was likely a holdover from the French, *la montre*.

32 *MB*, 1,184-1,185.

33 Ibid., 1,192. Edgehill was a Randolph property located slightly north and east of Monticello. Home to Martha (Jefferson) and Thomas Mann Randolph, it was eventually occupied by their son and his wife, Thomas Jefferson and Jane Hollins (Nicholas) Randolph.

34 Betts and Bear, *Family Letters*, 290.

35 *MB*, 1,192, 1,197.

36 *UB*, 19.

37 *MB*, 1,201. The next day, Thomas paid Randolph $2.00 for ferriages for Jesse Moore, who in 1802 had brought plank from Snowden to Monticello.

38 *UB*, 30. The 1810 Virginia census lists a J.H. "Fisbinder" in Richmond City. In 1820, he is enumerated as "J.H. Fasbender."

39 Ibid., 32 This may be Reuben B. Patteson, who served Buckingham in the Virginia House of Delegates 1821-1828. As Patteson was known far beyond Buckingham and in political circles in Richmond, Thomas Jefferson likely knew or certainly knew of Reuben B. Patteson, hence Randolph mentions him by name. See Whitley, *Genealogical Records of Buckingham*, 125.

40 *MB*, 1,297. In August 1813, Randolph had asked Thomas for a loan of $40.00, which he owed next court. See Ibid., 1,292.

41 *UB*, 52.

42 Ibid., 53.

43 Ibid. Varina was Thomas Mann Randolph's plantation, situated in Henrico County, close to Richmond. During these years, Thomas Mann Randolph was a

busy politician, frequently traveling to and from Edgehill and Varina. He served in the Virginia State Senate (1793-1794), the U.S. Congresses (4 March 1803 - 3 March 1807), the Virginia House of Delegates (1819, 1820, and 1823-1825), and as Governor of Virginia (1819-1822). During the War of 1812, he was a Colonel of the Twentieth Infantry.

44 Ibid., 55.

45 Stephen was Thomas Mann Randolph's hired man. By early 1815, he was living in the vicinity of North Milton, not far from Monticello, and, according to Thomas Jefferson, planned to go "down country" to live. See Ibid.

Chapter Twelve: The Second Mrs. Randolph Jefferson

1 William C. Hightower, III, "Jefferson Family Lineage," accessed April 2012, http://archiver.rootsweb.ancestry.com/th/read/JEFFERSON/1999-02/0917924860; Debra Reed, "The Descendants of Samuel Jefferson," accessed April 2012, http://www.tngenweb.org/hancock/DescendantsofSamuelJeffersonofWales.htm.

2 Buckingham County Personal Property Tax, 1800. In 1799, Randolph Jefferson paid tax for only one white male. The tax was high at $7.24; however, nothing unusual was noted. He paid for himself, fourteen slaves, and nine horses.

3 Hightower, "Jefferson Family Lineage; "Dist(s) of John Criddle the Elder vs. Admx of John Criddle the Elder, etc.," 1817-023, Cumberland County Chancery Causes, Vi, accessed April 2012, http://www.lva.virginia.gov/chancery. Named in the case are John Jefferson, Jr. and Sarah, his wife (formerly Sarah Criddle) and Nancy Ballowe of Kentucky, widow of Thomas Ballowe.

4 Cumberland County Deed Book 9, p. 88; Cumberland County Deed Book 9, p. 282. Concurrently, John Jefferson, Jr. of Buckingham County deeded John J. Reynolds another piece of property in Cumberland; this was 120 acres on little Grooms's Quarter Creek. The deed was written on January 29, 1803 and was received at court on September 26, 1803. See Cumberland County Deed Book 9, p. 260.

5 Buckingham County Personal Property Tax, 1802.

6 Buckingham County Personal Property Tax, 1803. John Jefferson, Jr. also may have sought a partnership with Hill & Rea, merchants in New Canton. In 1801, a John Jefferson of Cumberland sued Hill & Rea who were establishing a general store at New Canton, Buckingham County, which sits on the James River near the Cumberland County border. They consented to furnish John Jefferson with "such goods and articles of merchandize he might have occasion for . . . on condition that they would receive good bonds or accounts of their customers in payment of the debt which he might incur in consequence of such dealings." See Agnes E. Gish, "Project Report, Phase II: Searching for the Jefferson who Sired Campbell Jefferson," 2003.

Dr. Gish cites the "Bill of Injunction vs. Hill & Rea of New Canton," 24 March 1808, Cumberland County Chancery Causes, Vi.

7   Stinton, *Early Buckingham County, Virginia Legal Papers, Volume I, 1765-1806*, p. 80. In 2010, one historic currency converter states that "32 pounds in 1800 had the same buying power as $2501.51 current dollars." This was not a petty argument.

8   Virginia Chancery Case Digital Collection, Vi, accessed April 2012, http://www.lva.virginia.gov/chancery. John Jefferson, Jr. died c. March 1818. His will was written on March 1, 1818 and recorded in Cumberland County's April Court. In it, he named eight children, including Elizabeth and Sarah Suggett Jefferson. He appointed his widow, Sarah, as his Administratix. Sarah long outlived him, dying in Haywood County, Tennessee, on August 29, 1835. See Hightower, III, "Jefferson Family Lineage."

9   In 1804 and 1805, Zach Nevil paid personal property tax in Amherst County. In 1805, he maintained seven slaves and four horses on his Rockfish River plantation. Between 1804 and 1805, when Snowden's slaves increase by at least eight individuals, Zachariah Nevil was living in Buckingham. See Buckingham County Personal Property Tax, 1804, 1805.

10 Journals and ledgers of a Stony Point, Va. store, Journal and Ledger for 1808-1809, Small Special Collections, ViU.

11 Meaders, *Advertisements for Runaway Slave in Virginia*, 180. Gary is a very unusual slave name. It may come from the surname Gerry, which drifted to Gary. Robert Lewis Jefferson named his only son Elbridge Gerry Jefferson, after statesman and diplomat, Elbridge Gerry. He had a son, Elbridge Gerry Jefferson, Jr., called Gerry. In some records, it is written Gary.

12 Thomas Jefferson did not enter either marriage of his brother, Randolph, in the Jefferson Family Bible, which was in his possession in both 1781 and 1809.

13 Mitchie's name was likely pronounced "Mickey," as is the Virginia surname, Michie. Phonetic spellings of her name include McKey, MacKey, Mikey, and Mickey.

14 Abercrombie and Slatten, *Buckingham County, Virginia Publick Claims*, 21.

15 David Pryor's mother may have been a Patteson, a large family in northern Buckingham. His son, Zane, married Elizabeth Patteson and his granddaughter, Susan B. Pryor, returned to Buckingham from Hanover County to marry Charles Patteson of Buckingham. Nicholas B. Pryor's wife, Sarah M. Thomas, may also be a Patteson.

16 Whitley, *Genealogical Records of Buckingham*, 24.

17 Land Office Grants Digital Collection, Vi; Grundset, *Buckingham County Surveyor's Plat Book, 1762-1858*, 14, 16. For land descriptions also Buckingham County Land Tax, 1813.

18 Land Office Grants Digital Collection, Vi. This land is very likely on Rock Island Creek, sixteen to eighteen miles north of the courthouse, placing it close to Snowden. In 1813-1814, a James Stanton had his residence on Rock Island Creek, seventeen miles north of the courthouse. Members of the Couch family were also located on the creek about sixteen miles north of the courthouse. It is possible that one of the Pryor boys developed this land, located considerably closer to Snowden than Woodlawn, cementing a relationship between the families and, perhaps, introducing Mitchie Pryor to Randolph Jefferson.

19 The 1813 Buckingham County Land Tax shows 230 acres on Ripley's Creek in the David Pryor Estate, the residence of his widow. It is currently unclear who ultimately owned the Pryor dwelling house. After 1819, the farm was divided among the legatees. Over the years, two portions were sold to John F. Lightfoot of Nelson County; one to Charles Patteson; and 50 acres were sold to Watson B. Cobbs. In 1852, Cobbs transferred some land to Zane Pryor's widow, Elizabeth Pryor. Neighbor Rev. James H. Fitzgerald also ended up with some of the farm. Especially with no extant deeds to follow, it is extremely difficult to trace actual acres. Of course, as with so many homes in Buckingham from this period, the dwelling house may have burned, even shortly after the Pryor family dispersed in 1818-1819. See Ward, *Land Tax Summaries & Implied Deeds*, volumes 2 and 3.

20 Ward, *Land Tax Summaries & Implied Deeds*, 1:41, 249. In 1811, Boyd sold 300 acres to Charles Patteson, which is probably the same 300 acres.

21 Whitley, *Genealogical Records of Buckingham*, 106 -107; "The Pryor Family," *VMHB* (January 1900), 325-326.

22 Banister Pryor is also found in records as Baynton or Boynton Pryor.

23 Whitley, *Genealogical Records of Buckingham*, 37; Buckingham County Personal Property Tax 1804, 1806, and 1810.

24 Federal Population Census, Henrico County, Virginia, 1860.

25 Whitley, *Genealogical Records of Buckingham*, 106-107.

26 Lt. Langston Pryor served in the war from February 25, 1814 to March 26, 1814, attached to Capt. John Hays' company from March 26, 1814 to June 10, 1814. See Butler, *A Guide to Virginia Militia Units in the War of 1812*, p. 60; Carl Coleman Rosen, Sr., *200 Years of Freemasonry in Buckingham County, Virginia* (Westminster, MD, Family Line Productions, 1991), 6-8, 10.

27 "Bellews in Bibles," accessed December 2011, http://www.myplanet.net/gedmnds1/ballewsinbibles.htm. Thomas Ballow and Chloe Battersby were married in Richmond, Virginia, on March 17, 1757. Several Ballow/Bellew family histories state, but fail to prove, that Susannah Ballow was the daughter of Thomas Ballow and Chloe Battersby, daughter of William Battersby. This speculation fits nicely with the details of David Pryor's family in Buckingham County, placing her birth

in the early 1760s. However, "Susan" Pryor's Nashville, Tennessee obituary claims she was eighty-nine years old when she died in 1832. If her age is correct, her birth would be about 1743, putting her childbearing years between 1750 and 1775, which does not fit well with the estimated birthdates of the Pryor children. See Whitley, *Genealogical Records of Buckingham* 106-107; "The Pryor Family," *VMHB* (January 1900), 325-326; *National Banner & Nashville Daily Advertiser*, 25 April 1832. Additionally, the 1790 marriage of Micah Ballow of Buckingham County and Jesse Wood Thomas of Cumberland County reveals a connection between the Ballow family and John Jefferson, Sr. and John Jefferson, Jr. See "Thomas Bellew," accessed December 2011, http://genealogy.bellew-campbell.org/getperson.php?personID=I05960&tree=bellew-campbell.

28 In 1810, Nicholas B., "Zachius," John, and Langston Pryor are listed independently on the Buckingham County Personal Property Tax. That year, Mrs. Pryor paid taxes for three slaves above sixteen years old and for two horses.

29 The 1810 census lists the Jeffersons and Pryors spread around Buckingham County in numerous separate "households." They may or may not have been as widely dispersed as it appears, and it is virtually impossible to know where the various groups are actually living in the late summer of 1810. It is unclear whether or not the Randolph Jeffersons were residing at Snowden when the census was taken. Three young males are enumerated in the household with Randolph and Mitchie, and they are all recorded very near Mitchie's brother, Zane Pryor. Zane is enumerated adjacent to Jesse E. and Anne Couch, which suggests he was living on the Pryor's Rock Island Creek property.

30 William Wirt to John C. Pryor, 16 February 1816, William Wirt Papers, 1784-1864, MS.1011, MHi.

31 *UB,* 27, 30, 33.

32 Ibid., 44-45.

33 Thomas Jefferson deposition, 15 September 1815, Carr-Cary Papers, #1231, Small Special Collections, ViU.

34 The 1810 Buckingham County census records Mitchie as being between sixteen and twenty-five years old.

35 Joseph C. Cabell papers, 1796-1887, Accession #38-111, Small Special Collections, ViU; *UB,* 5.

36 Martha Randolph to her daughter, Ellen (Mrs. Joseph Coolidge), 1 September 1825. Accession 9090, Small Special Collections, ViU.

37 The 1812 muster records for Zach B. and Langston S. Pryor are certain due to their distinctive names. There is also a John C. Pryor in the Virginia Militia during the

War of 1812; he may or may not be Mitchie's brother. "War of 1812 Service Records," accessed December 2011, http://search.ancestry.com/search/db.aspx?dbid=4281.

38 Gordon-Reed, *The Hemingses of Monticello*, 311; 427-436. Gordon-Reed makes a very interesting comparison between Thomas Mann Randolph's marriage to Gabriella Harvie and Thomas Jefferson's alleged relationship with Sally Hemings, pointing out that maintaining a second illegitimate family typically does not threaten the property of the legitimate heirs. This is especially true if the second family is mixed-race.

39 This Pryor brother has not been absolutely identified. None of Mitchie's brothers are known to have died as early as 1809.

40 *UB*, 23.

41 Ibid., 33.

42 Ibid., 34.

43 Ibid.

44 Ibid., 44-45.

45 *PTJ:RS*, 7:534; "The Two John Nicholases: Their Relationship to Washington and Jefferson, *American Historical Review* (January 1940), 338-353; V. Dennis Golladay, "Jefferson's "Malignant Neighbor," John Nicholas, Jr.," *VMHB* (July 1978), 306-319. This John Nicholas is the son of the John Nicholas who was executor to Peter Jefferson.

46 *MB*, 1,302. Reviewing the memos from May 1 - August 27, 1814, there is no indication as to how or when Anna Scott Marks traveled to Snowden from Monticello.

47 *UB*, 58.

48 Ibid., 59. Randolph's original letter mentioned Charlotte County as the destination in December of 1809, yet in 1815 he recalled going to Prince Edward. Both counties lie to the south of Buckingham. The ill brother, who did not die, may have been William S. Pryor, who paid personal property tax in Prince Edward in 1810.

49 Patteson is a common surname in northern Buckingham. This Mr. Patteson could be one of several neighbors of the Pryors. Having spent considerable time at Woodlawn, Randolph may have gotten to know numerous Pattesons there. He could be Capt. Patteson, who had the grocery in Warren, or Mr. R. Patteson, who brought news from Richmond of Randolph's beloved watch in 1812. See Ibid., 24, 32.

50 Ibid., 24.

51 *MB*, 1,311. A footnote to Thomas Jefferson's memorandum offers a brief biography for Israel (b. 1800), who was the son of Ned and Jenny. "Israel was a scullion and

waiter for the Monticello household, as well as a driver and postilion for TJ's landau and a carder in the cloth factory."

52 *PTJ:RS* 8:636.

53 MB, 1,282. According to the editors of *Jefferson's Memorandum Books*, "There were two Flood's ordinaries on TJ's route to Poplar Forest...." One belonged to Noah Flood in Buckingham County, the other to "Maj. Henry Flood, who kept a tavern just northwest of Old Appomattox Courthouse on present State Route 24." Ibid., 1,161. Thomas' memos reveal that he was at The Raleigh tavern, just west of Buckingham Courthouse, on November 22, 1813. While he did not visit Snowden, at times like these the brothers may or may not have met with each other. Ibid., 1,304.

54 Ibid., 1,288.

55 Thomas Jefferson, deposition, 1815.

56 Ibid.

57 UB, 35. This is the only mention of a smith's shop at Snowden discovered thus far. Other notations indicate that Randolph had some work done by Monticello's smith, so the shop may be very old.

58 Ibid., 35-43.

59 Ibid.

60 Ibid., 44.

61 Ibid., 51.

62 *MB*, 1,293.

63 Even though Thomas Jefferson did not return to Snowden during Randolph's lifetime, transactions continued between the brothers' plantations. "James (b. 1796), son of Ned and Jenny, made an unsuccessful trip to the James River at Snowden and Warren to procure living carp for Tufton fishpond." [Ibid., 1,289] On August 27, 1814, Thomas' servant, Wormley, received $1.00 for ferriage to Snowden. [Ibid., 1,322] Also see letters dated between December of 1814 and June of 1815 in *Unknown Brother*.

64 UB, 21.

65 Ibid., 28; *MB*, 1,272. Thomas Jefferson paid Wormley $1.00 expenses for going to get Mrs. Marks on December 27, 1812. Nancy's husband, Hastings Marks, died shortly before that date.

66 UB, 42, 46; *PTJ: RS*, 7:518-519.

67 *MB*, 1,295.

68 UB, 57.

Chapter Thirteen: "My Brother Died This Morning"

1 *MB*, 1,312.

2 Ibid., 1,311. Tecumseh was likely purchased to replace Bedford, the horse that died after an attempted trip to Snowden on Tuesday, March 28, 1815.

3 *PTJ:RS*, 8:257-258; Albemarle County Order Book, 1813-1815, pp. 460-61; Albemarle County Order Book, 1815-1816, p. 266. The debt also involved Daniel Stone. Craven Peyton won the judgment in December of 1815.

4 Ibid., 1,307.

5 Betts and Bear, *Family Letters*, 409; *PTJ:RS*, 8:390-391. Ellen and Cornelia Randolph are younger sisters of Jefferson Randolph.

6 *UB*, 57.

7 Young Wilson Nicholas was the son of Wilson Cary Nicholas, Sr. of Warren. Thomas "Jefferson" Randolph married his sister, Jane Hollins Nicholas, on March 6, 1815, making the men brothers-in-law. See "Jane Hollins Nicholas Randolph," *TJE*, accessed December 2011, http://www.monticello.org/site/research-and-collections/jane-hollins-nicholas-randolph.

8 *UB*, 57. Editors Mayo and Bear transcribe Scott's middle initial as "S." However, an examination of the original letter indicates to this author that it is an "A," and this agrees with later records of the conveyance of the land.

9 Thomas Jefferson, deposition, 1815.

10 Ibid.

11 The "extra" taxable male is probably James Lilburne Jefferson, who returned home in 1815 after serving in the Virginia Militia during the War of 1812. Between August 19, 1814 and February 22, 1815, he was enlisted in the Virginia Militia. It is currently unclear where James Lilburne Jefferson resided during 1812-1814.

12 Buckingham County Land Tax, 1802, 1810, 1815, 1816.

13 "Carr Family," *TJE*, accessed December 2011, http://www.monticello.org/site/Jefferson/carr-family. This is likely Samuel Carr's wife, Eleanor B. Carr.

14 *UB*, 59.

15 *PTJ:RS*, 8:636.

16 Thomas Jefferson, deposition, 1815. The letters were likely submitted as evidence in the "Jefferson vs. Jefferson" case and are believed destroyed in the Buckingham Courthouse fire.

17 *MB*, 1,312.

18 Thomas Jefferson deposition, 1815.

19 Ibid.

20 *MB*, 1,312.

21 Anna Scott (Jefferson) Marks outlived both of her brothers, dying on July 8, 1828. At the end of her life she was practically blind, living with her not always patient niece, Martha (Jefferson) Randolph, who referred to her in one letter as "the old lady." In 1825, Martha wrote to her daughter, Ellen W. (Randolph) Coolidge, that Aunt Marks' newest fancy "now is a palsy in all her limbs to prove which she walks incessantly, if she had said of the *brain* she would have come nearer to the truth." See Correspondence of Ellen Wayles Randolph Coolidge, 1819-1861, Accession #9090, 38-58, Small Special Collections, ViU.

22 Wilson Cary Nicholas to Thomas Jefferson, 6 August 1815, *PTJ:* RS, 8:648. News of Randolph Jefferson imminent demise quickly reached Warren and W.C. Nicholas, perhaps indicating a relationship between him and Randolph. Indeed, Randolph would die the following day.

23 Thomas Jefferson to Wilson Cary Nicholas, 9 August 1815, *PTJ: RS*, 8:655-656.

24 Thomas Jefferson to Hardin Perkins, 11 August 1815, *PTJ: RS*, 8:661-662. Thomas Jefferson was seventy-two years old when he wrote to Perkins, addressing him as "Harding" Perkins, which is how his name was written in Randolph Jefferson's 1808 will. A day's ride from Snowden and even further from Buckingham Courthouse, distance would have made Thomas' job as executor difficult as well. Why a trained attorney would feel unqualified to serve as an executor is a mystery. Robert Craig, also named as an executor in the 1808 will, predeceased Randolph Jefferson.

25 *MB*, 1,312. Jefferson's slave, Gill (b. 1792), was the son of Ned and Jenny. He was a driver, postillion, and house servant at Monticello.

26 Betts and Bear, *Family Letters*, 409-410. Martha's reference to the "sick mending" concerns the epidemic of dysentery on the plantation.

27 Ibid., 410.

28 *PTJ:RS*, 8:676-677. The letter, written by one of Randolph's sons and received by Thomas Jefferson at Montpellier, has not been located.

29 Thomas Jefferson to Martha Jefferson Randolph, 31 August 1815, *PTJ:RS*, 8:697.

30 *MB*, 1,313.

31 Ibid.

32 Between 1802 and 1831, Chancery Cases were heard in Virginia's Superior Court of Law and Chancery. This Court met twice annually, in October and May. A formal bill of complaint was entered for "Jefferson vs. Jefferson," and the papers

in the case would have included the complaint, the answer, summons, the exhibits (i.e.: the wills, the letters Thomas Jefferson produced, etc.), depositions, and, finally, the judgment. All, excepting the 1808 will and Jefferson's copy of his deposition, are presumed lost in the Buckingham Courthouse fire.

33 Thomas Jefferson, deposition, 1815.

34 Randolph Jefferson will, 27 May 1808, file draft. Second copy: R. J., Carr-Cary Papers, #1231, Small Special Collections, ViU. Perkins' land was adjacent to Snowden, as was some of Robert Craig's land. On January 27, 1801, Craig ordered a survey of 7 acres, "joining lines of Randolph Jefferson, William O. Murrey (sic), and Robert Craig." Craig died in 1812, his large estate open for many years. In 1821, his 2,300-acre residence adjacent to Snowden sold to James Wiley Winfrey. This plantation was later purchased by Randolph's son, Field Jefferson, in 1854 and then by James Harris in 1858. See Ward, *Land Tax Summarized & Implied Deeds*, 3:149.

35 *West's Encyclopedia of American Law* defines hotchpot as, "The process of combining and assimilating property belonging to different individuals so that the property can be equally divided; the taking into consideration of funds or property that have already been given to children when dividing up the property of a decedent so that the respective shares of the children can be equalized."

36 It is unknown how Thomas Jefferson felt in general about the Pryor family, although he likely did not know them well. He was always cordial to Mitchie in his letters. Additionally, two letters from Mitchie's brothers survive in Thomas's papers. One from Nicholas B. Pryor, dated August 7, 1812, was written from Nashville, Tennessee, concerning his desire for a commission in the army. Thomas politely handled this request and responded to Nicholas. See *PTJ:RS*, 5:300-301, 327-328. Another letter came from John C. Pryor, dated December 10, 1813, asking Thomas' advice on how to obtain the office of tax collector in Prince Edward County. See Thomas Jefferson Papers, DLC; *PTJ:RS*, 7:45-46.

37 In 1816, Capt. John Harris of Albemarle acquired 178 7/8 acres from Charles A. Scott, which had been conveyed from R. Jefferson. See Ward, *Land Tax Summaries & Implied Deeds*, 2:185.

38 Thomas Jefferson, deposition, 1815.

39 Merrill, *Jefferson's Nephews*, 313-316.

40 Cabell to Cocke, 13 September 1815, Joseph C. Cabell Papers, 1796-1887, Accession #38-111, Small Special Collections, ViU. Cabell is undoubtedly writing about the September Court which Thomas Jefferson attended, not the August hearing. The dates for the September Court appear to run from Monday, September 11th, through at least Friday, September 15th. Jefferson's memo books indicate he is

in Buckingham from September 11-13; Cabell's letter is dated September 13th; Jefferson's deposition is dated September 15th. Presumably, this is the date it was entered, not the day it was presented.

41 The Oxford English Dictionary defines jade as "A term of reprobation applied to a woman. Also used playfully, like hussy or minx." In the novel, *Uncle Tom's Cabin*, Simon Legree calls the mulatto woman, Cassie, a "Jade." She controls Legree with both sex and cunning.

42 William Wirt had a relationship with the larger Cabell family. He, his first wife, Mildred Gilmer, and other members of the Gilmer family, including Francis Walker Gilmer, were guests at Judge Cabell's Buckingham County summer home, Montevideo. See John Pendleton Kennedy, *Memoirs of the Life of William Wirt: Attorney General of the United States* (Philadelphia, PA: Lea and Blachard, 1849), 1:319; The Cabell Family Papers, Small Special Collections, ViU. Francis Walker Gilmer spent the summer of 1813 at Montevideo, writing several letters to Wirt from Buckingham. See Francis Walker Gilmer Letters, 1812-1825, Small Special Collections, ViU.

43 "William Wirt." NNDB, accessed February 2011, http://www.nndb.com/people/112/000050959/.

44 "A Guide to the Microfilm Edition of the William Wirt Papers, 1784-1864," MHi.

45 Bruce Chadwick, *I am Murdered: George Wythe, Thomas Jefferson, and the Killing That Shocked a New Nation* (Hoboken, NJ: John Wylie & Sons, 2009), 132-147.

46 Interestingly, Randolph Jefferson's son, Isham Randolph Jefferson, named one of his sons Wirt Jefferson, born about 1841 in Kentucky. Was he named for William Wirt (d. 1834), despite his support of Mitchie B. Jefferson's case?

47 William Wirt to John C. Pryor, 16 February 1816.

48 Ibid.

49 At this point no more is known about the overseer named Goss or how long he was employed at Snowden. No Goss has been found in Buckingham on the 1810 census, 1815 personal property tax, or 1820 census, though a Goss family lived in Buckingham in the 18th century.

50 *MB*, 1,318.

51 Jeffersoniana, Bixby Collection, Missouri Historical Society.

52 James Lilburne Jefferson to Thomas Jefferson, 18 February 1816, Carr-Cary Acc. #1231, Small Special Collections, ViU. The man, "Guilley," may be Gill, who took Mrs. Marks to Snowden.

53 On March 25, 1816, Lilburne Jefferson charged five pairs of stockings, totaling £2.5, and fabrics totaling $25.66.

54 In 1818, James Lilburne Jefferson is found on both the Albemarle and the Fluvanna County personal property tax records. No earlier entry has been located for him. If he turned twenty-one years old in 1818, his birth was in 1797. A private in the Militia from at least August 1814 - February 1815, he would have been about seventeen years old when he enlisted. During the November Court of 1817, Lilburne was recommended to the office of Ensign in the second Battalion of the Fluvanna County Militia. See Fluvanna Count Order Book (1815-1819), 356. In the July Court of 1819, Lilburne was again recommended to the office of Ensign "in the room of William Woolling who was promoted." See Fluvanna Court Order Book (1815-1819), 599.

55 An Executor/Executrix is named in a will. An Administrator is appointed by the Court to handle the same duties as an Executor, particularly an estate's accounts – satisfying debts, dividing assets, etc. A Curator attends to all aspects of the estate while it remains open.

56 In 1813, the Pryor heirs sold their farm on Rock Island Creek to David Patteson. See Ward, *Land Tax Summaries & Implied Deeds*, 1:49. Situated much closer to Snowden, that farm was sixteen miles northeast of Buckingham Courthouse. An announcement for a Buckingham Chancery suit brought by Zane Pryor against the other heirs of his father, David Pryor, names "John Randolph Jefferson" as a legatee. To date, this is the only instance found where John R. Jefferson's full name is written. See *Richmond Enquirer*, 15 December 1832, p. 2.

57 James Lilburne Jefferson to Thomas Jefferson, 19 April [1816], Carr-Cary Acc. #1231, Small Special Collections, ViU. Thomas Jefferson continued to offer aid to his nephew. On June 13, 1816, he gave James Lilburne $10.00. See *MB*, 1,322. In 1817, he purchased a pair of shoes for the young man. See Ibid., 1,330.

58 Court receipt, 12 March 1818, frame 434, Microfilm reel 4615, Buckingham County Circuit Court Records, 1805-1869, Robert Alonzo Brock Collection, CSmH; Court receipt, 11 August 1818, frame 369, Ibid.

59 Court receipt, 8 February 1819, frame 313, Ibid.

60 In 1818, Susannah Pryor was taxed on 50 acres on Ripley's Creek and on another 166 acres. The 1819 land taxes indicate the remainder of the Estate was divided among Pryor's other heirs.

61 Wilena (Roberts) Bejach and Lillian Johnson Gardiner, *Williamson County, Tennessee Marriage Records, 1800-1850* (Memphis, TN: Privately published, 1957), 143; Mitchie and Josiah Johnson were married by Rev. Nicholas Scales. The bondsman was W.B. Hyde, who may have been related to Johnson's second wife.

62 Ibid. The transcription gives the marriage date as September 31, 1825. Since there is no September 31st, it may have been the 30th or a transposed 13th. The bondsman was H.B. Hyde.

63 Josiah Johnson Estate Inventory, Williamson County Will Book 12, p. 611.

64 Josiah Johnson Will, Williamson County Will Book 12, pp. 583-584.

65 "The Pryor Family," *VMHB* (January 1900), 326.

66 Mary Sue Smith, *Davidson County, Tennessee Deed Book T and W, 1899-1835* (Bowie, MD: Heritage Books, 1994), 21.

67 Martha (Jefferson) Randolph to Ellen W. (Randolph) Coolidge, 1 September 1825, MSS 9090, Small Special Collections, ViU. Martha includes her comments about Mitchie Johnson and Nancy Nevil in a "newsy" close to her letter, indicating the deaths were recent. It might have taken the news a bit longer to travel from Tennessee than from Nelson County.

68 Albemarle County Law Orders Book, 1809-1821; Albemarle County Law Orders Book 1821-1831.

69 Beginning in 1790, Robert Rives transacted increasing business with James Brown of Richmond; both were Virginia agents for the house of Donald & Burton, London. They became partners, styled as Brown & Rives, creating one of the foremost commission houses in Virginia. See Brown, *The Cabells and Their Kin*, 1895 edition, 221. Robert Rives married Margaret Cabell, daughter of Col. William Cabell; he died on March 9, 1845. See Ibid., 224-225.

70 Nelson County Will Book C, 315, 362, 424; Nelson County Will Book D, 414.

71 On the 1820 Albemarle County population census, Jefferson, Scott, and Moon are enumerated on consecutive pages.

72 Albemarle County Deed Book 26, pp. 178-179; Albemarle County Deed Book 25, pp. 148-150. Lower Plantation was also known as Belle Grove and as Valmont.

73 *The Scottsville Register*, 20 April 1861, Small Special Collections, ViU.

74 John Scott never knew his great-uncle, Daniel Scott, who had married Anna Randolph. She was the great-aunt of the Jefferson boys. Anna did not remain in the area after Daniel's death in 1754; she returned to her Fine Creek roots and eventually remarried.

75 Moore, *Scottsville on the James*, 51; Nicholas, *Scottsville*, 28; Vogt and Kethley, *Albemarle Marriages*, 1:278.

76 Nicholas Philip Trist Papers, DLC.

77 Nicholas, *Scottsville*, 28. Richard Nicholas shared the following thoughts concerning John Scott's quick exit from Scottsville: "I have always wondered why John Scott III vacated Virginia in such a hurry, so soon after coming of age, and inheriting so much from his grandfather. I was talking to a surveyor the other day and he confirmed what is obvious: that many of the lots laid out in the original plan of Scottsville in 1818 were "un-buildable" because of the steep topography. In fact, those lots have

never been built on to this day, and some of the streets never opened. Many of the people who bought those lots from Scott in the initial 1818 sale probably never saw the land before they bought. When they later learned that the lots were worthless, they probably blamed Scott. All things considered, Scott may have decided that it was a good time to take his money and run to Alabama!" Scott died in Alabama in 1829. See: Nicholas to Joanne Yeck, email, 29 January 2010.

78 Nicholas, *Scottsville*, 82. In 1850, Susan Scott was living in Charlottesville with her daughters, Pocahontas and Mary.

79 Nicholas, Scottsville, 72, 118.

80 Ibid., 72, 116-118, 122-123; Albemarle County Deed Book 21, p. 340; Albemarle County Deed Book 23, pp. 437-438. On December 9, 1818, James L. Jefferson purchased Lots 43 & 42. In 1830, he sold Lot 42 to George A. Scruggs. See Albemarle County Deed Book 28, p. 464. According to Richard Nicholas, "Lots 42, 43, 44 were on Jackson Street and were 'buildable.' But those lots east of there and on Holman Street were not, along with most of the Fluvanna lots." Prior to 1821, Randolph Turner sold Lot 3 to James L. Jefferson; however, the lot remained on the Albemarle tax rolls under Turner's name through 1833. See Fluvanna County Deed Book 7, p. 679. When James Lilburne Jefferson began investing in Scottsville real estate, he was residing at Scott's Ferry where he was postmaster from August of 1817 until 1819. See Edith F. Axelton, comp., *Virginia Postmasters and Post Offices, 1789-1832* (Athens, GA: Iberian Publishing Co., 1991), 6.

81 *Richmond Enquirer*, 15 October 1822, p. 4. According to www.nolo.com, "The administrator steps in either when a will fails to nominate an executor or the named executor is unable to serve."

82 In 1836, a revised assessment of Snowden indicated, "The improvements on this place are hardly worth mentioning consisting mostly of an old one story dwelling house with two rooms." It seems likely this is where Peter F. Jefferson lived; he was married with one child. Later it could have accommodated his brothers and his nephew.

83 During this transition, Robert Lewis Jefferson was in the process of purchasing his own farm on Buckingham County's Sharps Creek. See Ward, *Land Tax Summaries and Implied Deeds*, 2:185.

84 Ward, *Land Tax Summaries and Implied Deeds*, 2:158. In 1817, Harris purchased another 179 ½ acres from Charles A. Scott. In this case, the owner prior to Scott is unclear, but it is likely the property later known as Murray's. This land was adjacent to Snowden and once belonged to Richard Murray, then to Anthony Murray. See Buckingham County Land Tax, 1818.

85 Buckingham County Land Tax, 1826.

86 *Richmond Enquirer*, 3 February 1829, p. 4. Jesse Joplin of Nelson County was an Executor for Zachariah Nevil's estate.

Chapter Fourteen: *The Jefferson Brothers: A Study in Contrasts*

1 Thomas Jefferson, deposition, 1815. This summary may be slightly exaggerated, for Thomas' description of his brother had a very specific goal. Without completely misrepresenting Randolph, or perjuring himself, Thomas hoped to convince the Buckingham Court that his brother may have been unduly influenced by his young wife and her family in the re-writing of his last will. For a lengthy profile of Randolph Jefferson, see Yeck, "A Most Valuable Citizen," 1-37.

2 *MB*, 1,017.

3 *UB*, 27-30.

4 Woods, *Albemarle County*, 237-238.

5 Thomas Jefferson, deposition, 1815.

6 At the time of Thomas Jefferson's death, he was over $100,000 in debt, between $1,000,000 - 2,000,000 in today's currency. His family was faced with the sale of Monticello, other property, and Jefferson-owned slaves, which meant (in many cases) separating families who had served the Jeffersons for generations. For a thumbnail sketch of some of the reasons behind Thomas Jefferson's financial difficulties, see "Debt," *TJE*, accessed December 2011, http://www.monticello.org/site/research-and-collections/debt.

7 Randolph, *Domestic Life*, 39. In *The Domestic Life of Thomas Jefferson*, Sarah Nicholas Randolph (1839-1892) skillfully combined letters and commentary to present the Jefferson family.

8 Nathan Schachner, *Thomas Jefferson: A Biography* (New York, NY: Appleton-Century-Crofts, Inc., 1951), 88.

9 The bursar's records at the College of William and Mary are unclear as to whether Randolph Jefferson was enrolled in the Grammar School, the College, or both. In any case, it is clear that he was tutored for ten months by Rev. Mr. Gwatkins in mathematics and natural philosophy.

10 *UB*, 26.

11 Ibid., 27.

12 Ibid.

13 Thomas Jefferson, deposition, 1815.

14 Ibid.

15 Ward, *Land Tax Summaries & Implied Deeds*, 1:249.

16 Bear, *Jefferson at Monticello*, 22. It may be that Randolph Jefferson "hadn't much more sense than Isaac;" however, Isaac had enough sense to make the transition from slave to freeman in about 1822 and later to tell his memoirs and to have them published. Isaac was born into slavery and spent his childhood at Monticello, where he became a highly skilled tinsmith and blacksmith. How he gained his freedom is unknown. He left Monticello in the early 1820s. He recalled meeting and talking with Lafayette in Richmond in 1824. In 1847, he lived as a free man in Petersburg, Virginia and was still an active blacksmith at the age of seventy-two. See "Isaac Granger Jefferson," *TJE*, accessed December 2011, http://www.monticello.org/site/plantation-and-slavery/isaac-granger-jefferson.

17 Joseph J. Ellis, *American Sphinx: The Character of Thomas Jefferson* (New York, NY: Vintage Press, 1998), 171-180; Stanton, "Free Some Day," in *"Those Who Labor for My Happiness,"* 137; Lucia Stanton, "Monticello to Main Street: The Hemings Family and Charlottesville," in *"Those Who Labor for My Happiness: Slavery at Thomas Jefferson's Monticello* (Charlottesville, VA: University of Virginia Press, 2012), 215-231; *UB*, 17-18. Thomas Jefferson was declined to sell Mary Hemings' two older children, twelve-year-old Joe and nine-year-old Betsy, who were not Bell's children. See Stanton, "Free Some Day," in *"Those Who Labor for My Happiness,"* 189.

18 *UB*, 57.

19 Pearl N. Graham to Dr. Julian P. Boyd, 11 January 1958, p. 5, Hemings Family Genealogy, 1961-1966, Accession # 6636, 6636-a, 6636-b, 6636-c, Small Special Collections, ViU; Pearl N. Graham, "Thomas Jefferson and Sally Hemings;" *Journal of Negro History* 44 (1961), 89-103. Graham collected oral history concerning the Hemings family. Her conclusion that she had found the descendants of Harriet Hemings has since been challenged.

20 "Thomas Jefferson and Sally Hemings: A Brief Account," *TJE*, accessed December 2011, http://www.monticello.org/site/plantation-and-slavery/thomas-jefferson-and-sally-hemings-brief-account; Report of the Research Committee on Thomas Jefferson and Sally Hemings," accessed December 2011, http://www.monticello.org/site/plantation-and-slavery/report-research-committee-thomas-jefferson-and-sally-hemings; Annette Gordon-Reed, *The Hemingses of Monticello: An American Family*. For contrasting opinions, see Burton, *Jefferson Vindicated*; William G. Hyland, Jr., *In Defense of Thomas Jefferson: The Sally Hemings Sex Scandal* (New York, NY: St. Martin's Press, 2009); Karyn Traut, "Which Jefferson?," accessed December 2011, http://www.perihelionproductions.org/KarynTraut.htm; David N. Mayer, "The Thomas Jefferson - Sally Hemings Myth and the Politicization of American History," accessed December 2011; http://www.ashbrook.org/articles/mayer-hemings.html; Robert F. Turner, editor, *The Jefferson-Hemings Controversy* (Durham, NC: Carolina Academic Press, 2011).

21 Burton, *Jefferson Vindicated*, 52-60; Hyland, *In Defense of Thomas Jefferson*, 29-36.

22 There is no existing description of the route between Snowden and Monticello; however, the description of the effortful three-day route from Monticello to Poplar Forest is well documented in Thomas Jefferson's memoranda. See "The Route to Poplar Forest," *TJE*, http://www.monticello.org/site/research-and-collections/route-to-poplar-forest, accessed December 2011; S. Allen Chambers, Jr., *Poplar Forest & Thomas Jefferson* (Forest, VA: The Corporation for Jefferson's Poplar Forest, 1993), Chapter XVIII.

23 Randolph Jefferson's legacy as expressed through the success of his children has been overlooked as a significant part of his lasting contribution. In general, these paragraphs concerning the Jefferson children summarize a wide variety of primary sources, including Chancery Cases, death and marriage records, census material, personal property taxes, the Nevil family estate records, etc.

24 *Lynchburg Virginian*, 5 February 1849, p. 3.

INDEX

Page numbers in *italics* indicate illustrations. Page numbers with an italic *b* appended indicate a box feature. Abbreviations: RJ (Randolph Jefferson), TJ (Thomas Jefferson)

A
"Act of Exempting Dissenters" (1776), 137
Adams, Abigail, *134, 264*
agriculture. *See also* tobacco; wheat
  beef stock, appraisal of, 105–106, *121*
  crop rotation, 180–181
  crops, selling, timing 180
  Monticello, drought at, 179–180
  orchards, 182–184
  seed exchange between RJ and TJ, 180–181, 184–185
  at Snowden, 179–181
Albemarle County
  "Albemarle Declaration of Independence," 94*b*
  establishment of, 7–9
    challenges of living during, 9
  Old Courthouse
    military stores, 115–117, 119
    near Scott dwelling house, *8*
  seat location, 24
Alberti, Francis, 95, 343
alcohol, popularity of, 183
Allen, Samuel, 122
American Continental Congress, Association entered into by, 95
American Gardener (Gardiner and Hepburn), *273*
American Revolution, 93–100
  3rd Continental Light Dragoons uniform (Lefferts), *118*
  "Albemarle Declaration of Independence," 94*b*
  "Association entered into by the American Continental Congress," broadside of, *93*, 95
  Battle of Guilford Court House, 107
  beef stock, appraisal of, 105–106, 121
  certificates, 100–101, 105–107, 111, 117, 119–121, *122*
  contributions to, 100–101, 105–107, 117, 119
  Court Booklets, Revolutionary War, 100–101
  dissenters of Anglican Church, 133, 135, 137
  military stores
    Albemarle Old Courthouse, 115–117
    New London, 117, 119
    prisoners, Charlottesville, VA, *102–103*, 104–105, 107
  Public Service Claims, Revolutionary War, 100–101, 120–121
  at Snowden, 111, 115
  uniforms, 96–98, *97, 118*
Anderson, David, 155, 156
Anderson, Thomas, 123*b*
Anglican Church, dissenters of, 133, 135, 137
Armistead, William, 96–97
auction notice, land, slaves, and stock, *176*

B
Ballow, Susannah, later Pryor. *See* Pryor, Susannah, née Ballow
Ballow, Thomas, *14*, 119, 140, 389, 393–394n42
  dwelling house, value of, 141
  family lineage of, 267, 421n27
  selling farm, 183
Barracks, Charlottesville, VA, prisoners of war, *102–103*, 104–105, 107
bateaux, description of, 192
Bates, Isaac, 49–50
Beck, William, 58, 76
beer, brewing, 183–184
Bellew, Thomas, 8, 362n20
Ben Snowden or Snowden Ben (slave), 152, 155, 217–222, 394n50, 412n23
Beverley Town (a.k.a. Westham), 108, 366n52
the Black Doctor, identity of, 222–225
Bolling, Mary, née Jefferson, 30, 31*b*
Bolling, Susan Bathurst, later Scott, 329
Bolling, Thomas, diary, 145
Bolling, William
  Bolling Hall, 146, *147*
  portrait (Hubbard), *145*

social functions, diary on, 145–146, 148
Bondurant, Benjamin
  background on, 400n41
  "Bondurant v. Jefferson," 163*b*, 164–165, 167–171, 399n35
  residence, 400n42
Bondurant, Darby, 401n55
Bracken, John, 75–76
Branch, Mary, née Jefferson, 361n8
British soldiers, as prisoners of war, *102–103*, 104–105, 107
Brown, Betty (slave), 369n12
Brown, Rives & Co., 156
Brown, William, 197, 407n73
Bryan, Anderson, 95
Bryan, Martha, 375n20
Bryant, John, 403n11
Buck Island Creek, naming of, 38
Buck Island
  dwelling house, 39, *41*, 127
  Jefferson children, born at, 200
  RJ's education at, 41-45
Buckingham County
  churches, 136
  Court Days, 148
  courthouse, 148–150
  daily life in, 143–144, 148
  effects from establishing, 24
  politics in, 123*b*
  Tillotson Parish, 136, 393n30
Buckingham Militia, 104, 269–270
Burgess, William E., *139*

C
Cabell, Joseph, 123*b*, 310–311, 389n96
Cabell, William, 8, 117
Calvinistic Reform congregation, formation of, 136–137
Carr, Dabney, 128, 137, 313–314
Carr, Martha, née Jefferson. *See* Jefferson, Martha, later Carr
Carr, Peter, 128, 348
Carr, Samuel, 348, 369n12, 391n9, 425n13
Cary (slave), 152, 155, 216, 217–222, 347, 394n50, 411n18
Cary, Wilson Miles, 87
Chatsworth, 373n3
church buildings
  Buck and Doe Church, 136
  in Buckingham County, 136

Cove Presbyterian Church, 135
"Goodlings" Church, 393n31
Goodwins Church, 136, 393n31
Gooseberry Baptist Church, 138
"Jefferson's Church," Bristol Parish, 3
Maynards Church, 136
Old Bruton Church, Williamsburg, *75*
Churches
  Anglican Church, dissenters of, 133, 135, 137
  Baptist, 138, 204-205
  Presbyterian
    ministers preaching at homes, 135
    near Snowden, 135–136
Clay, Charles, 136–137
Clay, Green, 99, *100*
Cocke, Isabella, 375n20
Cocke, John Hartwell, 173, 311
Colbert, Brown (slave), 220–221, 411n18
"College Negroes," list of, 61–62, *63*
College of William and Mary. *See* William and Mary, College of
Continental Army, 95–96
Convention Army, encampment, *102–103*, 104–105, 107
Cornwallis, Charles, 111, 115–116, 117
Court Days, 148
Cove Presbyterian Church, 135
Criddle, Sarah "Salley" Smith, later Jefferson, 260
crops. *See* agriculture, *and specific crops*

D
David (Billy Logan; runaway slave), 210–211*b*
Dawson, Martin, 49, 138, 205, 207
Dawson's Meeting House, 138
"Diary of Arnold's Invasion and Notes on Subsequent Events in 1781," (TJ) 114–115
Diggs, Maria, accusations of neglect, 375n21, 376nn22–23
Dinah (slave), 235, *236*, 237, 238–240
dogs
  French sheepdogs belonging to TJ, 245-246, 248–249, 250–252, 417n13
  housedogs, 417n15
  hunting, 246
  sheep-killing, 246–247
  "Shepard's Dog" (etching), *248*

slave ownership, limits on, 246–247
Virginian's views on, 246
Donald, Scott & Co., 142, 219, 410n5
Dungeness, 7, 17, 28
Lord Dunmore. See: John Murray
Dwit, Phoebe, 61, 376n22

E
earthquake, 1774, Virginia, 86
Edgehill, *44,* 418n33
education. *See also* Sneed's English School;
William and Mary, College of
   costs of, 76–77, *77*
   Grammar School, William and Mary, 71-72
   plantation schools, 44–45, 84
   of RJ (*See under* Jefferson, Randolph)
   of RJ's children, 206
   Seven Islands school, 84–85
   of TJ, 55–56
"1816 Version of the Diary and Notes of 1781," (TJ), 115
Eppes, Elizabeth, née Wayles, 95, 129
Eppes, Francis, 95
Eppes, Francis Wayles, 392n21
Eppes, John Wayles, 146, 207, 249, 403n17
Eppes, Maria "Polly," née Jefferson, 129, 146, 199, 392n21, 408n1

F
Fan (slave), 49–50
Fanny (slave), 228–231
farming. *See* agriculture
fashions, women's, *134, 264, 283*
ferries
   business of, 13
   depiction of James River crossing, *16*
   ferriages, cost of, 361n08
   Jefferson family's business connection with, 366n52
   at Snowden, 193*b*
   Scott's Ferry, 17–18
   Scott's permission to operate, 364n35
   at Warren, 196
Ferry Road (Old Courthouse Road), 101, 148–149
Fine Creek, 3-4, 6, 361n7
fires
   Buckingham courthouse, 261, 289, 425n16, 427n32

Shadwell dwelling house, 26, 45, 47
Snowden dwelling house, 320, 322
Flat Hat Club (F.H.C.), 76
Flood, Henry, 281, 282, 424n53
flood. *See* "Great Fresh of 1771"
flood, Rivanna River, 86, 87, 383n12
Fluvanna River, James as, 382n1
Foland, Peter V., 112, 388n91
Fontaine, William, 382n5
Fry, Joshua
   appointments, Goochland County, 8
   background on, 22–23
   Fry-Jefferson map, *22*
   Viewmont, 362n20

G
Galt, John Minson, 79–80, 381n70
Garret, Margaret, 60, 375n21
Gary (slave), 210–211*b*, 262, 420n11
Gentleman Justices, of Albemarle County, 8
George family (slaves), poisoning of, 222–225
George (slave), blacksmith, 91–92
George (slave), murder of, 133
Goochland County
   appointed officials, 6–7
   government of, 8
Goode, Bennet, 4
"Goodlings" Church, 393n31
Goodwins Church, 136, 393n31
Gooseberry Baptist Church, 138
Grammar School, William and Mary, 71
"Great Fresh of 1771," 404n29
Greene, Nathaniel, 106
Greensville, VA, town of, 149
Guilford Court House, Battle of, 107
Gwatkin, Thomas, 73–74
   discipline of students and, 70
   England, return to, 74–75, 379nn54–55
   Flat Hat Club, 76
   payments to, 73, 76–77
   as tutor to RJ, 57, 71–77, 81, 82, 378n50

H
Hall, Thomas, 178, 403n12
Hannah (slave), death of, 49–50
Hardware tract, 174
harness, TJ's, 275–277, 290
Harris, John
   Snowden, purchaser of, 329, 330, 431n84

438 ~ Index

Snowden Trust Deed, securing, 333–334
Harvie, Gabriella, later Randolph, 270, 423n38
Harvie, John, 56, 373n3
Hemings, Eston (slave), 348
Hemings, Sally (slave), 347, 348
Henderson, Bennett, 130
Henderson, Elizabeth, née Lewis, 130
Henderson, Ned (runaway slave), 210–211*b*
Henley, Samuel, 379nn53–55
Henry Flood's tavern, 281, 282, 424n53
Hern, Lilly (slave), 222, 412n23
Hessian soldiers, prisoners of war, *102–103*, 104–105, 107
hogsheads
 defined, 262
 of tobacco, 363n25
Hopkins, Arthur, 366n51
horses
 Bedford, 286, 425n2
 Bony, 91
 Diamond, 90
 Hambleton, 186, *186*
 Jefferson family, 90–91
 Jenny Morris, 90
 Oroonoko, 90
 racing, 91
 RJ's impounded, 169–170
 Ryno, 90
 stud services on plantations, *186*, 186–187
 Tecumseh, 285, 425n2
Horseshoe Bend, James River, 13, 15, *139*
Hubbard, Cuthbert, 58–60
Hubbard, Jame (slave), 220, 411n18
Hubbard, William James, *145*
Hudson, Christopher, 155–156, 397n15, 397n19
Hughes, Wormley (slave), 275, 319, 412n23
Hyde, Fereby, marriage to Josiah Johnson, 325

I
inflation, 1781, 387n77
Innis, James, 376n23
intermarriages
 Jefferson family deficiencies from, 337
 Lewis-Jefferson, 40, 126–127
 Lewis-Henderson, 156
 Lewis-Randolph, 132, 392n21
 Lewis-Woodson, 202–203, 408n4

Israel (slave), 277, 423n51

J
"Jade of genuine bottom," Mitchie B.
 Jefferson as, 269, 311, 428n41
James, Richard, 154–155, 397nn11–12
James (slave), identity of, 412n33
James (slave), searching for carp, 225–227
Jane (slave), *158*, 158–160
Jefferson, Anna Scott "Nancy," later Marks, 34–35
 birth of, 23
 Jefferson, Mitchie B., showing kindness to, 274–275
 naming of, 18
 siblings of, 31*b*
 Snowden, visits to, 282, 284, 286
Jefferson, Anna Scott "Nancy," later Nevil, 206–207, 303, 327
Jefferson, Anne "Nancy," née Lewis
 birth order of, 390n2
 death date, 203–204
 family background, 408n4
 fashions, current, *134*
 marriage to RJ, 125–127, *126*
 TJ, relationship with, 131
"Jefferson v. Bates," 49–50
Jefferson, Elbridge Gerry, 138
Jefferson, Elizabeth, 30, 31*b*, 32
 death of, 86–87
Jefferson, Field, 6
Jefferson, Isaac (slave)
 daguerreotype of, *345*
 freedom gained by, 433n16
 memoirs of, 345–346
 on RJ playing fiddle, 343*b*, 345
Jefferson, Isham Randolph, Jr.
 RJ's will, controversy over, 290–291
 Snowden, after leaving, 263
Jefferson, James "Lilburne"
 Fluvanna County Militia, 429n54
 Monticello, offer to study at, 207–208
 on postponement of "Jefferson v. Jefferson," 323
 on Snowden fire, 320
 in Virginia Militia, *209*, 425n11
Jefferson, Jane, née Randolph
 background on, 26
 children of, 30, 31*b*, 32–35
 entertaining at Shadwell, 21–22

estate of, 88
marriage to Peter Jefferson, 7, 28–29
personality of, 28
will, last, of, 87–88
Jefferson, Jane, Jr., 7, 31*b*, 32–34
Jefferson, John, suit against Richard James, 154–155, 397nn11–12
Jefferson, John "Garland," 168
Jefferson, John, Jr.
   RJ, work relationship with, 260–261
   will, last, of, 420n8
Jefferson, John Randolph
   after Mitchie B. Jefferson's death, 326
   birth of, 318, 322
Jefferson, Lucy, later Lewis, 31*b*, 34–35, 127
   death of, 128
   Kentucky, move to, 267
   TJ, relationship with, 130–131
Jefferson, Maria "Polly," later Eppes, 129, 146, 199, 392n21, 408n1
Jefferson, Martha, later Goode, 4
Jefferson, Martha, later Carr, 31*b*, 32
   death of, 338–339
   traditional silhouette of, *338*
Jefferson, Martha "Patsy," later Randolph
   to Ellen on deaths of Mitchie B. (Pryor) Johnston and Nancy Jefferson Nevil, 327, 430n67
   inheritance, change in Randolph family, 270
   on Jupiter's poisoning, letter, 222–223
   on TJ attending court for RJ's will, 297, 298
Jefferson, Martha, née Wayles, then Skelton, 31, 58, 140, 415n64
Jefferson, Mary, later Turpin, 4
Jefferson, Mary, later Bolling, 30, 31*b*
Jefferson, Mary, later Branch, 361n8
Jefferson, Mary "Polly" Randolph, née Lewis, 207
Jefferson, Mitchie B., née Pryor, 185
   death of, 325–326
   debts owed, 324
   fashions, current, *264, 283*
   ill brother, visit to, 423n48
   infidelity, accusations of, 315, 318
   inheritance from second will, 309
   marriage to Josiah Johnson, 325
   marriage to RJ, 263, 265
   move to Woodlawn, 320, 322, 323

   name pronunciation, 265, 420n13
   pregnancy at time of RJ's death, 308–309, 318
   RJ's financial skills, criticism of, 279, 306
   spending habits, 287, 306–307, 308
   TJ, gift from, 274
   TJ, letter on RJ's illness, 290
   unpopularity of, 269
Jefferson, Nancy (Anna Scott). *See* Jefferson, Anna Scott, later Marks
Jefferson, Nancy (Anna Scott). *See* Jefferson, Anna Scott, later Nevil
Jefferson, Nancy (Anne). *See* Jefferson, Anne, née Lewis
Jefferson, Peter
   appointments, Albemarle County, 8
   appointments, Goochland County, 5*b*, 6–7
   background on, 3
   children of, 30, 31*b*, 32–35
   Fluvanna lands, 13
   as frontier hero, 9–11, 363n26
   Fry-Jefferson map, *22*
   inventory at Shadwell, 21–22
   land, acquirement of, 4, 6, 7, 361n7, 362n24, 363n32
   library of, 20*b*
   marriage to Jane Randolph, 7, 28–29
   nicknames used in Jefferson family, 25
   sheriff, bond as, 362n14
   Spencer, Edward, meeting, 233–234
   survey of land, *12*
   as surveyor of Goochland County, 5*b*
   Tuckahoe, move to, 11–12, 17
   will, last, of, 23–24
Jefferson, Peter Field, 210–211*b*
Jefferson, Randolph (RJ). *See also* Snowden
   accounts, in full control of own, 92
   American Revolution
      contributions to, 100–101, 105–107, 111, 115, 117, 119
      Nelson's Corp of Light Horse, 95–99
      Public Service Claims, 120–121, *121*
   birth of, 23
   Bondurant case, 163*b*, *166*, 167–171
   at Buck Island, 40–42, 44
   in Buckingham Militia, 122
   children of, 200–203
      birth order, naming practices by, 200–203

education, 206
relationship with, 349–350
"undutiful and disrespectful conduct" of sons, 308
clothing costs for, 86
coming of age, 89–92
"Continental Association of 20 October 1774," signature on, *93*, 95
death of, 285, 289
death, reflections on, 339
debts owed, 152–153, 155
doctor, visits to, 79, 380n68
education of
  Buck Island, 42–44, 371n21, 371n23
  Gwatkin, Thomas, as tutor, 57, 71–77, 81, 82, 378n50
  Seven Islands school, 84–85
  Sneed's English School, 42–44
  William and Mary, 55, 57–58, 60, 71, 76-77, *77*, 81-82
estate, ("Jefferson v. Jefferson"), 298–299, 309, 310–311, 314–316, 323, 326, 330–331, 426n32
as fiddler, 343*b*
finances
  frugality, 172, 337
  loans between TJ and, 108–109, 151–152, 155, 172, 231–232
  transactions between TJ and, 92
Hemings, Eston, as possible father of, 348
humor, corn and squirrel joke, 337
intelligence level of, 336–339
inventory, taxed, 408n2
Jefferson, John, Jr., work relationship with, 260–261
kidney stones attack, 250, 339
land holdings, 173–174
legacy of, 349–350, 434n23
letter from nieces on Lucy Jefferson Lewis' death, 128
letters to TJ, 122, 390n190
  on carp, 226–227
  final, 289
  gig axle repair, needing, 190–191*b*
  gig harness, 275–277, 290
  invitation to Snowden, 271
  Marks, Anna Scott, safe arrival, 286–287
  silver watch, requesting, 254–256
Lewis, Anne, marriage to, 125–127, *126*

Lewises of Monteagle, relationship with, 129, 132–133
map, "Randolph Jefferson's Virginia," *54*
marriage
  Lewis, Anne, 125–127, *126*
  Pryor, Mitchie B., 265
mental clarity, 280–281
"Moon v. Jefferson," 157–161, 162*b*, 398n23–24
naming of, 18
Nevil, Zachariah, work relationship with, 261–263
personality described by TJ, 335–336
personality differences between TJ and, 338–339, 341
political orientation, 95–96, 123*b*
Pryor, Mitchie B.
  attempts to stop overspending by, 287
  marriage to, 265
  religious views, 135–138
Snowden, overseers, 175, 177–179
sheepdogs, requests for, 248–249, 251–252
siblings of, 31*b*
silver watch
  maintenance of, 255–257
  purchasing, 253–255
slaves
  attitudes toward, 347
  distribution of Jefferson family slaves, 48–49
  impoundment of, 158–160
  list of, 216–217
  Snowden, family members at
    after Mitchie B. Pryor's arrival, 263
    before Mitchie B. Pryor's arrival, 262–263
  spinning jenny, from TJ, 227–231
  summons to Prince Edward County District Court, 165, *166*
  surveys of RJ's land holdings, 174–175
  taxes on (*See under* taxation)
will
  controversy over second, "Jefferson v. Jefferson," 289–293, 295–296, 306–307
  first, copies of, 302–303, 304–305b
Williamsburg, traveling to, 58
writing in family *Book of Common Prayer*, 46

Jefferson-Randolph-Lewis "dynasty," 34, 370n17
Jefferson, Robert Lewis, 138
  runaway slaves, advertisements for, 210–211*b*
  Snowden, managing, 332–334
Jefferson, Sarah "Salley" Smith, née Criddle, 260
Jefferson, Thomas (TJ)
  accounts of RJ, managing, 92, 385n43
  as attorney, 374n13
  carp, attempt at acquiring, 225–227
  cave, legend of hiding in, 112, 114
  "Continental Association of 20 October 1774," signature on, *93*, 95
  crop rotation advice, 180–181
  death of RJ, reaction to, 285
  debts owed, 337, 432n6
  deposition on RJ's wills, 299–302
  education, 55–56
  engraving (Longacre), *278*
  European trees, introduction of, 246
  as executor for RJ, 306
  fire at Shadwell dwelling house, 45, 47
  French sheepdogs
    introduction of, 246, 249
    for RJ, 248–249, 251–252
    virtues of, 250
  garden vegetables, cultivating, 184–185
  as Governor of Virginia, 108–109
  Hemings, Sally, 348
  inheritance of, 47–49
  Jefferson, Lucy, relationship with, 130–131
  Jefferson, Lilburne
    interest in, 207–208
    letter on attending court RJ's will, 297–298
  Jefferson, Anna Scott "Nancy," interest in, 206–207
  Jefferson, Thomas, Jr., interest in, 206
  letters
    to Lafayette, 111, 388n88
    to Lilburne Jefferson offering clothing and shelter, 319
    from Martha Jefferson Randolph on attending court for RJ's will, 297
    to Martha Jefferson Randolph on attending court for RJ's will, 298
    from Mitchie B. Jefferson on RJ's illness, 290
    from nieces on Lucy Jefferson Lewis' death, 128
    from Paris, 199–200
    from RJ requesting silver watch, 254–256
  letters to RJ, 122, 390n190
    borrowed harness, 275–277, 290
    with gift to Mitchie B. Jefferson, 274
    Monticello invitation, 270–271
    from Paris, 408n1
    requesting carp, 226–227
    on sale of Dinah (slave), 235, *236*, 237
  Lewis, Anne, later Jefferson, relationship with, 131
  Lewis, Isham, helping, 133
  Lewises of Monteagle, relationship with, 129–131
  loans between RJ and, 108–109, 151–152, 155, 172, 231–232
  Monticello, flight from, 111–112, 114–115
  Nicholas, Wilson Cary
    letter on Randolph's second will, 293, *294*, 295
    relationship with, 406n57
  paying for education of RJ, 76–78, *77*
  Perkins, Hardin, Jr., letter on RJ's will, 295, 426n24
  portrait of, *59*
  Randolph, Thomas Jefferson, letter on RJ's health, 286
  religious views, 136–137
  request for aid, Revolutionary War, 106–107
  silver watch, assisting RJ with purchase of, 253–256
  slaves
    attitudes toward, 219–220, 347
    distribution of Peter Jefferson family slaves, 48–49
    purchasing and selling, 217, 381n76
    work, by age, 219
  Snowden, visits to, 278–282
  spinning jenny, to RJ, 227–231
  surveys of RJ's land holdings, 174–175
  transactions between RJ and, 92
  violin, importance of, 47

at Westham during the Revolutionary
  War, 387n78
will of RJ, controversy over, 290–291
in Williamsburg with RJ, 58
Wren Building, improvements to, 70,
  378n42
Jefferson, Thomas, II
  background on, 3
  property of, 361n7
  will, last, of, 4
Jefferson, Thomas, Jr.
  runaway slaves, advertisements for,
    210–211*b*
  special attention from TJ, 206
"Jefferson v. Jefferson," 298–299, 309, 310–
  311, 314–316, 323, 326, 426n32
"Jefferson's Church," Bristol Parish, 3
"Jefferson's Landing," 6
Johnson, James Monroe, 325
Johnson, Josiah, 325, 326
Jouett, Jack, 111
Jupiter (slave), 58, 222–223

K
Kinsolving cemetery, 242–243
Kinsolving, Elizabeth "Betsy," née Leigh,
  238, 240
Kinsolving, George W., 242
Kinsolving, James
  horses, 186, *186,*
  industries owned by, 238
  slaves, inventory, 416n80
  slaves purchased, 237–238

L
Ladd family, Horseshoe Bend land owned
  by, 13, 15, 364n36
Lafayette, Marquis de, 111
  Cornwallis, blocking military stores from,
    115–116
  portrait (Court), *116*
Leigh, Elizabeth "Betsy," later Kinsolving,
  238, 240
Lewis, Anne, later Jefferson. *See* Jefferson,
  Anne, née Lewis
Lewis, Charles, Jr., 37–39, 40, 135
Lewis, Charles Lilburne, 127
  Jefferson brothers, relationship with,
    129–133
  Kentucky, move to, 267

religious views, 133, 135
Lewis, Elizabeth, later Henderson, 130
Lewis-Henderson marriages, 156
Lewis, Isham, 133
Lewis, Jane Woodson, later Jefferson, 132,
  408n4, 409n4
Lewis, John and Elizabeth Warner, family
  lineage of, 202–203
Lewis, Lucy, née Jefferson. *See* Jefferson,
  Lucy
Lewis, Mary, née Randolph, 37–38
  children of, 40
  entertaining at Buck Island, 39–40
Lewis, Mary Randolph "Polly," later
  Jefferson, 207
Lewis, Nicholas, 105, 386n63
Lewis, Robert, 202, 408n4
Lindsay, Reuben, 109–110, 388n82
Logan, Billy (David; runaway slave),
  210–211*b*

M
Madison, Dolley, portrait of, *264*
Madison, James, 208, 296
maps
  Buck Island plantation, *44*
  Edgehill, *44*
  Fry-Jefferson, *22*
  "Randolph Jefferson's Virginia," *54*
  Shadwell, *22, 44*
  Sneed's English School, *44*
  Snowden, *22*
Marks, Anna Scott "Nancy," née Jefferson.
  See Jefferson, Anna Scott "Nancy"
Marks, Hastings, 282
Marrs' Road, 148–149
Martin, Hudson, 104–105, 126
Mattocks, Joseph, 361n08
Maury, Walker, 64–65
Mayo, William, as surveyor of Goochland
  County, 5*b*
Maysville, VA, origin of, 149
Mazzei, Filippo "Philip," 152, 396n5
Merino sheep, 187, *188*
military stores
  Albemarle Old Courthouse, 115–117
  New London, 117, 119
Miller, Lewis
  every day life in Virginia (sketch), *178*
  evoking road to Snowden (sketch), *280*

ferry crossing (sketch), *16*
"Negro Dance" (sketch), *342b*
slave auction, Christiansburg,
  Montgomery County (sketch), *160*
spinning wool (sketch), *227*
Miller, Robert, 79, 379n61
Monteagle, 129–133, 394n49
Monticello
  blacksmith operations at, 91–92
  drought at, 179–180
  dwelling house, value of, 141
  European trees, introduction of, 246
  illustrations of
    early 20th century engraving, *205*
    "View from Monticello Looking Toward Charlottesville" (Peticolas), *351*
    "View of the West Front of Monticello" (Peticolas), *351*
  nailery, 219-220
  overseers, need for, 177
  peach trees at, 182
  textile industry at, 187–188
Montpelier (*View of Montpelier;* Thornton), *296*
Moon, John S., 328–329
Moon-Scott family connection, 329
Moon, William, Jr.
  "Moon v. Jefferson," 157–161, 162b, 398n23–24
  Scottsville, establishment of, 328–329
  Stony Point store, 189, 192
Moore, Edward, 130
Moore-Henderson family connections, 130, 392n16
Moore-Lewis family connections, 130, 392n16
Mount Warren, 194
Murray, John (Lord Dunmore), *69*, 70
Murray, Richard, 18, 122, 366n54

N
naming practices, birth order, 200–203
Nelson, Thomas, Jr., 95–96, 98–99, 385n51
Nelson's Corp of Light Horse, uniforms for, 96–98, *97*
Nevil and Jefferson (company), 261–262
Nevil, Lafayette, obituary of, 350
Nevil, Anna Scott "Nancy," née Jefferson, 206–207, 303, 327
Nevil, Zachariah, 207, 389n102

as administrator of RJ's estate, 327–328, 330–331, 429n55, 431n81
debts owed, 328
RJ, work relationship with, 261–263
residences, 420n9
Nicholas, John, 82–84,
Nicolas, John, Jr., 274–275, 423n45
Nicholas, Wilson Cary, 194
  industries owned by, 195
  portrait (Stuart), *292*
  TJ, letter on RJ's second will, 292–293
  TJ, relationship with, 406n57
Nicholas, Wilson Cary, Jr., 194, 286, 425n7

O
Oglesby, Ann, née Perkins, 400n45
Oglesby, David, 168, 171, 400n44, 400n45
Oglesby, Ware, 117, 389n99
Oglesby, William, 400n45
Old Bruton Church, Williamsburg, 75
Old Courthouse Road (Ferry Road), 101, 148–149
Orange County, origin of, 233
Orange (slave), 233–237
ordinaries
  Albemarle County, colonial, 16–17
  Anderson's tavern, 13, 59–60
  Brown's, 197
  Buckingham County, colonial, 149
  cost of services, 17
  Henry Flood's, 278-279, 281-282, 298, 424n53
  Orange County, colonial, 233
  Raleigh, 171
  Snowden, 15, 18, 122, 193b, 261, 365n44
  Warren tavern, *195*, 196–197
overseers
  Goss, 315, 317–318, 428n49
  occupation description, 175, 177
  Snowden, 175, 177–179, 260
  Stanly, accused of stealing, 234–235

P
Page, Carter, 65
Page, John, 65
Pasteur, William, 380n68, 381n70
Patterson, John, 197, 407n73
Patteson, Reuben B., 255, 418n39
peach trees, Monticello, 182
Peasely, William, 136

Perkins, Ann, later Oglesby, 400n45
Perkins, Hardin, Jr., 122
   Hermitage, dwelling house of
      sketch by Pennington of, *143*
      value of, 140
      note from RJ to Benjamin Bondurant, 164–165, *165*
      TJ, letter on RJ's will, 295, 426n24
Perkins, Hardin, Sr., last will, bequests of slaves in, 224
Perkins, Stephen, 177, 402n10
Perkins, Walker, 84
Perkins, William, 122
Personal Property tax (1815), 309
Peter (slave), 48
Peyton, Craven, 129, 144, 391n7
Philadelphia, engraving of, *253*
plantations. *See also* Buck Island; Monticello; Shadwell; Snowden
   Chatsworth, 373n3
   Dungeness, 7, 17, 28
   farm industries on, 185–189, *186*, 194–197
   Fluvanna lands, 13, 15
   Mount Warren, 194
   Poplar Forest (Bedford), 111-112, 114–115, 227, 278–279, 414n61, 434n22
   schools, 44–45, 84
   Seven Islands, 83, 84
   Stony Point, 189, 192, 261–262
   Temple Hill, 238, *241*, 241–243
   Tuckahoe, 11–12, 29, *29*
   Westham (a.k.a. Beverley Town), 108–109, 387n78
   Woodlawn, 265–266, 344, 421n19
Presbyterian Churches
   ministers preaching at homes, 135
   near Snowden, 135–136
Prince Edward County Court, *156*
   "Bondurant v. Jefferson," 163*b*, 164–165, 168, 399n35
   execution book, Prince Edward County District Court, 162–163b
   "Moon v. Jefferson," 161, 162*b*, 398n23–24
   summons for RJ, 165, *166*
Pryor, Banister S., 266
Pryor, David, 420n15
   children of, 266
   land, acquirement of, 265–266

Pryor, John C., 267
Pryor, Langston S., 267, 421n26
   in Buckingham Militia, 269–270, 422n37
Pryor, Mitchie B., later Jefferson. *See* Jefferson, Mitchie B., née Pryor
Pryor, Nicholas Ballow, 267
Pryor, Susannah, née Ballow, 184, 265–268, 324–325
   children of, 266
   family lineage of, 421n27
   RJ, relationship with, 267–268
Pryor, William S., 266
Pryor, Zane, in Buckingham Militia, 269–270, 422n37

R
Rachel (slave), 51
Randolph, Anne, later Scott, 17–18
Randolph, Beverley, 64
Randolph, Cornelia, 286, 425n5
Randolph, Dorothea, later Woodson, 361n10, 365n50
Randolph, Ellen, 286, 425n5
Randolph, Gabriella Harvie, 270, 423n38
Randolph, Isham, 26–27, *27*
Randolph, Jane, later Jefferson. *See* Jefferson, Jane, née Randolph
Randolph, Jane, née Rogers, 28, 365n50, 369n6
Randolph, John, of Roanoke, 64–65
Randolph, Martha, née Jefferson. *See* Jefferson, Martha, later Randolph
Randolph, Mary, later Lewis. *See* Lewis, Mary, née Randolph
Randolph, Peter, 56, 373n3
Randolph, Sarah Nicholas, 337
Randolph, Thomas Jefferson, 42, 286, 425n7
Randolph, Thomas Mann, Jr., 91, 128, 144, 194, 220–221, 248, 249, 253, 257, 270, 418–419n43
Randolph, Thomas Mann, Sr., 12, 18, 270, 423n38
Randolph, Virginia J., later Trist, 329
Randolph, William, last will of, 11–12
Reynolds, Sir Joshua, *69*
Revolutionary War. *See* American Revolution
Richardson, Isham, 174–175
"Rives & Brown," Zachariah Nevil's debts

paid by, 328, 430n69
roads, "new-cut" road to Warren, 196
Rock Island Creek, residences at, 421n18
Rogers, Jane, later Randolph. *See* Randolph, Jane
Rose, Hugh, on military stores, Albemarle Court House, 117
Rose, Robert, 22, 367n66

S

Sam (slave), as possible Black Doctor, 224
Samuel Shelton & Co. distillery, 197, 407n75
schools. *See* education
Scott, Anne, née Randolph, 17–18
Scott, Daniel, 17–18, 362n22
Scott, Edward, *8,* 9
Scott, John, III
　marriage to Susan Bathurst Bolling, 329
　move to Alabama, 329, 430n77
　Scottsville, establishment of, 328
Scott-Moon family connection, 329
Scott, Susan Bathurst, née Bolling, 329
Scott's Ferry, 8–9, 17–18, 24, 362n21
"Scott's warehouse," 192
Scottsville, *139,* 192, 328, 405n55
Seven Islands, 83, 84
Shadwell, 21–22
　alcohol, stock of, 183
　description of, 7, 21
　expansion of, 21
　fire at dwelling house, 26, 45, 47
　floor plan, 19
　improvements to, 18–19
　inventory of, 21–22
　location on map, *22, 44*
　origin of name, 28
sheep
　Merino, 187, *188*
　wolves, danger of, 247
Shelton, Samuel, 197
Skelton, Martha, née Wayles, later Jefferson, 31, 58, 140, 415n64
slaves. *See also specific slaves by name*
　"A Typical Negro Cabin," *346*
　auction, Christiansburg, Montgomery County (sketch; Miller), *160*
　auction notice, *176*
　bequeathal of Peter Jefferson's, 23–24
　blacksmiths, Monticello, 91–92

impoundment of, *158,* 158–160
Jefferson brothers' attitudes toward, 347
marriages of, 222, 234–237
Perkins, Hardin, Sr., list of slaves owned by, 224
as personal property, 49–50
redistribution of Jefferson family slaves, 48–49
runaways, advertisements for, 210–211*b*
at Snowden, 216–217
spinning wool (sketch; Miller), *227*
taxes on, 154, 396n9
TJ purchasing and selling, 381n76
training on spinning jenny, 228–231
value of, 222, 234, 399n31, 412n23, 416n78
William and Mary, belonging to College of, 61–62, *63*
work assigned by age, 219
Small, William, 56, 373n4
Sneed, Benjamin, 42
Sneed's English School, 206
　location on map, *44*
　payments for RJ at, 42–44, 371n21, 371n23
Snowden Ben or Ben Snowden (slave), 152, 155, 217–222, 394n50, 412n23
Snowden Ferry House, 15
Snowden
　Administrator, court appointed, 327, 429n55, 431n81
　American Revolution at, 111, 115
　auction of, 330–331
　beef stock, appraisal of, 105–106
　crops, 179–181
　curator, court appointment of, 320–321
　description of, 15
　dwelling house
　　fire, 320, 322
　　value of, 140
　family, visits from, 284
　ferry and ordinary, 193*b*
　furniture, 142–143
　Hannah (slave), death of, 49–50
　horse advertisement, 85
　industries at, 189
　inventory of items, 141–142
　Jefferson, Thomas, visits to, 278–282
　land patent, 363n32
　leasing part of, 18, 366n54

location on map, *22*
Marks, Anna Scott Jefferson, visits to, 282
naming of, 364n42
neighbors, 140
orchards, 182–184
ordinary, 193*b*
overseers, 175, 177–179
photograph, northern tip, *139*
purchased by John Harris, 329, 330, 431n84
road conditions to, 280–281
as "Snowdon," 83, 382n1
stud services, 186–187
survey of, *14*
tobacco as cash crop, 181
Trust Deed, John Harris securing, 333–334
social functions, Buckingham County, 145–146, 148
Spencer, Edward, Peter Jefferson, meeting, 233–234
spinning jenny, 187–188, *227*, 227–231, *229*, 282
spinning wheel, *33*
Squire (slave), 231–232
"still house creek," 407n75
Stony Point general store, 189, 192, 261–262
Stuart, Gilbert, *278*, *292*
"Subscription to Support a Clergyman in Charlottesville," 136–137

T
Tarleton, Banastre, 111, *113*, 115
taverns. *See* ordinaries
taxation
capitation tax, 382n2
Jefferson, Randolph
on Buckingham County tax list, 83
house at Snowden, value of, 140–141
personal property tax record, 1815, 141, 309
on slaves and self, 408n2
taxes paid, 1810 to 1815, 288
personal property tax record, 1815, 141, 309, 394n48
poll tax, 382n2
on slaves, 154, 396n9, 408n2
tithables, 83, 382n2
War of 1812, rising taxes from, 288
Temple Hill, 238, *241*, 241–243

Terrick, Richard, 73–74
3rd Regiment of the Light Dragoons, 115, *118*, 119
Thomas, John, 168–169, 400n50
Thompson, Joseph, appointments, Albemarle County, 8
Thornton, Presley, 119
Thornton, William, 249
Three Notch'd Road, 22
Tillotson Parish, 136, 393n30
tobacco
as cash crop, 181
hogsheads of, 363n25
inspectors, salaries of, 192, 196
on Snowden, 181
warehouses, 196
wheat, switch from tobacco to, 194–195
Trist, Virginia J., née Randolph, 329
Tuckahoe, 11–12, 29, *29*
Tucker, St. George, 64, 377n32
Tucker, Harry, on life at William and Mary, 72
Turpin, Mary, née Jefferson, 4
Turpin, Thomas, 4, 6, 12, 361n7, 363n32

V
Viewmont, 362n20
Virginia, "Randolph Jefferson's Virginia," *54*
Voigt, Henry, 253, 418n29

W
Wales, Jefferson family connection to, 364n42
Ware, Thomas, 177–178, 260, 402n9, 403n11
Warren
distillery, Samuel Shelton & Co., 197, 407n75
general store, 190–191*b*
mill, 225
tavern, 195
town of, establishment of, 192, 194–197, 405n55
Wayles, Elizabeth, later Eppes, 95, 129
Wayles, Martha, later Skelton, then Jefferson, 31, 58, 140, 415n64
Westham, loan between TJ and RJ, 108–109, 387n78
wheat, as crop
profitability of, 194–195
switch from tobacco to, 194–195

White, Hugh, 204–205
William and Mary, College of
   accommodations at, 62
   board, cost per term of, 78
   Bursar's Books, payments in, 60, 70,
      76–79, *77*, 375n16
   conditions at, 57
   curriculum, 73–74
   discipline at, 68, 70–71
   employees of, 60, 375nn20–21
   Flat Hat Club, 76
   four components of, 374n7
   Grammar School, 71
   housekeeper, Wren Building, 60
   prayer schedule, 68
   RJ at, 55, 57–58, 60, 71, 76, 81, 82
   salaries of employees, 61
   student viewpoint from, 72
   students enrolled during RJ's years at,
      64–65
   tuition payment reform, 70, 78–79
   Wren Building, 60, 65, *66–67*, 70,
      375n19, 378n42
Winkfield (slave), 61–62, *63*, 376n25,
   376n26
Wirt, William, attorney, 310–311
   Cabell's, friendship with, 428n42
   case against Jefferson family, 316–318
   engraving, *312*
   letter to John C. Pryor, 314–316
   reputation of, 311, 313
women's fashions, *134, 264, 283*
Wood, Valentine, 95
Woodlawn, 265–266, 344, 421n19
Woodson, Dorothea, née Randolph, 361n10,
   365n50
Woodson, John, 361n10, 364n48
Woodson, Tucker M., 46, 58, 408n4
Wren Building, College of William and
   Mary, 60, 65, *66–67*, 70, 375n19, 378n42

About the Author

Photo by Andy Snow

After earning her doctorate in cinema studies at the University of Southern California, Joanne Yeck taught and wrote about film history for many years. She is the author of numerous articles concerning Classic Hollywood and American Popular Culture and is the co-author of *Movie Westerns* and *Our Movie Heritage*. Since 1995, her interest in Virginia history has become a full-time occupation. Years of research resulted in *"At a Place Called Buckingham" . . . Historic Sketches of Buckingham County, Virginia* (Slate River Press, 2011). Her work concerning Randolph Jefferson and Snowden resulted in a Jefferson Fellowship at the International Center for Jefferson Studies in 2010. When she's not driving the back roads of Virginia, she resides in Kettering, Ohio.